CRUCIAL EVENTS
Why are Catastrophes Never Expected?

STUDIES OF NONLINEAR PHENOMENA IN LIFE SCIENCE*

Editor-in-Charge: Bruce J. West

*For the complete list of titles in this series, please go to
http://www.worldscientific.com/series/snpls

Studies of Nonlinear Phenomena in Life Science – Vol. 17

CRUCIAL EVENTS
Why are Catastrophes Never Expected?

Bruce J West
Army Research Office, USA

Paolo Grigolini
University of North Texas, USA

World Scientific

NEW JERSEY · LONDON · SINGAPORE · BEIJING · SHANGHAI · HONG KONG · TAIPEI · CHENNAI · TOKYO

Published by

World Scientific Publishing Co. Pte. Ltd.
5 Toh Tuck Link, Singapore 596224
USA office: 27 Warren Street, Suite 401-402, Hackensack, NJ 07601
UK office: 57 Shelton Street, Covent Garden, London WC2H 9HE

Library of Congress Cataloging-in-Publication Data
Names: West, Bruce J., author. | Grigolini, Paolo, author.
Title: Crucial events : why are catastrophes never expected? / Bruce J. West, Paolo Grigolini.
Description: Hackensack, New Jersey : World Scientific, [2021] |
    Series: Studies of nonlinear phenomena in life science, 1793-1428 ; vol. 17 |
    Includes bibliographical references and index.
Identifiers: LCCN 2021012581 (print) | LCCN 2021012582 (ebook) |
    ISBN 9789811234095 (hardcover) | ISBN 9789811234101 (ebook for institutions) |
    ISBN 9789811234118 (ebook for individuals)
Subjects: LCSH: Computational complexity. | System theory. | Fractals. |
    System failures (Engineering)
Classification: LCC QA267.7 .W48 2021 (print) | LCC QA267.7 (ebook) | DDC 003/.75--dc23
LC record available at https://lccn.loc.gov/2021012581
LC ebook record available at https://lccn.loc.gov/2021012582

British Library Cataloguing-in-Publication Data
A catalogue record for this book is available from the British Library.

For any available supplementary material, please visit
https://www.worldscientific.com/worldscibooks/10.1142/12203#t=suppl

Contents

Preface

This book was motivated, in part, by the positive response to our review article concerning maximizing information exchange between complex networks [587]. Such strong positive reactions are often the result of fortunate timing, but what cannot be so easily brushed aside are the numerous applications that have been made of the ideas discussed in that article in the decade since its publication. We have followed up on much the research left open-ended in the review and can now document what has developed into a new area of information theory. An area that we believe takes on the mantle of a principle within Complexity Science, one that does, in fact, determine how information is handed back and forth between complex networks. We discuss the unexpected nature of this switching of information in a conversational way and leave a large part of the mathematics to appendices and citations.

The review demonstrated that there exists a subtle connection between information exchange within and between networks and the complexity (nonsimplicity) of those networks. We replace the term complexity, with nonsimplicity in this essay/book, for many of the same reasons that the term nonlinear was introduced into the dynamics of physical systems. In large part because we understand such phenomena by the properties they lack, rather than by those they possess. This essay presents a generalization of the complexity matching effect discussed in the review and provide a rationale for a Principle of Optimal Information Exchange (POIE), which quantifies the transmission of information in terms of the information imbalance between networks and interrelates information and nonsimplicity, as well as the control of nonsimple networks.

In a complex system there is shown to be a mix of crucial and non-crucial events, with very different statistical properties and it is the crucial events that determine the efficiency of information exchange between complex networks. It has been determined that in a large class of complex systems it is the crucial events which determine catastrophic failures, from heart attacks to stock market crashes. This essay outlines a data processing technique that separates the effects of the crucial from those of the non-crucial events in complex time series. An informal, almost conversational style is adopted at many places in the manuscript, without sacrificing the clarity necessary to explain, understand and use such concepts as fractals, complexity and randomness in interpreting what the mathematics is telling us about the science.

Nomenclature

AMI: Acute Myocardial Infarction
AR: Allometry Relation
ASM: Altruism-selfishness Model
ATA: All-to-All
BA: Barabási–Albert
BOLD: Blood–Oxygen Level Dependence
BRA: Breath Rate Analysis
CBF: Cerebral Blood Flow
CI: Citation Index
CLT: Central Limit Theorem
CME: Complexity Management Effect
CWR: Complexity Without Renewal
CTRW: Continuous Time Random Walk
DEA: Diffusion Entropy Analysis
DFA: Detrended Fluctuation Analysis
DMM: Decision Making Model
EEG: Electroencephalogram
EIE: Empirical Information Exchange
FBM: Fractional Brownian Motion
FC: Fractional Calculus
FDT: Fluctuation–dissipation Theorem
FLM: Fractional Lévy Motion
FPE: Fokker–Planck Equation
FFPE: Fractional Fokker–Planck Equation
FLE: Fractional Langevin Equation
FRE: Fractional Rate Equation
FM: Fibromyalgia
fMRI: Functional Magnetic Resonance Imaging
GLE: Generalized Langevin Equation
GLRT: Generalized Linear Response Theory
GME: Generalized Master Equation
GOP: Generalized Onsager Principle

GCM: Global Circulation Model
HRV: Heart Rate Variability
IT: Information Theory
IEI: Inter-event Intervals
IPCC: Intergovernmental Panel on Climate Change
IPL: Inverse Power Law
KAM: Kolmogorov–Arnold–Moser
LFLE: Linear Fractional Langevin Equation
LRT: Linear Response Theory
MDEA: Modified Diffusion Entropy Analysis
MEG: Magnetoencephalography
ML: Mittag–Leffler
MLT: Mittag–Leffler Theorem
MoS: Method of Stripes
MRI: Magnetic Resonance Imaging
NPR: Non-Poisson Renewal
NER: No-ergodic Renewal
NERP: Non-ergodic Renewal Process
NME: Nonsimplicity Management Effect
NPRP: Non-Poisson Renewal Process
OGM: Ontogenetic Growth Model
OP: Onsager Principle
PCM: Principle of Complexity Matching
PDF: Probability Density Function
POIE: Principle of Optimal Information Exchange
RAC: Renewal Approach to Complexity
RF: Radio Frequency
RW: Random Walk
SCPG: Supercentral Pattern Generator
SCLT: Stochastic Central Limit Theorem
SHM: Stochastic Habituation Model
SNR: Signal-to-Noise Ratio
SOC: Self-organized Criticality
SOTC: Self-organized Temporal Criticality
SR: Stochastic Resonance
SRA: Stride Rate Analysis
SOGM: Stochastic Ontogenetic Growth Model
TBI: Traumatic Brain Injury
TBM: Total Body Mass
WH: Wiener Hypothesis
WWW: World Wide Web

Acknowledgment

The opinions expressed herein are those of the authors and do not necessarily reflect those of DEVCOM Army Research Laboratory. However, PG thanks the U.S. Army Research Office (ARO) for financial support of the research through Grant W911NF1901 on which much of the authors' opinions are based.

Chapter 1

Information is Fundamental

I often say that when you can measure what you are speaking about, and measure it in numbers, you know something about it; but when you cannot express it in numbers, your knowledge is of a meager and unsatisfactory kind; it may be the beginning of knowledge, but you have scarcely in your thoughts advanced to the state of science. — Lord Kelvin [285]

The Industrial Age began its transition into the Information Age with the end of the Second World War, which is probably as good a road sign as any other to mark its beginning, and just as arbitrary. The war research literally threw together the world's best minds in the physical, social, life and mathematical sciences to work on the most challenging scientific, technological and theoretical problems imaginable and all with a common purpose, to overcome the technological advantage of the German military. These scientists broke codes, built hardware, initiated a new vision of what it means to be human, and proposed a new understanding of how humans communicate with one another and with their machines. To capitalize on these and other advances in science President Franklin Deleno Roosevelt wrote a memo to Vannevar Bush, a physics professor on leave from MIT, who was serving as the Director of the Office of Scientific Research and Development (1941–47), asking him how the substantial research that was being accomplished during the war could be carried over to peace time and applied within the civilian sector.

V. Bush's response to the president's question was the legendary report *Science, The Endless Frontier* [126], which he wrote synthesizing the input from multiple committees he convened to examine the president's question. In the report he argued for how the United States could retain the scientific advantage it had achieved during the war years and laid out the reasons for building a civilian-controlled organization for doing fundamental research to support national needs, with close liaisons with the Army and Navy. Unfortunately FDR died before the report was finished, but he would have no doubt been gratified by the result, since the report became and remains the founding justification for a country's government to financially support basic science for the good of its society.

It was in this heady atmosphere of the release of research from the security constraints of world conflict that scientists began thinking and writing about applications of that research to the post-war world, or in some cases bringing into daylight what they had been secretly thinking outside the confines of weapons research. In regard to the nascent field of information science, two American scientists stand out: the mathematician Norbert Wiener, who synthesized his collaborations with a substantial number of scientists into *Cybernetics* [611], and the engineer/mathematician Claude Shannon, who did the same with *Information Theory* (IT) [488], both investigators using the physical concept of entropy to define information. It was the scientific concept of information that molded the new vision of what it is to be human, which in turn is based on the necessity of probability theory for understanding the world of humans and machines, along with their interactions.

This new scientific vision of the transformation taking place in the world around them was lucidly expressed by Wiener in *The Human Use of Human Beings* [613]:

> ..physics now no longer claims to deal with what will always happen, but rather with what will happen with an overwhelming probability....It is true that the books are not yet quite closed on this issue and that Einstein (as well as others) ... still contend that a rigid deterministic world is more acceptable than a contingent one; but these great scientists are fighting a rear-guard action against the overwhelming force of a younger generation.
>
> ..In control and communication we are always fighting nature's tendency to degrade the organized and to destroy the meaningful; the tendency...for entropy to increase.
>
> ..Information is a name for the content of what is exchanged with the outer world as we adjust to it, and make our adjustment felt upon it.

Wiener identified information with disorder, Shannon associated information with uncertainty, and the Russian mathematician Kolmogorov thought of it in terms of complexity. Understanding of richly textured phenomena in the social and life sciences have significantly benefited from the order/disorder distinction, the certainty/uncertainty separation and the simple/complex characterization. History has endorsed all three interpretations, each one in a variety of distinct venues, using entropy as the measure of information. On the downside, this is not unlike Saxe's 1872 poem about three blind men of Indostan who wanted to know the shape of an elephant and each exploring different parts of the elephant drew very different conclusions about its true nature. The moral of the poem guides us here [470]:

> So oft in theologic wars,
> The disputants, I ween,
> Rail on in utter ignorance
> Of what each other mean,
> And prate about an Elephant
> Not one of them has seen!

The science/engineering community adopted at least three different orientations of Shannon's IT shortly after it was introduced: (1) it was directed towards the generation, storage, transmission and processing of information; (2) it was aligned with the statistics of signal detection in the presence of noise; (3) it was steered towards the study of communication and control of organisms and machines as in cybernetics. Shannon strongly advocated the first of these potential uses of IT, but it was cybernetics that first quantified scientific discussion of the interface between humans and machines, and made explicit Wiener's belief that the social and life sciences are as lawful as the physical sciences. The failure of science to find such social laws was a consequence of the complexity of the phenomena being studied, not a justification for not seeking them out. Shannon shied away from speculation on the use of IT outside a strict engineering context, and indeed often ridiculed those that made them. On the other hand, Wiener's cybernetics embraced the human potential of the new discipline. It is one of Wiener's world-changing speculations, its subsequent proof and the understanding it has provided about the increasing complexity in modern society that we address in this essay.

1.1 The Wiener Hypothesis

In a 'popular'[1] 1948 lecture Wiener observed [612]:

---

[1] In this context, popular is interpreted to mean non-mathematical.

> We have a system of high energy coupled to a message low in energy, but extremely high in amount of information, i.e., of great negative entropy. This unlike the usual interaction in thermodynamics, where all the coupled systems enjoy high entropy. But it may happen in the development of such a system that the internal coupling causes the information, or negative entropy, to pass from the part at low energy to the part at high energy so as to organize a system of vastly greater energy than that of the present instantaneous input.

In a less formal style he was saying that according to thermodynamics, two complex networks, under normal circumstances, one high in energy (hot) and another low in energy (cold), when brought in contact with one another transfer energy from the hot to the cold network. This is, of course, the second law of thermodynamics in which it is implicitly assumed that both networks are physical and high in entropy. The second law is therefore predicated on the networks' dynamics being energy-dominated processes in which the control of behavior is determined by the network with more free energy.

Wiener goes on to conjecture that if the hot network is low in information and the cold network is high in information, then information can be transferred from the cold network to the hot network. Schrödinger [480] introduced the term negentropy to quantify how a living being extracts order from the environment and discards disorder. Schrödinger explained that it is necessary to locally violate the second law of thermodynamics by exchanging entropy with the environment and in so doing maintain the ordered state of an organism's life, while disrupting its immediate environment. Wiener identified information with negentropy and saw these kinds of processes as being information-dominated and assumed that they need not be physical in nature. In such situations, unlike the familiar thermodynamic energy-dominated processes, in which the hot network governs things, it is the information-dominated processes wherein cold networks can control the behavior of more energetic networks as suggested in Fig. 1.1.

He speculated that the complex networks in the social and life sciences behave differently from, but not in contradiction to, those in the physical sciences, with control emanating from the flow of information, not from the flow of energy. The significance of this observation cannot be overstated. In the physical world: cars roll down hill, traveling from higher to lower potential energy; an apple pie cools off, heat radiating from the hot apples to the cooler room; hot air rises, being buoyed from the region of more dense to the less dense air. The force laws and therefore control in physical phenomena are a consequence of the negative gradients of energy potentials.

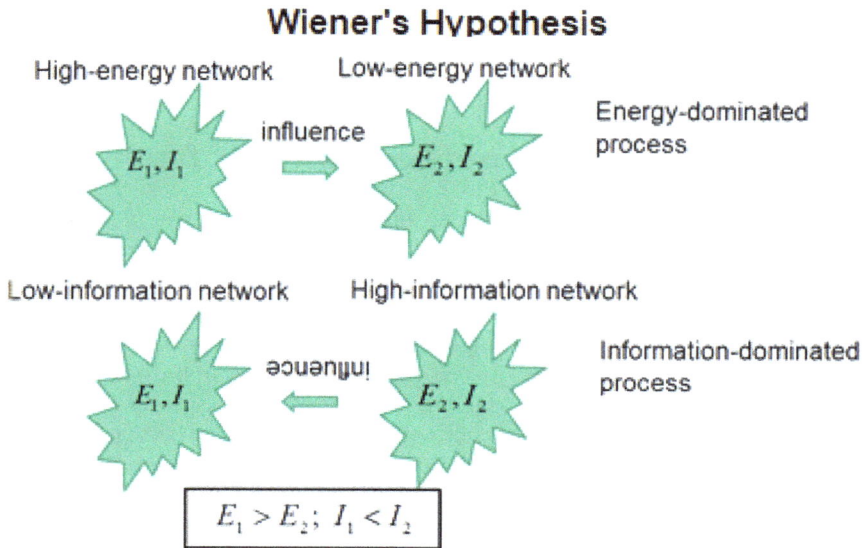

Wiener's Hypothesis

High-energy network   Low-energy network

influence

E_1, I_1 E_2, I_2

Energy-dominated process

Low-information network   High-information network

influence

E_1, I_1 E_2, I_2

Information-dominated process

$$E_1 > E_2;\ I_1 < I_2$$

Figure 1.1: Wiener's Hypothesis: The upper panel denotes the familiar thermodynamic situation of an energy-dominated interaction. The lower panel depicts the counter-intuitive information dominated interaction. This is emphasized by the influence in the lower panel being turned on its head. Adapted from [586].

He did not explicitly make the point, but Wiener's speculation implied that the force laws and therefore control of social phenomena typically do not follow the negative gradients of energy potentials, but rather they follow gradients due to imbalances of information. The force laws that control social phenomena need not follow the negative gradients of energy potentials (even if they could be defined), but instead follow gradients produced by information imbalance.

    The story of how this information force comes into being is a long one and is the thread that connects that various aspects of this essay into a coherent whole. Since most physicists would dismiss the assertion of an information force out of hand we proceed slowly, so that even the most volatile dispositions have an opportunity to cool. The point of departure is a recasting of the Wiener quote into the form of a hypothesis that we prove in stages, what we honorifically call the Wiener Hypothesis (WH):

> Given the proper conditions the force between two complex systems, produced by an energy gradient acting in one direction

between the systems, can be overcome by the force produced by an information gradient acting in the opposite direction between the systems.

1.2   Networks, Complexity and $1/f$-noise

The transportation networks of planes and highways; the economic webs of global finance and stock markets; the social mesh of governments and terrorists; the physical wicker of the Internet and climate change; the bionets of gene regulation and the human body; the ecowebs of food networks and species diversity, all bear a striking similarity. As the networks in which we are immersed become increasingly complex, a number of apparently universal properties begin to emerge. One of those properties is a version of the WH having to do with how complex networks, perhaps involving phenomena assigned to very different disciplines, exchange information with each other.

The first order of business in establishing the WH requires that we settle on a working definition of complexity. We note that empirically the dynamic behavior of a complex network is often expressed in terms of a power spectrum, wherein the amount of energy in each frequency interval is recorded. A simple process has all the energy concentrated at one or a few frequencies, such as in a harmonic, or in a chord, of music. As the process becomes more complicated, the number of frequencies involved becomes greater, approaching a continuum. However, complexity is not just a consequence of the number of frequencies involved in the dynamics, but also results from the relative amount of energy within each frequency interval as measured by the mode amplitude. In addition, the relative phase between modes can be of equal importance. Complexity often arises when the power spectrum $S_p(f)$ takes on an inverse power-law (IPL) form:

$$S_p(f) \propto \frac{1}{f^\alpha}, \tag{1.1}$$

with the IPL index α in the empirical interval $0.5 \leq \alpha \leq 1.5$ and goes by the technical name of $1/f$-noise, or $1/f$-variability. In fact $1/f$-noise (variability) is taken by many scientists to be the signature of complexity and appears in a vast array of dynamic phenomena including the human brain [284], body movements [166], music [418, 516, 553], physiology [585], genomics [325], sociology [518] and many more that are discussed by, for example, West et al. [587].

Historically, West and Shlesinger [575] discussed noise in natural phenomena pointing out that a completely random process, such as white noise is a colorless, hiss as in the sound of running water. We experience this as the static between stations on the radio and the 'snow' on TV without programming. White noise has a broad spectrum, but the energy content is the same for each frequency interval, that is, $\alpha = 0$ in Eq. (1.1) and therefore, although random, it is a relatively simple dynamical process. The non-zero IPL index is related to the dimension of the underlying process, which we now know need not be integer, that is, the process can be fractal with a non-integer dimension. The IPL index indicates how what happens in here and now is related to what happened in there and then. Consequently, the color of the spectrum ($\alpha \neq 0$) determines how what happened in the past influences the present and how what is happening now may change the future. So the scaling indicated by the IPL index is a measure of the memory in the phenomenon. Consequently, $1/f$-variability is a ubiquitous property of complex networks, whose origin is still the topic of debate and controversy. However, there seems to be no doubt that $1/f$-noise spreads upwards from molecules to the macrocosm and that spreading from the microworld to the macroworld is facilitated by the ease of communication among networks that share $1/f$-variability [587].

For the purpose of proving the WH we adopted $1/f$-variability as one definition of complexity and the IPL index as its quantitative measure. In doing this, we recognized the difference between continuous and discrete random processes. We focus on discrete random processes, since these are the more common form of experimental data, particularly the time series recording the occurrence of an event, e.g., heart beat, stride interval, earthquake of a given magnitude, solar flare eruption, starting and stopping of traffic flow, and so on. A process $s_1(\tau)$ is depicted in Fig. 1.2 and the events are the transitions between $+1$ and -1; recorded as the time intervals between successive events $\tau_1, ..., \tau_j, \tau_{j+1}, ..., \tau_N$, a time series with N data points.

In the time domain a complex network has a waiting-time probability density function (PDF) $\psi(t)$, with a corresponding IPL index given by $\mu = 3 - \alpha$, such that the asymptotic PDF is

$$\psi(t) \propto \frac{1}{t^\mu}, \tag{1.2}$$

and $\psi(t)\,dt$ is the probability of an event being generated by the network dynamics in the time interval $(t, t + dt)$. Consequently, networks characterized by $1/f$-variability have long-time memory whose extent increases with decreasing IPL index μ. In the next chapter we show that the average time between events for a hyperbolic PDF, which is an IPL PDF asymptotically, is:

$$\langle t \rangle = \int_0^\infty t\psi(t)dt = \begin{cases} \frac{T}{\mu-2}; & \mu > 2 \text{ ergodic} \\ \infty; & \mu < 2 \text{ non-ergodic} \end{cases}, \tag{1.3}$$

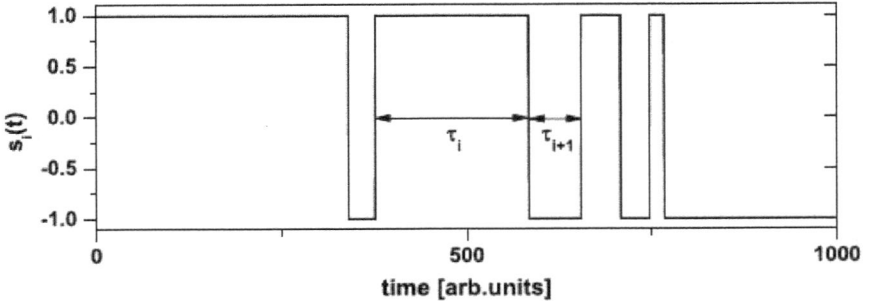

Figure 1.2: The time series for a two-state process is depicted in which the switching time between states is determined by an exponential PDF. The time interval between successive events, say between events j and $j + 1$ is τ_j, as shown.

where T is characteristic parameter of the waiting-time PDF. Note that for $\mu > 2$ there is a well-defined average time between events, but for $\mu < 2$ the average time between events diverges, so that if this were a communications network the rate of message transmission would cease to have meaning and the process is said to be non-ergodic. An ergodic time series is one for which a time average is equal to an average over the PDF and a non-ergodic time series is one for which this is not true. We shall have more to say about non-ergodicity, renewal processes and crucial events, subsequently.

The value $\mu = 2$ separates the ergodic and non-ergodic dynamic regions and this separation value occurs when there is true $1/f$-noise, that is, for $\alpha = 1$ in Eq. (1.1). Note the empirical range of this parameter when dealing with real data and for which the name of $1/f$-noise is still used.

1.3   Complexity Management Effect

We have, so far, restricted discussion to complex networks that can be quantified by IPL PDFs and consequently the IPL index is used to quantify our working notion of complexity. However, because of the ubiquity of such IPL networks, the imposed restriction is not overly severe. Using this concept of complexity we can determine how information is exchanged between two such networks, or in a more restricted sense, how information is transferred from one complex network to another complex network and use this to establish reasonable bounds for the WR. The more complicated situation of mutual information exchange is considered in due course.

Consider two complex networks interacting with one another and through that interaction they exchange information. For example, two people talking to one another face-to-face; a patient's body 'talking' to a physician during a physical examination; the music of a symphony orchestra exciting an audience member. Each complex network, talking, moving, music, etc., has its own characteristic exponent and the efficiency of the information transfer is determined by the relative values of the inverse power-law indices of the sender and receiver [586]. This an important point, since it contradicts the assumption made by Shannon in his characterization of information, that being that information is independent of the properties of the sender and receiver [488]. He apparently made this assumption because he was more interested in the engineering properties of the message being transmitted through a channel than he was in the properties of the persons at either end of the message. He explicitly assumed that the message was independent of the sender and receiver.

One measure of the information transfer between two complex networks is the cross-correlation between the output of a complex network P (stimulating network) and the stimulation of a complex network S (responding network) being perturbed by P. Denoting the normalized output of network P by $\xi_P(t)$, that of network S by $\xi_S(t)$ and the averages over the time series by brackets, the cross-correlation function is defined:

$$C(t) \equiv \langle \xi_P(t)\xi_S(t) \rangle . \tag{1.4}$$

If the output of P increases and the response of S tracks that increase, the two networks are positively correlated. The tracking could be perfect, in which case the cross-correlation function would be one. However, the response may be tempered in some way, yielding a less than perfect response with a correlation function less than one. The other limit is that the network S does not react to the changes in output of P and the two networks are said to be uncorrelated. The cross-correlation function is the simplest measure of the information transferred from P to S.

In Fig. 1.3 the asymptotic cross-correlation function is normalized to one and graphed as a function of the IPL indices of the two networks to form a cross-correlation cube. For a given level of complexity of the stimulating network P denoted by the IPL index μ_p the response of network S depends on its own level of complexity denoted by the IPL index μ_s. The values of the two IPL indices define a plane and the value of the cross-correlation function at each point on the plane defines a third dimension, so that the three together give rise to the cross-correlation cube. Note that this cube denotes the asymptotic values of the cross-correlation function.

Cross-correlation cube

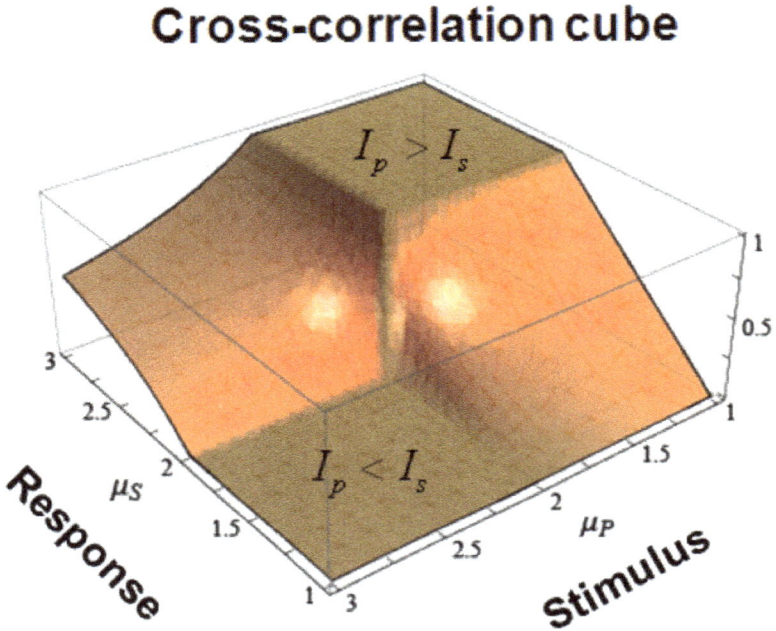

Figure 1.3: The cross-correlation cube depicts the asymptotic cross-correlation function, graphed as a function of the two IPL indices of the responding network S and the stimulating network P. Modified from [43].

This cube displays a number of remarkable properties:

(1) When the two IPL indices are equal to 2 there is an abrupt jump from zero correlation to perfect correlation of the dynamics of the two networks. This is a singular point at which the spectra of the two networks display $1/f$-noise. Of course, whether the abrupt jump is up from zero to one, or down from one to zero, depends on the value of the IPL indices just before they converge on 2.

(2) The upper plateau indicates that when the perturbing network P is non-ergodic $1 \leq \mu_p \leq 2$, and the average time between stimuli $\langle t \rangle_p$ diverges to infinity. The responding network S is ergodic $2 \leq \mu_s \leq 3$, and the average response time separating events $\langle t \rangle_s$ is finite. The time intervals between stimulating P-events are very long and the time intervals between unperturbed S-events are much shorter. Consequently, the S-events have more than enough time to adjust to a given P excitation, that is, to transfer the influence of the perturbation throughout the S network before the

next perturbation occurs. In this way, the S network relaxes to the P network perturbation. The greater information in the non-ergodic output of the P network I_p dominates the ergodic statistics of the S network having information I_s and produces complete correlation between stimulus and response. We discuss the mathematics leading to this result subsequently.
(3) The lower plateau indicates that when P is ergodic $2 \leq \mu_p \leq 3$, the average time between stimuli $\langle t \rangle_p$ is finite, and S is non-ergodic $1 \leq \mu_s \leq 2$, the average response time $\langle t \rangle_s$ diverges. The time intervals between P-events are therefore much shorter than those between the naturally occurring S-events. The responding S-events are much less numerous than those of the stimuli, which interfere with one another and their influence is lost before the responding network has time to transfer the perturbation even a short distance within its network. Consequently, there is no detectable response asymptotically. The information-rich S network, $I_s > I_p$, is seen to be unresponsive to the influence of the stimuli, due to the significant difference in time scales.
(4) In the two regions in which the IPL indices are in the same domain $1 \leq \mu_s,\, \mu_p \leq 2$ and $2 \leq \mu_s,\, \mu_p \leq 3$, both mean times either diverge or are finite, respectively. The information content of the stimulus and response behaviors are the same, so that the value of the cross-correlation function depends in detail on their respective values, as seen in the figure. The analytic expressions for the size of the cross-correlation in these two regions is derived subsequently.

We conclude from studying the cross-correlation cube that the manner in which one complex network responds to perturbation by another complex network is determined by which of the two networks has the greater information according to the statistics of their respective dynamics. The measure of this content it given by the IPL index. This is the one-way version of the complexity management effect (CME), which is to say, the transmission of information from the more complex to the less complex network.

The WH is described by the influence of the stimulus as it appears on the upper plateau region of the cross-correlation cube, where the information in the stimulus exceeds that in the response. In all regions except the lowest one, a weak stimulus significantly modifies the properties of the responding network. In the upper plateau region the stimulus not only influences, but actually dominates the properties of the response and reorganizes it, just as Wiener predicted. The CME is embodied by the cross-correlation cube that incorporates the WH into this larger principle.

1.3.1   Habituation

Let us consider the phenomenon of habituation as an exemplar of the information transfer from one complex network to another. This transfer can be explained by means of the CME, using the cross-correlation cube. Habituation is a ubiquitous and simple form of learning through which animals, including humans; learn to disregard stimuli that are no longer novel, thereby allowing them to attend to new stimuli [562]. One of the more interesting aspects of habituation is that it can occur at different levels of the nervous system. For example, with strong odors, sensory networks stop sending signals to the brain in response to repeated exposure to the stimuli. But odor habituation has been shown in rats to also take place within the brain, not just at the sensor level. The statistical habituation model (SHM) [591] hypothesized that $1/f$-noise, characteristic of complex networks, arising as it does in both single neurons and in large collections of neurons, is the common element that explains suppressing signals being transmitted to the brain and inhibiting signals being transferred within the brain.

A repetitious stimulus of unvarying amplitude and constant frequency content, such as a strong odor, a persistent hum, or the gentle sway of a boat, all induce responses that fade over time even though the stimulus persists, since no new information is being presented. This is the situation captured by the region with $I_s > I_p$, or region II in Fig. 1.4. The habituation response to the lack of new information allows the brain to shift its focus from the more to the less familiar, the latter providing new information that may have survival value.

However, we know the brain does not habituate to all external stimuli so let us consider two distinct kinds of stimulation; one ergodic $2 \leq \mu_s \leq 3$ and another non-ergodic $1 \leq \mu_p \leq 2$. The perturbation in the ergodic regime can be expressed as a simple spectrum, which allows us to generalize a previously established result for a periodic stimulation of a complex network [505, 571]. The non-ergodic stimulus can be drawn from a number of sources; here we choose for contrast the sequence of splashes from a dripping faucet [425] and certain pieces of classical music [553].

Consider first the case of the sound of waves crashing on the beach coming in through the window as you lay in a vacation motel bed at night, after a long day of reading on the beach. This naturally generated noise is typically a broadband of uncorrelated frequencies with random amplitudes and phases, like the static mentioned earlier. Consequently, the time series has a time average and ensemble average that are equal, a condition with the technical name of ergodicity. Most people habituate to this pleasant noise and after a short time they no longer hear it and fall asleep. The hippie

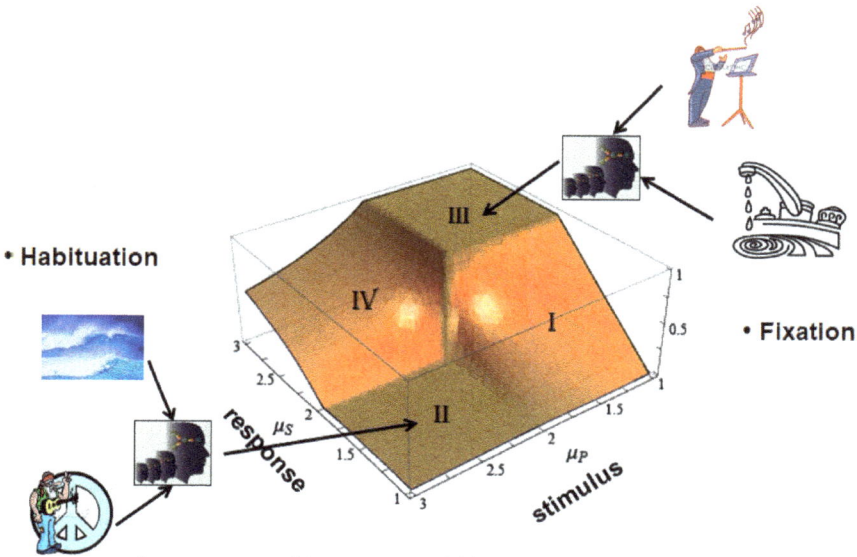

Figure 1.4: Examples of stimuli that habituate asymptotically in region II and those that are in 'resonance' with the complexity of the human brain and consequently fixate in region III, like the melody you cannot get out of your head.

playing the guitar in Fig. 1.4 is meant to represent simple, or uncomplicated, music, such as a ballad, which would facilitate rather than disrupt the onset of sleep. In the present context, it is possible to prove using SHM that the response of a model brain to such stimuli fades as the inverse power law in time $1/t^{2-\mu_s}$.

The plateau region II in Fig. 1.4 is the parameter region $2 \leq \mu_p \leq 3$, and $1 \leq \mu_s \leq 2$ where the dynamics of the model brain and the external stimulus are asymptotically statistically independent of one another. Brain activity in the non-ergodic regime $1 \leq \mu_s \leq 2$ is asymptotically unresponsive to ergodic and/or periodic stimuli; the complexity of the neuron network essentially swallows up simple signals through its complex dynamic interactions and the response fades in time as described by the IPL. The asymptotic suppression of a periodic stimulation of a complex physical network using linear response theory (LRT) was previously demonstrated by a number of investigators [505, 571]. In the present context we used generalized LRT to determine the asymptotic suppression of the stimulus to explain the phenomenon of habituation [591]. But even when the brain is in the ergodic regime $2 \leq \mu_s \leq 3$ its response, depicted in region IV, habituates to a constant value less than the full response value of the plateau region III, but it does not vanish altogether.

The second stimulus we consider is the sound of water from a leaky faucet splashing in the sink below. This sequence of water drops can set your teeth on edge and leads some to toss and turn throughout the night. The leaky faucet stimulus has been determined experimentally to have a distribution with an IPL index in the domain $1 \leq \mu_p \leq 2$, leading to the conclusion that the statistics of the time intervals between slashes are non-ergodic [425]. The brain's response ramps up from zero to one as the index μ_s increases from 1 to 2 as shown in region I of Fig. 1.4. Over the interval $2 \leq \mu_s \leq 3$ the response to the intermittent splashes given by the cross-correlation is maximal. Plateau III depicts the parameter domain where the brain is ergodic and records the sound of every intermittent drop of water, just as the WH anticipated.

It was determined experimentally [425] that the intermittent sequence of water drops from a leaky faucet is described by a Lévy stable PDF that is asymptotically an IPL with index $\mu_p \approx 1.73$. Consequently, the power spectral density has an index $\alpha_p = 1.27$, very close to the value where the WH predicts the maximum response of the brain to the stimulus. A discussion of the significance of Lévy statistics and its importance in understanding complexity is given subsequently.

Of course it is not just annoying stimuli that refuse to fade away. Classical music has been shown to manifest $1/f$-behavior [418, 553] and to resonate with the human brain, leaving strains of melody running through your head long after the music stops. The influence of the more pleasant stimuli also resides on plateau III of Fig. 1.4. West and Deering [577] review the occurrence of the $1/f$-variability in classical western music, classical ragas of India, traditional music of Japan and jazz, as well as the spatial variability in successful paintings. They speculated that the aesthetic judgments we make regarding music and art may well have a biological origin in that the stimuli resonate with the complexity of the human brain. This speculation is now supported by experiment, as well as being explained by a generalized LRT and the extension of WH to the CME.

It should be noted that the concept of energy has not been used in the above discussion of how one network influences the other. The discussion of the cross-correlation cube and habituation serve as a surrogate for information transfer from one dynamic network to another. But we have not yet examined how information has been expressed in terms of entropy, so let us do that now.

1.3.2   Entropy

The thermodynamic concept of entropy has recently been applied in a biological context by Demetrius *et al.* [169] to explain Darwinian evolution. They introduced *evolutionary entropy* to provide a context in which allometry relations, which describes the dependence of functionality, such as metabolic rate and maximal life span, on body size can be predicted. This is one of the more recent contributions in an ongoing strategy to offer an alternative to energy as the fundamental tool for the understanding of complexity, order and organization in bio-systems. Here we stress the need for creating new tools with which to analyze complex systems and thereby gain an understanding of complexity. The tools presented herein have been, in part, generalized from non-equilibrium statistical physics, as well as, others stemming from new applications of the fractional calculus. But more about that later.

In 1953, one of the first meetings was held to explore the importance of entropy to quantify the information content of living systems, from macroscopic biodynamics down to protein structure. Some three decades later, entropy was offered as a unifying principle for biological evolution by Brooks and Wiley [118]. Similar attempts have been made to explain social organizations using theories of social entropy [43, 44]. All these efforts and many more went unaware of Wiener's prescient contribution to the discussion. But now a theory of the stochastic dynamic basis for allometry is available [603].

The WH remained a provocative speculation for over sixty years. It was only with the recent activity to develop a science of networks that an extension of the WH was proven to be true. The implicit idea necessary for the existence of a network science is that a network's macroproperties are determined by emergent behavior, not by its microdynamics. It is this emergent behavior that entails commonality in the macroproperties of social, biological and physical networks. The behavior that is common to networks is a consequence of their complexity.

The proof of the WH consequently relies on generalizing some of the fundamental ideas of non-equilibrium statistical physics, in particular LRT to non-stationary phenomena [43, 44]. The science of thermodynamics explains the movement of heat and other irreversible phenomena in the physical world and statistical physics seeks to explain the thermodynamic formalism, using the microdynamics of physical systems. On the space/time scales we live our lives the laws of thermodynamics and statistical physics are familiar and seem to encapsulate our experience into a handful of intuitive principles. The network dynamics of an individual or of a group are very different, however. Predicting the ups and downs of the stock market may have an analogy with Brownian motion [351], where a heavy particle is buffeted about by a fluid of lighter particles, but we cannot predict the average behavior of a stock's price, the way we can the likely position of the Brownian particle.

As the networks in which we are immersed become increasingly complex, a number of apparently universal properties begin to emerge. One of those properties is a generalized version of the WH, one version of which we expressed as the CME in discussing the cross-correlation cube. Another generalization has to do with how complex networks, perhaps involving phenomena from different scientific disciplines, exchange information with one another; this is the Principle of Optimal Information Exchange (POIE). As mentioned earlier, this essay presents a generalization of the CME discussed in WGG [586] and provides a rationale for a POIE, which quantifies the transmission of information in terms of the information imbalance between networks and interrelates information and nonsimplicity, as well as the control of nonsimple networks. The efficiency of information transfer is dependent on the relative complexity of the two networks, and the complexity gradient gives rise to a complexity-induced information force, which we subsequently discuss.

On a larger stage, Karl Marx talked about class conflict as being the driver of social evolution [361], whereas Adam Smith invoked an invisible hand to visualize the unintended social benefit resulting from individual actions of self-interest [499] and Freud argued for instinct as the primary driver of human behavior [201]. At both levels of the individual and the group these historical arguments exemplify what could be included under the general heading of information forces; non-physical forces resulting from gradients in the complexity of the phenomena being studied [602].

Chapter 2

Empirical Information Exchange

> *You can have data without information, but you cannot have information without data.* — D.K. Moran [387]

Let us now turn our attention to the often subtle, but nonetheless significant, empirical connections between information exchange within and between networks, as well as, the complexity of those networks. We offer observational and experimental evidence from a variety of naturally occurring networks, as well as social networks, that directly support the POIE hypothesis, before becoming involved with the mathematics developed in subsequent chapters.

We explore some of the implications of the POIE by introducing a context that is perhaps more familiar than is the non-stationary, non-Poisson, non-ergodic stochastic processes of crucial events, that being, $1/f$-phenomena, or $1/f$-noise. The existence of such phenomena was recognized by Daniel Bernoulli [88], as indicated by his introduction of the utility function, which he used to characterize the social well-being of an individual in the 18th century. He reasoned that a change in some valuable, but unspecified, quantity f, denoted by Δf, elicits different responses from different people, depending on how much of f they already possess (utility). He concluded that people respond to the fractional change (marginal utility) $\Delta f/f$, rather than to the magnitude of the change, Δf, itself. Over nearly three centuries since Bernoulli first thought and wrote about these things, his ideas have been transformed in a number of different ways — each way attempting to capture the complexity of the network to which his ideas were applied.

Consider the observable represented by the dynamic variable $X(t)$ that at a time t could represent the position, or velocity, of an object, the profit of an investment stock, the electrical recording of brain activity, the size of an earthquake, or strands of music coming from your neighbors' window. If the phenomenon is complex, or even simple but complicated, the dynamic variable is typically stochastic and data comes in the form of a time series. The analysis of such time series in terms of moments and the PDF determines what we can know about the phenomenon being observed. The lowest-order moments that determine the dynamics of the underlying process are the average $\langle X(t) \rangle$ and variance $\sigma^2(t) \equiv \langle X(t)^2 \rangle - \langle X(t) \rangle^2$, if the statistics satisfy the bell-shaped curve of Gauss and the brackets denote the average over this PDF. For example, the Gaussian process with the second moment:

$$\langle X(t)^2 \rangle \propto t^H; \quad 0 < H \leq 1, \tag{2.1}$$

H being the Hurst exponent, has been related to random walk (RW) processes that are persistent for $H > 1/2$, in which the walker after taking a step has a bias to continue without changing direction in his next step, and is anti-persistent for $H < 1/2$, in which the walker after taking a step has a bias to change direction in her next step.

The time variation of the first two moments are determined by the system's dynamics, which can be characterized, at least in part, by another second-order quantity, the autocorrelation function:

$$C(t, \tau) = \langle \Delta Z(t + \tau) \Delta Z(t) \rangle, \tag{2.2}$$

where we introduce the normalized zero-centered variable:

$$\Delta Z(t) \equiv \frac{X(t) - \langle X(t) \rangle}{\sigma(t)}, \tag{2.3}$$

and $\sigma(t)$ denotes the standard deviation of $X(t)$. Note that, in general, the correlation function depends on both t and τ when the underlying time series is non-stationary, but is independent of t when the time series is stationary. The function given by Eq. (2.2), also called the correlation function, determines how the deviation in the observable from its average value at one time influences that deviation at a later time; it is a measure of the memory of the process. Note that this memory is not the same as the persistence that arise from the statistical bias in the steps of a random walk process in Eq. (2.1). We stress that power-law autocorrelation functions for stationary time series:

$$C(\tau) \propto \tau^{\alpha-1}, \tag{2.4}$$

have IPL spectra, as defined by the Wiener–Khinchine theorem in terms of the Fourier transform:

$$S_p(f) \equiv \mathcal{FT}\{C(\tau); f\} \propto \frac{1}{f^\alpha}, \tag{2.5}$$

as is easily shown using a Tauberian theorem. Note that this expression for the spectrum was previously given by Eq. (1.1). The importance of this observation is emphasized here because $1/f$-noise is ubiquitous in biological, social and physical networks and analysis suggests that the scaling identified in complex networks may be causally related to these familiar phenomena.

There is widespread conviction that $1/f$-noise is the signature of complexity, where the spectrum of the phenomenon is given by Eq. (2.5). Here we briefly discuss a number of complex processes described by $1/f$-spectra, emphasizing how these networks may communicate among themselves. The examples assist in formulating the question of what may be the underlying reason for the transfer of information between two or more complex networks.

2.1 Physiological Variability

We begin our discussion of empirical complexity by partitioning the phenomena of interest into discipline categories. In some cases the disciplinary assignment is arbitrary because the phenomenon of interest spans two or more disciplines, but that need not detract from the discussion.

2.1.1 The brain

Electroencephalograph (EEG) time series: The mammalian brain generates a tiny, but measurable electrical signal; first measured by Caton in 1875 in small animals and in people by Berger in 1925. The trace left by an ink stylus on a moving strip-chart by this amplified signal was called an electroencephalograph and the term EEG has subsequently been used to characterize the electrical signal. The power associated with the EEG signal is distributed over the frequency interval 0.5 to 100 Hz, with most of it concentrated in the interval 1 to 30 Hz. A typical EEG signal looks like a random time series with contributions from every part of the spectrum appearing with random phases. This aperiodic signal changes throughout the day and changes clinically with sleep, that is, the *alpha* (8–14 Hz) rhythm, which dominate the EEG signal of an awake individual, with eyes closed, is almost completely absent when a person is asleep.

In the over one hundred years since the discovery of EEGs, there have been a variety of methods used in attempts to establish the taxonomy of

EEG patterns in order to delineate the correspondence between brain wave patterns and brain activity. Wiener mistakenly believed that his *generalized harmonic analysis* [614] would provide the mathematical tools necessary to penetrate the mysterious relations between the EEG time series and the functionality of the brain. More recently, nonlinear dynamics, chaos and fractals have lead the parade of methodologies hoping to accomplish this task [577]. The progress along this path has been slow and the understanding and interpretation of EEG signals remain elusive. However, along the way, certain properties of EEG signals have revealed themselves.

The erratic behavior of the EEG signal is so robust that it persists through all but the most drastic situations, including near-lethal levels of anesthesia, several minutes of asphyasia and the complete surgical isolation of a slab of cortex [198]. It has been suggested that cognitive attractors exist, as demonstrated using attractor reconstruction techniques, and that the observed fluctuations in EEG time series were the result of chaotic dynamics within the brain [46]. The EEG time series of awake, alert individuals were shown to have non-stationary statistics [315] and are of high, perhaps infinite, dimension, suggesting that the signals are not generated by strange attractors, but are random and not chaotic. However, these observations were made before the properties of intermittent chaotic maps were widely used.

By way of contrast, it was shown [45] that the various stages of sleep can be characterized by quite different fractal dimensions, implying that if a cognitive attractor does exist it is not static; the attractor and its corresponding fractal dimension varies with sleep level. One of the strongest arguments for this latter perspective was made using an explicit set of coupled nonlinear differential equations, with interconnections that are specified by the anatomy of the olfactory bulb, the anterior nucleus and the prepyriform cortex of a rat [199]. The solution to these equations yields an attractor that is qualitatively very similar to that obtained from the observed EEG time series for an induced epileptic seizure in the rat's brain, using the attractor reconstruction method.

The brain's electrical signals originate from the interconnections of the neurons through collections of dendritic tendrils interleaving the brain mass. These collections of dendrites generate signals that are correlated in space and time near the surface of the brain, and their propagation from one region of the brain's surface to another can be followed in real time. The EEG signal is consequently the field generated partly by a superposition of spike trains from individual neurons and partly the result of such trains triggering additional neurons along their path in the neuronal network. According to some neurophysiologists, the statistics of neural spike trains are renewal [544], and they are remarkably non-Poisson [47]. The mathematics of such point processes are presented in the excellent monograph of Lowen and Teich [335].

The experimental evidence of *in vitro* observations coupled with the analysis of *in vivo* spiking patters indicate that single neurons generate non-Poisson time series [188, 255, 465, 466, 467, 485, 492, 512]. The implicit assumption made in much of the neurophysiology literature is that even if the spiking activity of a single neuron is not Poisson, the activity of a network of many neurons is Poisson. However, the implied Poisson nature of the human brain has been proven to be invalid [133]; a conclusion reached by others [94, 96], as well. Consequently, we anticipate that brain activity can be interpreted as a non-stationary, non-Poisson, renewal process and implement a number of techniques entailed by this interpretation.

Another interesting study, concerning the human brain, indirectly confirms the existence of $1/f$-properties [222]. These authors investigate patterns of collective phase synchronization in brain activity in awake, resting humans with eyes closed. They find that the alpha range of the human EEG activity is characterized by changing patterns and that these fluctuations generate renewal events. The time interval between consecutive EEG events is described by a waiting-time IPL PDF, with IPL index $\mu = 1.61$. We have seen that $1/f$-noise, yields Eqs. (1.1) and (2.7). Thus, the work of Gong *et al.* [222] generates $\alpha = 1.39$, well within the widely accepted range for $1/f$-noise. We consider these experimental observation as additional evidence that multiple activities of the brain generate $1/f$-noise.

In order to understand the noise property in typical auditory evoked potentials, Peng *et al.* [424] analyzed spontaneous EEG data obtained from auditory brainstem response studies from which 20 randomly selected EEG epochs were used. The spectra determined by these epochs were averaged and are depicted in Fig. 2.1 to represent the typical frequency characteristics of stimulus-free EEGs. The IPL index in this case is determined to be $\alpha \approx 1$ or $\mu \approx 2$, the boundary between ergodic and non-ergodic signals.

Measures of Dynamics: The brain is perhaps the most interesting example of a complex dynamic network. Thus, it is a good place to begin our discussion of experimental research that registers, either directly or indirectly, the brain's dynamic activity. The results of recent psychological experiments [284] have been used to argue that $1/f$-scaling is a consequence of the coordination within the cognitive function. This interpretation is in keeping with the main conclusions of a review [559], that being, that cognitive tasks are the source of $1/f$-noise in EEG signals.

The emergence of $1/f$-noise can be detected by converting the underlying time series, such as the EEG, into a diffusion process. This conversion of time series data enables the determination of the corresponding Hurst

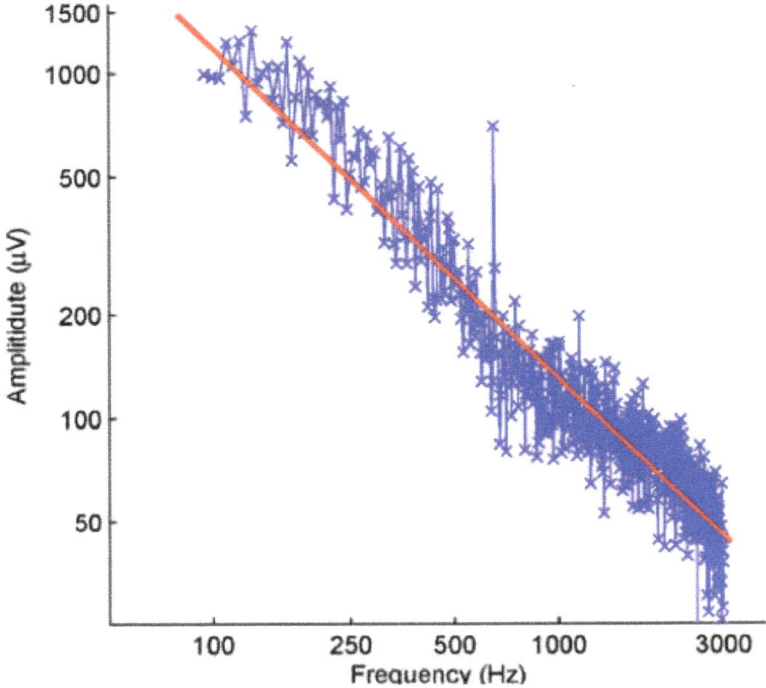

Figure 2.1: Stimulus-free EEG spectrum averaged over 20 sweeps for auditory brainstem response within 100–3,000 Hz. The straight line is the least-squares fit to these data: $2H - 1 = 0.97$, $R^2 = 0.90$ and $p < 0.001$. Adapted from [424] with permission.

coefficient H. One method for converting a time series into a diffusion process is detrended fluctuation analysis (DFA) [421]. Thus, since the autocorrelation function scales as t^{2H-2}, it is possible to obtain for the scaling index of the spectrum in terms of the Hurst exponent

$$\alpha = 2H - 1, \qquad (2.6)$$

or in terms of the IPL index of the waiting-time PDF:

$$\alpha = 3 - \mu. \qquad (2.7)$$

According to the convention adopted in the literature that a phenomenon is $1/f$-noise when the empirical spectral index is in the interval $0.50 < \alpha < 1.50$, we obtain that the Hurst exponent in the interval $0.75 < H < 1.25$ is equally a signature of $1/f$-noise. Note that either index exceeding the value 1 is a sign of aging, as pointed out by Kalashyan *et al.* [280] to explain EEG signals. We return to this concept of aging subsequently.

The work of Buiatti *et al.* [125] affords additional evidence that the brain is a source of $1/f$-noise. Notice that some interest is also devoted to the condition $H < 0.50$, which generates an IPL spectrum such as given by Eq. (1.1) in the low-frequency limit [258] that actually increases with frequency since $\alpha = 2H - 1 < 0$. Some authors [542] study how space, time, and physical constraints affect planning and control process when aiming at targets. They find that the aiming tasks generate values of H leading to $1/f$-noise. A recent analysis [150] on reaction times to stimuli indicates that the more challenging the task the less intense the $1/f$-noise produced. This may not conflict with the conviction that cognition generates $1/f$-noise, if we take into account that $\alpha > 1$ implies $\mu < 2$ and consequently, the rate of event generation vanishes asymptotically in time: the more extended the observation time, the less intense the $1/f$-noise becomes. This conjecture, of course, rests on the assumption that a challenge of increasing difficulty increases the time necessary to complete the task.

2.1.2 Body movements

Experiments that stimulate regular tapping, or the regular oscillation of a joystick by means of a metronome, have provided significant insight into the control of body movements by Deligniéres and collaborators [165, 166]. After switching off the metronome, the fluctuating departures from the induced periodicity are monitored, and it is found that the spectrum has the IPL property given by Eq. (1.1) with $\alpha = 0.97$ and 1.19, with and without the metronome, respectively. Other scientists [278] studied the human movements in synchronization with external events and found values fluctuating in the vicinity of $\alpha = 1$, with measurement errors compatible with the emergence of perfect $1/f$-noise.

Note that some authors analyze the movements produced by physical activity [413], using the Hurst coefficient H determined by DFA [421] and the relation between the indices given by Eq. (2.6). The typical values found [413] are $\alpha = 0.70$ and 0.60 for healthy subjects, as well as, 0.50 and 0.47 for patients with chronic pain. Note that the pathological patients are on the lower boundary for $1/f$-noise. In this field, as well, the convention has been adopted by researchers that $1/f$-noise represents a healthy physiological condition.

An intriguing role played by $1/f$-noise is in the relief of chronic, intractable pain for which transcutaneous electrical nerve stimulation has been found to be beneficial [390]. This treatment was used clinically as early as 1855 and has a history dating back to 46 AD. In experiments, the use of random stimuli, having various spectra in such treatments, were compared. The electrical signal was applied in such a way as to intercept and disrupt pain messages to the brain. The spectrum of frequencies in the intercepting signal determined the success of the disruption. In the majority of cases, a $1/f$-spectrum for the applied signal was preferred and reduced pain levels caused by trauma, cancer, lumbago and back pain for significant periods after the stimulation were stopped.

The most familiar body movement is, of course, regular walking. However, the regular gate of everyday experience is not very regular. The variability in stride interval, the time interval between successive heel strikes, was first recognized in the 19th century [549], but thought to be inconsequential. That is probably the reason why follow-up experiments, to quantify the degree of irregularity in walking [241], took nearly 120 years. The inter-stride interval time series was determined to manifest scaling behavior, using a number of different methods of data analysis, to obtain consistent results. The average fractal dimension for normal healthy adults is determined to be approximately 1.25 [242, 585], yielding $\alpha = 0.50$.

Scafetta *et al.* [476] point out that walking is the result of the two-way interaction between the neural networks in the central nervous system plus the intraspinal nervous system on one side and the mechanical periphery consisting of bones and muscles on the other. The muscles receive commands from the nervous system and send back sensory information that modifies the activity of the central neurons. The coupling of these two networks produces a complex stride interval time series that is characterized by fractal and multifractal properties, which depend on several biological and stress constraints, such as walking faster or slower than normal. The IPL power spectral density for the time fluctuations in stride intervals determined by Deligniéres and Torre [167], as you might expect, does not look very different from the EEG power spectral density depicted in Fig. 2.1.

The fractal and multifractal natures of the stride interval fluctuations become slightly more pronounced under faster or slower paced frequencies relative to the normal paced frequency of a subject, as depicted in Fig. 2.2. The randomness increases as subjects are asked to synchronize their gait with the frequency of a metronome, or if the subjects are elderly, or suffering from neurodegenerative disease, such as Parkinson's disease (PD). The supercentral pattern generator (SCPG) model of West *et al.* [581] was able to reproduce these known properties of walking, as well as, to provide physiological and psychological interpretations of the model parameters.

Moreover, people basically use the same control network when they are standing still, maintaining balance, as when they are walking. This leads one to suspect that the body's slight movements around the center of mass

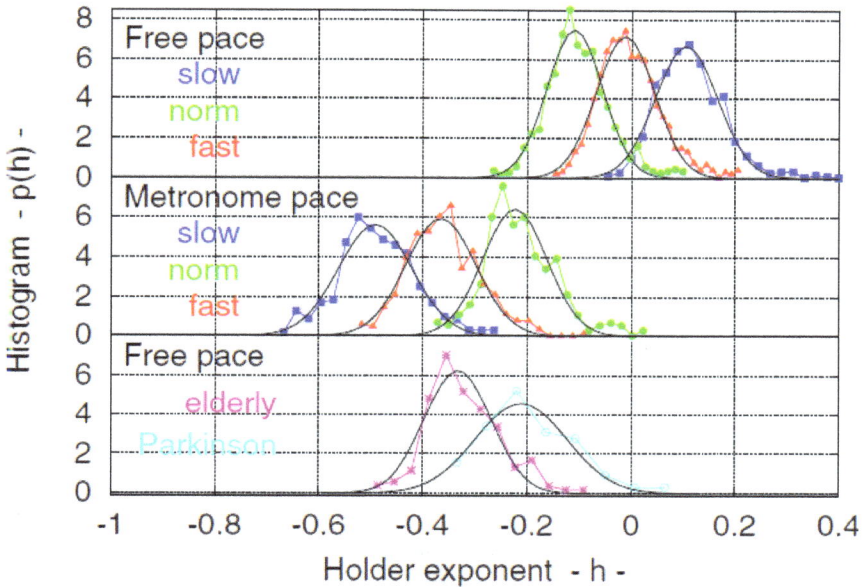

Figure 2.2: Typical Hölder exponent $h = H - 1$ histograms for the stride interval series in the freely walking and metronome conditions for normal, slow and fast paces, for elderly, and for a subject with PD. The histograms are fitted with Gaussian functions. From [476] with permission.

of the body, while standing still, postural sway, has the same statistical behavior as that observed during walking and this is found to be the case [143].

Another aspect of walking is the phenomenon of synchronization, or coordination, of two individuals walking side-by-side, which is a commonly observed phenomenon on any sidewalk, in any city, in any country, during any period. Almurad *et al.* [33] investigate this effect using a number of alternative frameworks and provide statistical tests to distinguish among alternative hypotheses to account for the coordination. As explained in the discussion of information transfer between complex networks, the maximum information is transferred when interacting networks share similar complexities. It is also the case that interacting networks tune their complexities to increase their level of coordination. They applied a variety of statistical analyses to time series collected from individuals undergoing such synchronized walking and concluded that the data is consistent with complexity matching and which we suggest is also consistent with the POIE.

In the sequel to this study these investigators [34] conducted the first experiment that determined a potential therapeutic benefit from complexity matching. The complexity of locomotion systems is reflected in the $1/f$-variability in the stride intervals of normal gait in young healthy individuals and is an indication of both stability and adaptability. By way of contrast, walking tends to be more disordered in older people, and this loss of regularity is correlated with a greater incidence of falling. Almurad *et al.* [34] experimentally determined that if an elderly person walks in close synchrony with a young companion, the complexity matching effect restores once lost complexity to the gait of the elderly. Synchronization within the dyads is dominated by the CME and a restoration of complexity is observed in the elderly after 3 weeks, and this effect persisted for at least 2 weeks after the training session ended.

Motor control and multifractality: It has been noted that the CME [586] entails the transport of global properties from one complex network to another, relying on the crucial role of criticality and ergodicity breaking. This agreement suggests that a connection exists between multifractality and ergodicity breaking, in spite of the fact that current approaches to detecting multifractality are based on the ergodicity assumption [275], as we subsequently show. Note that multifractality is defined as a property of a time series having a spectrum of Hurst exponents (fractal dimensions), which is to say the scaling index of the statistical fluctuations of the time series changes over time [184], resulting in no single fractal dimension, or scaling parameter, characterizing the physiological process.

In psychophysical experiments, for example, a person is asked to synchronize a responding tapping finger to a chaotic auditory stimulus, and a generalization of the complexity matching effect is interpreted as the transfer of scaling from the fractal statistical behavior of the stimulus, to the fractal statistical response in the brain of the subject. Note that this extends the complexity matching effect observed in dyadic walking, since it involves information transfer between very different complex networks. The brain response, in such a motor control task, when the stimulus is a multifractal metronome, has been established [510]. Additional experimental results [168] confirm this interpretation, based on the transfer of global properties from one complex network to another. In these latter experiments the multifractal metronome generates a spectrum of fractal dimensions $F(\alpha)$ as a function of the average singularity strength of the excitatory signal and it is this dimensional spectrum that is captured by the brain response simulation, as depicted in Fig. 2.3. The multifractal behavior manifest in the uni-modal distribution provides a unique measure of complexity of the underlying network. It is worth noting that the same

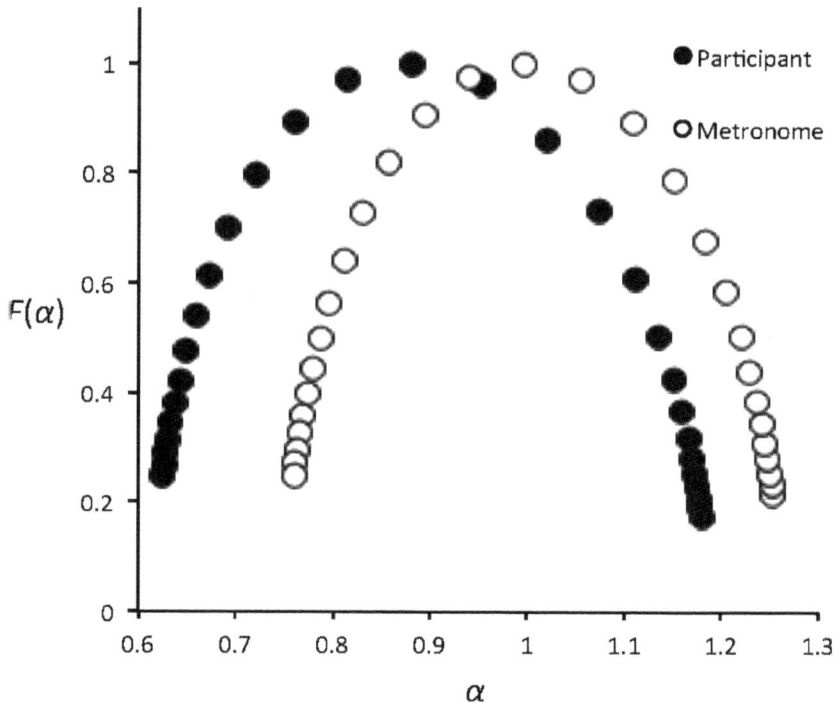

Figure 2.3: Walking in synchrony with a multifractal metronome. Multifractal spectra for the participant (black circles) and the metronome (white circle). This figure is derived with permission from the left panel of Fig. 3 of [168].

displacement of the metronome spectrum, from the body response spectrum, is observed for walking in response to a multifractal metronome.

Mahmoodi *et al.* [345] show that a complex network at criticality generates a distinct multifractal spectrum and describe how to detect the renewal events hidden within the dynamics of a multifractal metronome. In addition they establish the equivalence between the multifractal metronome and a complex network at criticality. This analysis is extended to the perturbation of one complex network by another, numerically establishing that a network with a broad multifractal spectrum perturbing one with a narrow multifractal spectrum distinctly broadens the spectrum of the responding network. This result suggests that, as done by Deligniéres *et al.* [168], it may be convenient to measure the CME by observing the correlation between the multifractal spectrum of the perturbed network S and the multifractal spectrum of the perturbing network P, rather than the cross-correlation between the crucial events of S and the crucial events of P. This method, although more closely related to the occurrence of crucial events is made difficult by ergodicity breaking [427], even when the crucial events are visible, not to mention the fact that usually crucial events are hidden in a cloud of non-crucial events typically having Poisson statistics [14].

Mahmoodi *et al.* [346] use another technical tool, that being subordination theory, which may be understood as the difference between the objective time measured by a clock and the subjective time experienced by the brain, to establish that when two networks of differing complexity levels interact, the multifractality, see Fig. 2.4, of the more complex network attracts the less complex, thereby inducing an increase of complexity in the latter, see also [343]. Almurad *et al.* [33] comment that the latter theoretical framework inspired the experiment that tested the restoration of complexity in the walking of elderly people.

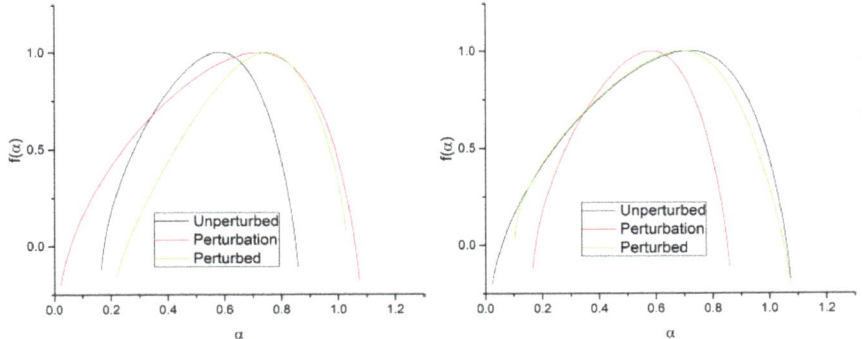

Figure 2.4: Effects of perturbation between multifractal spectra. Left: Complex network with at IPL index $\mu = 2.4$ (red line) perturbing a complex network with an IPL index $\mu = 3.4$ (black line) producing the green line. Right: System with $\mu = 3.4$ (red line) perturbing a system with $\mu = 2.4$ (black line) producing the green line. The multifractal spectrum of the less complex perturbed network (narrower) is strongly affected by the more complex perturbing network (broader) in the first case, in contrast to the second case.

2.1.3   Heart rate variability (HRV)

The time interval between consecutive heart beats (the time interval between consecutive R peaks in an electrocardiogram) is not constant and its fluctuations generate $1/f$-noise [282, 585, 596]. Research [249] confirms the $1/f$-noise nature of the RR interval PDF. In fact, it has been known for some time that 'normal sinus rhythm' is not strictly periodic, but is characterized by a broadband, IPL spectrum [218, 292]. The heart rate is modulated by a complex control network consisting of respiratory, sympathetic and parasympathetic regulators. Akselrod *et al.* [9] show that individual suppression of these regulators alters the interbeat interval power spectrum in healthy individuals dramatically, but a broadband spectrum persists. The importance of heartbeat (heart rate) variability was finally established by a task force formed by the *North American Society of Pacing and Electrophysiology* and the *Board of the European Society of Cardiology* in their published report in 1996.

There are at least 16 different methods that have been developed to assess HRV, a few of which will be touched on in this essay, but most of which do not concern us here. The number of techniques is mentioned only to indicate that the scaling nature of the beat-to-beat interval time series has an extensive literature and has been established from a number of different analytic perspectives. The scaling nature of the RR interval PDF was first recognized by Peng *et al.* [420] along with the fact that the increments in the human heartbeat time series were thought to possess the limit distribution of Lévy and not that of Gauss. Using the DFA on the differences between successive time intervals, they found that $H \approx 0$ for healthy individuals, whereas for those with severe heart failure have $H \approx 0.50$; but the statistical distribution for the two groups is the same. They concluded that the difference in the observed scaling is a consequence of the ordering of the heartbeat interval differences and not their statistics, that is, the pathology appears in the correlation of the time intervals in the underlying time series.

The analysis of the variability of RR-intervals for healthy individuals and those with heart disease were shown to have the same PDF, that being Lévy stable [420]. This non-Gaussian behavior of the HRV intermittent time series has, a quarter of century later, been shown to be more subtle than originally believed. The scale invariant fluctuations in the healthy human heartbeats, as well as, those for certain pathologies were reviewed by West [598]. In Fig. 2.5 the curves denote the HRV statistical PDFs from a study [240] of a collection of 670 post-AMI (acute myocardial infarction) patients using 24-hour Holter monitor datasets. In this study, a number of individuals suffered a cardiac death, others died by non-cardiac causes and

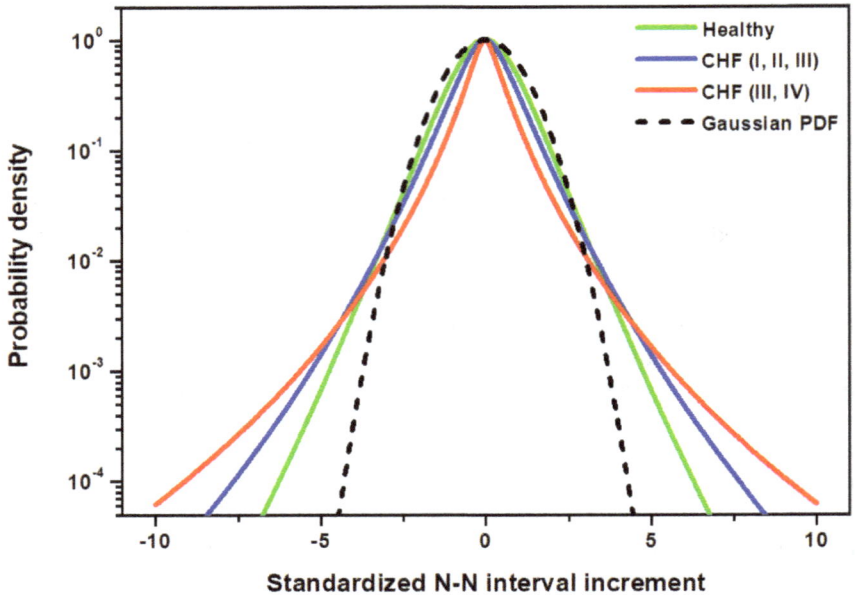

Figure 2.5: The HRV PDFs are indicated schematically from 24 h RR interval time series for a group having suffered a trial myocardial infarction. The patients are separated into those that suffer cardiac death (red) from congestive heart failure (CHF), another with non-cardiac death (blue) and a third consisting of survivors (green). Adapted from [604] with permission.

some survived. The PDFs determined by histograms constructed from the time series for the three groups are indicated schematically.

Notice that even though the PDFs on death and survival in Fig. 2.5 are only sketched, it is clear that no group has Gauss statistics. The latter would coincide with the dashed curve, which is purely theoretical. The survivors (green curve) and those succumbing to non-cardiac death (blue curve) have essentially the same variability as estimated by eye from the widths of their respective PDFs. On the other hand, the PDF for the cardiac death patients (red curve) is very different from those that survive, the difference between the PDFs for those that survive and those that do not (cardiac death) had been modeled using a truncated Lévy PDF [240, 598], and more recently a theoretical justification of this truncation based on a fractional control processes has been hypothesized [604]. In Sec. 7.2 we review a method based on the detection of crucial events that enables us to discriminate between individuals with healthy HRV time series and those with pathological HRV time series.

2.2 Variance in the Arts

2.2.1 Music and noise

The mathematician, George Birkhoff, argued that the aesthetics of art have a quantitative measure. His book on aesthetics [100] discussed the structure of various art forms, particularly music, and was largely overlooked by other scientists until near the end of the 20th century when Mandelbrot introduced the scientific community to fractals [354]. Mandelbrot's colleague Voss applied the fractal concept to the mathematical analysis of music. Voss and Clarke [553, 554] used stochastic, or $1/f$-music, in which notes are selected at random and the frequency use of a particular note is determined by a prescribed distribution function, to gain insight into the structure of more conventional music.

There is experimental evidence that one complex network, the brain, a $1/f$-network, is sensitive to the stimulation exerted by another $1/f$-network, as in a number of well-designed experimental tasks. The hypothesis that the matching of the exponents in separate $1/f$-networks facilitates the transmission of information between them, provides scientific support for the observation that music can, through the POIE, entrain the brain. Given that the brain is a $1/f$-noise network [72, 587], as mentioned in the previous section will be subsequently discussed, it is expected to be especially sensitive to $1/f$-noise. The fascination exerted on the brain by music has been explained as its being due to the fact that music is a source of $1/f$-noise [553] as depicted in Fig. 2.6. There is general agreement on the $1/f$-nature of music [260], although other research work [110], resting on a dynamical approach to music, suggests that the IPL index α may be significantly larger than 1.

West and Deering [577] explain that music with $1/f$-spectra falls midway between that of uncorrelated white music, with its flat spectrum, and brown music, with its steep $1/f^{-2}$-spectrum. $1/f$-music, with its blend of regularity and spontaneous change was selected to be more aesthetically pleasing by a substantial majority, when presented to a number of people, along with white and brown music [553]. This is more than a curious result regarding stochastic music, but carries over into the analysis of classical music and jazz, both of which reveal an underlying $1/f$-spectrum of variations.

It is worth mentioning that the fractal dimension ($D = 2 - H$) of music can be evaluated with such great precision as to make this a reliable way to classify music [98]. Music is a familiar example of an external stimulus that can generate emotion [238]:

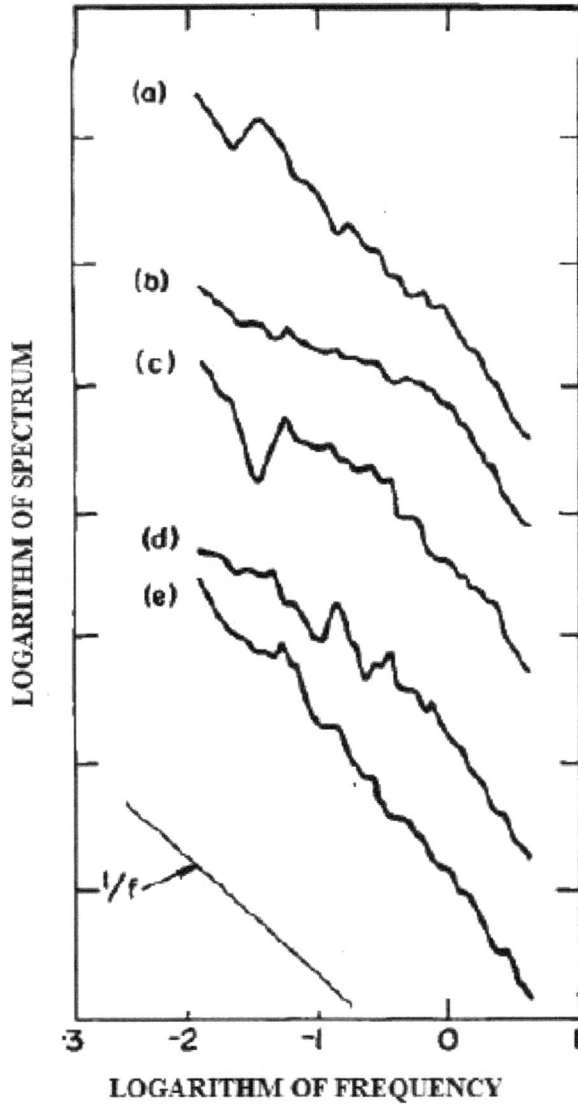

Figure 2.6: Frequency fluctuations spectral densities: (a) Davidovsky's Synchronism I, II and III; (b) Bach Spring Quartet number 3; (c) Jolas' Quartet number 2; (d) Carter's Piano concerto in two movements; (e) Stoonhauser's Moments. Adapted from [555] with permission.

In every culture music is used to communicate, entertain, express and elicit emotion. Yet underlying questions remain. Why
do people respond to music? Why are we overcome with melan-
choly when hearing one tune and overjoyed when hearing another? Why do some melodies cause chills up and down one's
spine and others evoke calm, peace and well-being? How can
combinations of sounds produce such diverse emotional and
physical reactions?

It has been argued that music triggers a hierarchical cooperation [238],
between the two hemispheres of the brain. This cooperation is analogous
to the phase transition phenomena realized by the synchronization of many
elements in a physical system. It is the condition behind the picture advocated by Gong *et al.* [222], which we subsequently discuss in some detail.

On the basis of the correspondence between music and the $1/f$-nature of
many physiological processes [585], one may be tempted to interpret music
as an expression of an evolutionary homeodynamic feedback-mechanism
more than as a social function [560]. However, we notice that the principle
of *emotion transfer* is not confined to the influence of music on the brain.
Rather, as we show subsequently, the same $1/f$-noise principle leads to
social networks, with the self-organizing elements as human beings, whereas
the brain's EEG, has the neurons as the self-organizing elements.

2.2.2   Visual arts

Human beings do not only make judgments concerning the variability and
regularity within music. We similarly discriminate in examining photo-
graphs and paintings. A study of the distribution of the breakup of space
in a number of great works of art found $1/f$-phenomena in the paintings by
Dürer, Munch, Rembrandt and Picasso [403]. More recently, the method of
scaling analysis has been applied to the paintings of Pollock [35, 526, 527].
Analysis establishes a connection with the $1/f$-noise issue and also suggests
how the visual inspection of a painting can be converted into an *equivalent*
time-dependent stimulus, of a nature comparable to musical stimuli. In
fact, these authors [35] imagine an observer looking at a painting from
left to right at a fixed level. This corresponds to exerting on the eye a
time-dependent stimulus that is an effective one-dimensional stochastic trajectory. They evaluate the fractal dimension D of this stochastic trajectory,
thereby yielding information through the well-known formula $D = 2 - H$.
Using the earlier relation for the IPL spectral index $\alpha = 2H - 1$, we obtain
$\alpha = 3 - 2D$. In this way, Alvarez-Ramirez *et al.* [35] find typical values of
the fractal dimension $D = 1.10$ and $D = 0.90$, corresponding to $\alpha = 0.80$

and $\alpha = 1.20$, respectively, suggesting an excursion of $1/f$-noise into the non-ergodic regime. Forsythe *et al.* [194] have argued that visual complexity has been known to be one of the more significant predictors of how people discriminate between works of art. They conclude that the fractal dimension provides a quantitative measure of perceived beauty.

Here we see two contrasting ways $1/f$-phenomena enter our lives. First, through the aesthetic judgments we make regarding music and visual art; second, through the perception of pain in our bodies [577]:

> Apparently, the same $1/f$-signal that can soothe the savage beast, can in another way also relieve the beast of his pain. Perhaps in a way we do not yet understand, a pain-free life and a joyfully aesthetic one may be related in some biologically fundamental way.

2.2.3 Architecture

The architect as scientist/engineer investigates one kind of hierarchical ordering in design and that is in the distribution of subunits according to their size. This consideration in the design is distinct from the more obvious geometrical ordering of forms through symmetry — as, for example, arranging elements along a line or in a reflected pattern — yet it is perhaps even more important for the eventual coherence of the structure. This ordering, although necessary to maintain the integrity of the structure, is also required to satisfy the aesthetic perception of fellow human beings.

As Salingaros and West [468] observed, architectural scales are defined by the same elements repeating within a structure across a multitude of sizes. Independent scales arise from the structure, functions, and even material used in construction, which distinguish specific scales that are necessary to the structure's character. For example, a building might have several obvious levels of scale X_j. Each level j is denoted by the size X_j of similar, repeating elements. A structure begins with an overall scale and each subsequently smaller scale is generated by a connection to the largest scale, say a reduction in a characteristic diameter of a structural subunit.

Scales play a major, even if on a subconscious level, role in design, because they stimulate human cognition. The mind of the observer collects and orders similar-sized objects into a single level of scale. This process, not unlike that of digital image compression, reduces the amount of information presented to the observer by a complex structure. Humans apparently also unconsciously estimate the number of similar-sized objects on each scale, that is, their relative multiplicity. We do this to compare these numbers to what we have mapped in our brain regarding the complexity of naturally occurring structures, such as trees, rock formations and even mountains. If the distribution of scales and the relative multiplicity of elements correspond to some experientially generated internal standard, we perceive the structure as coherent.

A structure's immediate psychological impact on a person depends to a large extent on the distribution of its subunits. This first impression is essentially independent of other features such as shape, form, and proportion. It is a matter of answering the old question of what makes a complex structure interesting to human beings. The answer, at least in part, is that it has to be a balance between two extremes: too regular is monotonous, which is boring; too incoherent, which is disruptive. Its features must be sufficiently divergent that the eye never rests too long at one location, but sufficiently convergent that the eye pauses at each feature, never wanting to overlook anything. Artifacts and buildings are as important for the pleasure they give as for any purely utilitarian function they serve. Just as in music, we enjoy a building or city because it offers a mixture of regularity, or comfort, and surprise in a certain ratio [575]. Comfort and surprise are complementary qualities that depend on how subdivisions are distributed.

2.3 Social Diversity

2.3.1 Income

The IPL PDF was first observed in the systematic study of data on income in Western societies as analyzed by the engineer/economist/sociologist Vilfredo Pareto (1848–1923) at the end of the 19th century. Subsequently, beginning a century later with [59], scientists recognized that phenomena described by such IPLs do not possess a characteristic scale and referred to them collectively as scale-free, in keeping with the history of such distributions in social phenomena, see e.g. [580].

Pareto was an economics professor at the university in Lausanne, Switzerland. His analysis of income/wealth data collected from various countries led to his being the first to recognize that the distribution of income/wealth in a society was not in fact random, but followed a consistent pattern. An exemplar of such an income distribution, drawn from the United States at the beginning of the 20th century is depicted in Fig. 2.7. This pattern was and is described by an IPL PDF, which now bears Pareto's name, and which he called: The *Law of the Unequal Distribution of Results* [414]. He referred to the inequality in his distribution more generally as a "predictable imbalance" which he was able to find in a variety of other complex social phenomena. This imbalance is ultimately interpretable as an implicit unfairness found in complex networks. The income imbalance was somewhat

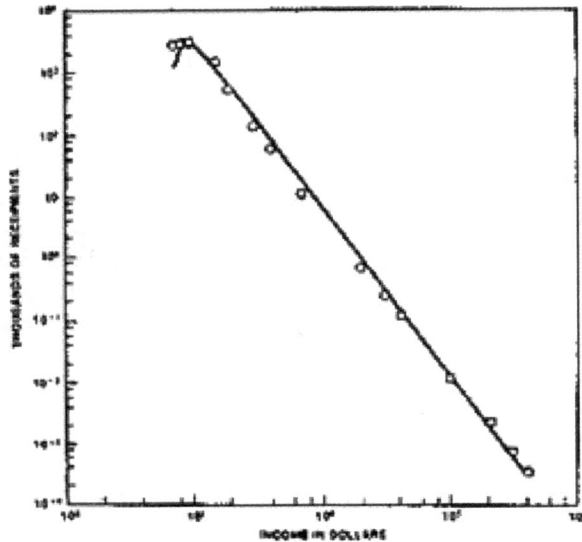

Figure 2.7: The distribution of income in the United States circa 1920 plotted on log–log graph paper. The point at which the distribution turns over is when the 'wolf is at the door' and survival is a concern. Taken from [381].

mitigated by the growth of the middle class in the 20th century, which resulted in a lognormal distribution of income for approximately 97% of the population and an IPL for the top few percent that persists in the Western world today.

So how do we go from random networks with their average values and standard deviations with its implicit fairness, to complex networks that are scale-free and explicitly unfair? [601] More importantly, how does this relate to the assessment of the research quality of laboratories, programs and individual scientists and engineers?

An in-depth understanding of IPL PDFs in the context of complex social networks began with small-world theory; a theory of social interactions in which social ties can be separated into two primary kinds: strong and weak. Strong ties exist within a family and among the closest of friends, those that you call in case of emergency and contact to tell when you get a promotion. Weak ties connect the many colleagues at work with whom you chat, but never reveal anything of substance, friends of friends, business acquaintances and most of your teachers.

Clusters form among individuals having substantial interactions, developing closely knit groups; clusters in which everyone knows everyone else. These clusters are held together by strong ties, but then the clusters are coupled to one another through weak social contacts. The weak ties provide contact from within a cluster to the outside world. It is the weak ties that are all-important for interacting with the world at large, say for getting a new job. A now classic paper by Granovetter, "The strength of weak ties." [225], explains how it is that the weak ties to near strangers are much more important in getting a new job than are the stronger ties to one's family and friends. In this 'small-world' these are the shortcuts that enable the connections between one tightly clustered group to another tightly clustered group very far away. With relatively few of these long-range random connections, it is possible to link any two randomly chosen individuals with a relatively short path. This short path has become known as the six-degrees of separation phenomenon [59, 514, 564]. Consequently, there are two basic elements necessary for the small-world model, clustering and random long-range connections.

Recent research into the study of how networks are formed and how they grow over time reveals that even the smallest preference introduced into the selection process has remarkable effects. It has been assumed by a number of investigators that one mechanism is sufficient to obtain the IPL distributions that are observed in the majority of complex phenomena. That mechanism is contained in the principle that: *the rich get richer...* [294]. In a computer network context, this principle implies that the individual with the greater number of connections attracts new links more strongly than do individuals with fewer connections, thereby providing a mechanism by which a network grows as new individuals join.

The IPL nature of complex networks affords a single conceptual picture spanning scales from those in the WWW to those within a single organization. As more people are added to an organization, the number of connections between existing members depends on how many links already exist. In this way the status of the oldest members, those that have had the most time to establish links, grows preferentially. This mechanism has been called *preferential attachment* leading to an IPL index of 3 [59] for the Barabási–Albert model. However, the model had been developed by Yule [624] over a half century earlier in the context of speciation in a somewhat more general form and was put in a citation index (CI) context by de Sola Price [439], who called the CI mechanism *cumulative advantage* a quarter century before the tag preferential attachment captured the imagination of the scientific community.

Thus, some members of an organization have substantially more connections than do the average, many more than predicted by a typical bell-shaped distribution. These are the individuals out in the tail of the IPL distribution, the gregarious individuals that seem to know everyone and to be involved in whatever is going on. In a research context these are the individuals that members of the laboratory seek for scientific dialogue and collaboration, but with IPL indices having a variety of values [600].

2.3.2   Cybernetics

The POIE emphasizes the condition necessary to most efficiently transport information between complex networks and the mechanism can be traced back to the 1957 *Introduction to Cybernetics* by Ross Ashby. Unlike this earlier work we argue, both here and elsewhere, that complexity can be expressed in terms of crucial events, which are generated by the process of spontaneous self-organized temporal criticality (SOTC) [342], which we explain in detail in subsequent chapters. Complex phenomena, ranging in disciplines from anthropology to zoology, and all those in between, must satisfy the homeodynamic condition and host crucial events that have been shown to drive information transport and information exchange between complex systems.

It has been over 60 years since Ashby alerted the scientific community to the difficulty of regulating biological systems, and that "the main cause of difficulty is the variety in the disturbances that must be regulated against". This insightful observation led some scientists to the conclusion that complex systems cannot be regulated. But others reasoned that it is possible to regulate them as long as the regulators share the same degree of complexity (variability) as the systems being regulated; the *requisite variety*. Herein we replace Ashby's term requisite variety with complexity matching, or information matching. The CME has been widely identified [1, 33, 141, 191, 360], since its introduction [586].

It is important to stress that there exists further research directed toward the foundation of social learning [164, 173, 202, 627] that is even more closely connected to the ambitious challenge made by Ashby. In fact, the latter research aims at evaluating the transfer of information from the brain of one player to that of another, by way of the interaction the two players established through their avatars [202]. The results are exciting in that the trajectories of the two players turn out to be significantly synchronized. But even more important than synchronization is the fact that the trajectories of the two avatars have a universal structure based on the shared EEGs of the human brain. Mahmoodi *et al.* [346] provide a theoretical explanation of the universal structure representing the brains of the two interacting individuals. In addition, the theory can be adapted to the communication, or information transfer, between the heart and the brain [432] of a single individual.

2.3.3 Linguistics

A surprising connection exists between music and sociology, established by a well-known linguistic principle, Zipf's law [352]. This law establishes that the relative frequency of occurrence of a written word is proportional to $1/r$, where r is the rank of that word in the given language. The connection with $1/f$-noise is obtained by identifying the rank r with the frequency f. Manaris *et al.* [352] found that the $1/f$-noise properties of music can be interpreted as a manifestation of Zipf's law in music, for instance, that the chromatic-tone distance, namely, the time interval between consecutive repetitions of chromatic tones, obeys Zipf's law. Simultaneously, Zipf's law is closely connected to Pareto's law [484], which establishes that the probability of a randomly chosen individual from within a society having an income larger than a specified income level X, no matter how large, is proportional to $1/X^{\beta}$. In this way, Zipf's law and Pareto's law have been shown to be equivalent [5, 484] representations of the same information, although in terms of the distribution of different quantities.

If we denote by α the IPL index of Zipf's law, it is straightforward to prove that $\alpha = 1/\beta$. It is remarkable that the PDF $\psi(\tau)$, where τ now denotes the chromatic tone interval, with the IPL index:

$$\mu = 1 + \beta = 1 + \frac{1}{\alpha}. \qquad (2.8)$$

We point out that α must be very close to 1, so as to fit the condition of realizing the $1/f$-condition [352].

The equivalence between linguistic laws and sociological laws suggests that a process of interaction exists for the transfer of information, simultaneously reflecting the synergy of the brain's neurons and the cooperation among the individuals within society. We can state that language may, in much the same way as does music, be considered to satisfy both neurobiological and sociological constraints [267], if we accept the hypothesis of PIE, according to which the communication between complex dynamic networks rests on them sharing the same $1/f$-property.

2.3.4 Psychosociology

The cooperation among neurons within the human brain is a problem of general interest, but in addition the behavior of the brain has become a paradigm for many forms of cooperative behavior within complex networks including those of sociological interest [518]. Consider neuroeconomics [287], which is a relatively new discipline based on this paradigm. Decisions involving trade-offs between costs and benefits occurring at different times are important and ubiquitous in economics [200]. The influence of neuroeconomics on decision making rests on an interdisciplinary research activity, involving the intervention of behavioral psychologists [444], as well as, that of neurophysiologists [620] and health care providers. The psychological experiments on how individuals discount delayed rewards, which is to say individuals prefer smaller rewards immediately after a decision, rather than larger rewards delayed for some time, are used as a way to evaluate the influence of drugs and other addictions on the functioning of the brain [620].

Research work in neurophysiology [75, 320, 429] plays a central role in complexity science and is expected to shed light on the origins of complexity itself. For example, the relatively new concept of an neuronal avalanche (brain quake) is a set of elements showing simultaneous enhanced activity. In neurological experiments [75, 429] the elements are electrodes, monitoring the global activity of a local set of neurons, and in numerical work [211] the elements are integrate-and-fire model neurons. The avalanche size is the number of elements in the avalanche and is distributed according to an IPL PDF with IPL index $\mu = 1.50$ for [74, 75, 429] and in the interval $2 \leq \mu \leq 1.5$ for [320].

There seems to be general agreement on the IPL PDF of the avalanche size, which fits the theoretical prediction of SOC theory [51]. The adoption of the SOC perspective prompted the conclusion [158] that the time interval between consecutive avalanches is also distributed as an IPL. The IPL PDF of the time intervals, between consecutive events, is a property shared by some theoretical models for decision making processes [60, 597], thereby yielding the same time statistics as the results of psychological experiments [524], which have been shown to mirror brain activity [620].

However, the stable recurrence over time of an avalanche's spatiotemporal patterns remains an elusive issue requiring clarification [429]. This required clarification is of importance not only for neurophysiological processes, but for the foundation of non-equilibrium statistical physics, as well. The recent discovery of the statistical aging and non-ergodicity in the fluorescence of single nanocrystals [117], named blinking quantum dots (BQD), triggered the interest of many researchers. A number of them adopted the CTRW perspective [113, 373, 501], from which an important approach to the study of non-ergodic processes has emerged [450]. Bianco *et al.* [94] discovered quakes in the brain that seem to share the same non-ergodic properties as the BQD, thereby making it even harder to account for the stable time recurrence. Note that earlier numerical results indicate that the emergence of an IPL is entailed by strong cooperation, which generates values of μ significantly smaller than 2, and consequently very far from the $1/f$-noise condition. This is in keeping with the empirical fact that accomplishing a difficult task may produce a significant deviation from the $1/f$-noise condition.

2.3.5 Conversation turn-taking

Human conversation is a highly complex phenomenon. In order to facilitate smooth communication, speakers coordinate their behavior alternating between speaker and listener roles. The turn-taking behavior in a casual conversation follows a norm of *minimal-gap-overlap*, where speakers do not pause to begin their turn and tend to avoid gaps and overlaps in the conversation. In this way conversation is considered a form of cooperation. Normal conversation is generally accepted to be the speech exchange between two or more parties, where there is no externally imposed procedure on the flow of talk. The order of speaking, length of time for a given turn, or the content of what is said is managed locally by the participants, turn by turn, in terms of who speaks when and for how long. Speech is exchanged nearly continuously for speakers taking turns, with minimal gaps in the talk and with minimal overlap.

A fundamental part of the infrastructure for conversation turn-taking, or the apportionment of who is to speak next and when, is that in English conversations, speakers do not wait for pauses to begin their turn, but avoid gaps and overlaps. A significant body of research has modeled the time at which a person begins to speak in terms of their anticipating when the speaker will stop talking based on semantics, or through the identification of various cues [256].

Abney *et al.* [1] set up a study of dyadic conversations and found evidence for a new kind of coordination they identified with the CME of West *et al.* [587]. Without going into the details of the experimental protocol, suffice it to say that the investigators recorded the conversation between a number of carefully selected dyads of undergraduate students. The experiment consisted of ten-minute conversations that were independently prepared under the condition of either affirmative or argumentative topics. The events were the onset and offset of individual speaking events and the logarithmically binned data of the time interval between successive events is displayed in Fig. 2.8. They [1] remark:

> ..leads us to predict complexity matching in the temporal clustering of acoustic onset events, However, West *et al.* [587] showed that complex systems in general are expected to exhibit complexity matching when their inter-event intervals (IEIs) are IPL distributed with an exponent near two, $P(\text{IEI}) \sim 1/\text{IEI}^\gamma$, where $\gamma \sim 2$. West *et al.* analysis suggests that we test IEIs for the predicted IPL.

An Alan Factor Analysis was carried out on the logarithmically transformed event data to independently measure the scaling hypothesis [1] and

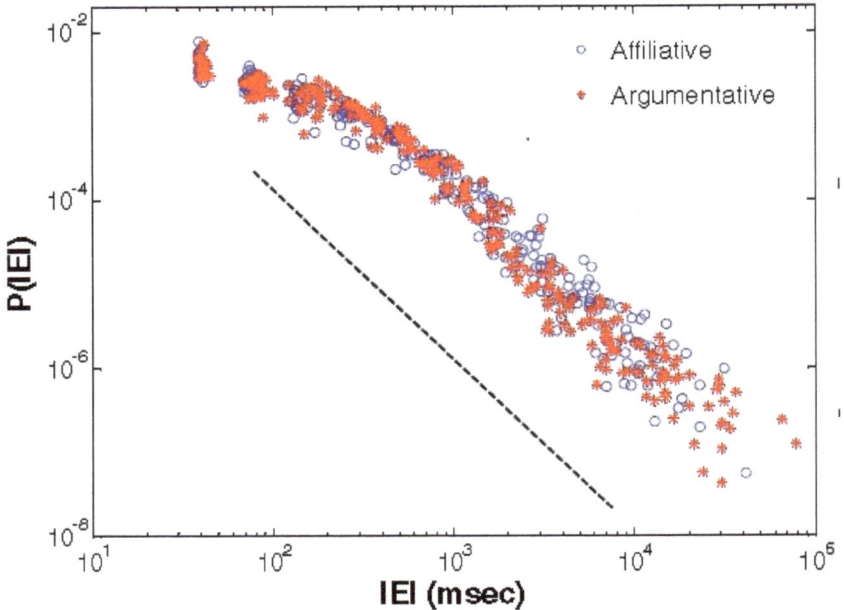

Figure 2.8: Interevent interval (IEI) PDF for individual interactions in dyadic conversations using logarithmic binning of events. Dashed line shows idealized slope of -2 as suggested by the theory in [587]. From [1] with permission.

the matching condition. A statistical measure of the scaling index was applied to the Allan Factor under two separate conditions and they found the level of support for the interpretation of two multilevel, complex systems are most responsive to each other when their power laws converge to be remarkably strong. This is the point at which the IPL indices are both at the center of the range of the cross-correlation cube depicted in Figs. 1.3 and 1.4, so that each time series of the onset of speaking satisfies a $1/f$-condition.

In the sequel to the above work, Abney *et al.* [2] point out the generalization to the complexity matching condition in which if the two systems do not match in scaling patterns, information transmission would not be maximum, but the reduced rate could be written as an explicit function of the degree of matching. In this way, information matching gives way to information management and is distinguished from a number of related phenomena [2]:

For infant-caregiver interactions [3] there is also evidence for matching of multiscale clustering in prelinguistic vocalizations produced by infants with those produced by their caregivers. Notably, there is more complexity matching for speech-related vocalizations (speech, non-word babble, and singing) relative to non-speech-like vocalizations (laughing, crying, burping and coughing), suggesting that the matching process is sensitive to different types of vocal behavior.

2.4 Science of Science

The phrase *science of science* was introduced into the scientific lexicon in 1935 [407]. The wife and husband team argued that the science of science would subsume such earlier disciplines as *epistemology*, the *philosophy of science*, the *psychology of science*, and the *sociology of science*. In the 21st century the nascent field of network science has developed some of the tools necessary for science to carry out this inward looking investigation using quantitative measures. However, we need to have at least a preliminary understanding of how human networks operate in order to determine how to evaluate the quality of the research carried out within laboratories, supported by research programs, or conducted by individual scientists or engineers. In order to achieve even this primitive level of understanding, let us sketch out how members of a community might choose from a large number of options.

Suppose a large cohort group has a hypothetical set of choices, say a large set of nodes on a computer network, to which they may connect in order to perform a function. If we assume that the choices are made independently of the quality of the nodes, or without regard for the selections made by other members of the group, the resulting distribution in the number of times a given node is selected has the familiar bell-shaped form. In this case the selection process is completely uniform with no distinction based on personal taste, peer pressure or aesthetic judgment, resulting in a network of random links between humans and nodes, or more generally between individuals by means of computers.

The bell-shaped distribution describes the probable number of links a given node has in a random network. Barabási and Albert (BA) [58] determined such distributions to be unrealistic. Using real-world data they showed that the number of connections between nodes on networks as diverse as the WWW or citation patterns in science deviate markedly from the bell-shaped distribution. In fact they, along with others [59, 514, 564], found that complex networks in general have IPLs, rather than Normal PDFs.

2.4.1   Publications

It is surprising that the tools developed within the various science disciplines, to quantify the subtle variations in natural variables, have not been systematically applied across those same disciplines to quantify the success of research organizations. Nor have they been applied to the many research programs within such organizations, nor to the government programs funding that research. Finally, such application has been made in fairly rudimentary form to the individual scientists contributing to these programs.

Some may protest that there are a variety of disciplines that have developed statistical measures of the efficacy of various measures of quality and that we will concede. But these efforts have been discipline specific and lacked the universality hoped for when the science of science was first introduced. A contemporary of sociologist Pareto was the biophysicist Lotka, who in 1926 [332] graphed the fraction of a cohort of investigators versus the number of papers published N, see Fig. 2.9. We will have a great deal more to say about the IPL form of the PDF and the general interpretation of the IPL index μ, but for the moment we list a number of other empirical IPLs that have been named after those that discovered them and which are discussed elsewhere in terms of complex networks, see [577] for a review:

Pareto's Law (1896): income
Auerbach's Law (1913): city sizes
Willis' Law (1922): speciation
Lotka's Law (1926): scientific publications
Murray's Law (1927): size of limbs
Rosen–Rammler Law (1933): fragmentation of ore
Zipf's Law (1948): frequency of words
Gutenberg–Richter Law (1954): size of earthquakes
Rall's Law (1959): neuron branching
Richardson's Law (1960): size of wars
Price's Law (1963): scientific citations
Goldberger–West Law (1987): size of bronchial airways

However, before we can make sense of these empirical laws, it is necessary to review what the traditional statistical measures tell us about such datasets. What relation do averages have with these empirical laws and where does the central limit theorem enter the picture, if at all? It turns out that knowing how these empirical laws deviate from simple statistical theory is very important, because we have developed our intuition about uncertainty in the world using the concepts underlying the simple theory. However, all too often, when these same concepts are applied to

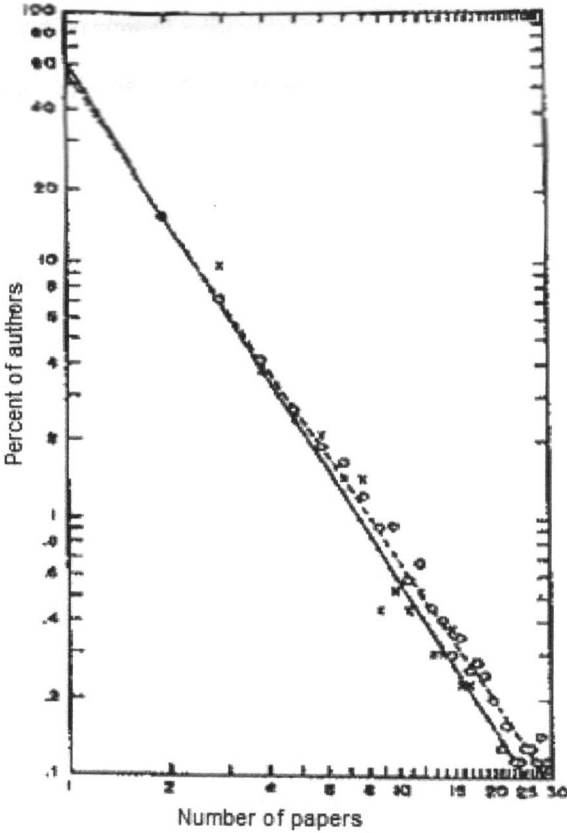

Figure 2.9: *Lotka's Law*: The number of scientists publishing exactly N papers, as a function of N. The open circles represent data taken from the first index volume of the abridged *Philosophical Transactions of the Royal Society of London* (17th and 18th centuries), the filled circles are those from the 1907–1916 decentenial index of *Chemical Abstracts*. The solid curve shows the exact inverse-square law of Lotka; the dashed line is the fit to data. All data are normalized to a basis of exactly 100 investigators publishing a single paper [332].

empirical data gleaned from complex phenomena we encou... tion and paradox, therefore, before we apply what we know a... contradic- to data relating to the quality of a research activity, we revie... statistics tinguishes complex from simple phenomena, and interpret the m... dis- a bell-shaped PDF in Chapter 5.

The connectivity of complex networks is IPL [59, 514, 597], physiol... ical time series generated by such complex networks, such as heart rate variability [259, 585], breathing variability [521, 584] and stride interval variability [549, 579] all have IPLs distributions and these among others are reviewed in [596]. There is a large literature on the interpretation of IPL PDFs that has developed over the past quarter century, much of which has been applied to uncovering the cooperative mechanisms present in complex adaptive networks. The reliance of the IPL PDF on the complexity of the network being examined is discussed and the relation of the complexity to the quality of the measures is examined in Chapter 5.

Phenomena described by IPL PDFs have a number of interesting properties, particularly when applied to science of science phenomena. In the case of publications, it is possible to say that most scientists publish fewer than the average number of papers. It is definitely not the case that half the scientists publish more than average and half publish less than the average, as they would in a Gauss distribution, but rather that the vast majority of scientists publish less than the average number of papers appearing in an interval of time. This is referred to as Pareto' Law or the 80/20 rule, which in the present context would say that 20% of the scientists publish 80% of the papers.

Consequently, if the number of papers were the metric of quality used by management to evaluate an individual, a laboratory, a university, or even a country, most such entities would rate far below average. However, this would be a distortion of the true quality of the individual, lab, university or country, and the research being done, since this application of the metric presupposes a linear additive world as the basis for comparison with other entities, or among members within a research group within those aggregate entities.

Another PDF of interest in this context is the time interval between successive scientific publications by a scientist. The PDF of publication intervals is depicted in Fig. 2.10 and it too is found to be IPL [274]. This PDF indicates that it is natural to have periods of high publication activity interspersed within intervals of low activity, the phenomenon of bursting. Therefore an individual's publication process is bursty and is reflected in the fact the interevent time interval is an IPL and the event is the publication of a research paper. As a matter of fact, Jia *et al.* [274] show that the IPL form of the waiting time PDF does not change when the data are shuffled,

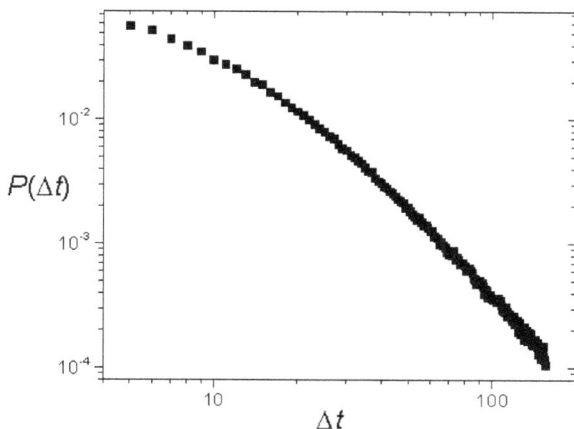

Figure 2.10: The time interval between a scientist's successive publications follows an IPL, documenting the burstiness in scientific publication. The time interval Δt is measured by the unit of month. Taken from [274].

indicating that the publication events are crucial. This is discussed in a network science context in Chapter 3.

The empirical laws for complex systems that we listed above are all IPL PDFs and the theoretical reasons for this are addressed in Chapter 5 and subsequent chapters. The IPL form may not be the case universally, but even when it is not true across the range of variation of the statistical variable, it does appear to be true asymptotically. We mentioned the central limit theorem (CLT) above. An assumption on which this mathematical argument (theorem) is based is that the second moment of the dataset being analyzed is finite, otherwise the CLT breaks down. However, in the 1920s the mathematician P. Lévy showed the existence of a generalized CLT, valid for time series having diverging second moments. A qualitative comparison of a Normal and a Lévy PDF is given in Fig. 2.11, and the asymptotic form of this new limit PDF is IPL in the tails of the Lévy distribution. It is in these tails that the Normal and Lévy distributions are the most divergent and when you think about it, this is where most catastrophes happen, at the very large and the very small scales.

In Chapter 7 we return to the question of violation of the CLT and its consequences. The argument presented in this chapter has to do with the visibility of crucial events and the resulting form of the CLT when the crucial events are masked by a fog of much more frequent Poisson events. The argument develops into a stochastic CLT (SCLT) leading to a

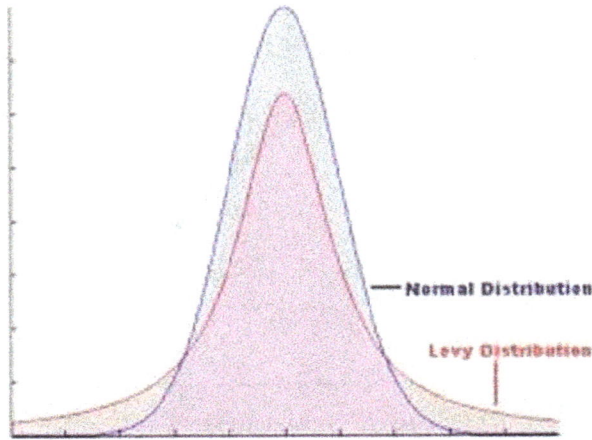

Figure 2.11: Normal PDF is depicted in blue superimposed on a Lévy PDF in red. The tails of the red clearly have non-zero weight for large deviations from the mean of the PDF. It is the tail region that produces the divergence in the second moment.

Mittag–Leffler function (MLF) probability for the measurable, that is to say visible, crucial events

2.4.2 Citations

We apply what has been developed in the above-mentioned literature to determining how the scientific quality of a paper can be quantitatively measured in terms of citations. The number of citations to papers across science are found, as you may have guessed, to be IPL and consequently the average does not characterize the PDF, nor can it be used to compare the relative quality of papers without additional interpretation. If the average is not the quantity of interest, what is the proper measure of the quality of a paper? Assuming that such a measure of quality for an individual paper can in fact be identified, the question remains concerning how this measure can be aggregated to determine the overall quality of an individual scientist, of a research program, or of a research organization. Put more generally, can science itself be measured? This question was answered with a tentative yes by van Raan [445], through the use of bibliometric analysis.

Contrary to the romantic myth of the brilliant but socially inept scientist working in isolation to bring their scientific investigations to fruition, science is, in point of fact, a social activity. Newton's comment about 'standing on the shoulders of giants" was an acknowledgment of this social interaction with scientists, both living and dead, whose contributions were necessary for the work of the individual to develop and be completed. This is not just a parsing of words, but addresses the deeper matter of the sociology of scientists and science. The importance of acknowledging how prior research guides and supports contemporary investigations cannot be over stated. Such acknowledgment not only puts the new work in context by connecting it to what is already known, but assists in the clear communication of what new knowledge is being uncovered through the research.

In today's world citations constitute more than just the tipping of one's hat to those whose work has been found necessary to carry out their own research, but is a recognition of how the present work fits into the landscape of established knowledge. Whether the earlier work is background, providing motivational context for the research, addresses a mathematical challenge that had to be dealt with, clarifies an obscure technical point, or any other of the dozens of problems that typically have to be addressed or circumvented to carry out and subsequently communicate the research are all accounted for through the cited literature. The citations are lampposts lighting the path across a rugged landscape, revealing the knowledge gaps that the research is attempting to bridge.

The number of times a research paper is cited is a measure of the paper's luminosity and here again, as depicted in Fig. 2.12, the number of scientists having a given number of citations is an IPL (fat-tailed) PDF with an index $\mu > 2$. It is evident that the number of citations is being used as a proxy measure of a paper's scientific quality or impact on a focused area of research. Consequently, any patterns, over and above the citation PDF, are useful in further assessing the quality of research.

The number of citations to all science papers published within a given year is schematically depicted by IPL in Fig. 2.12. In this figure, 35% of all papers published in the sciences within a typical year have no citations; 49% of all such publications have 1 citation; 9% have 2 citations; 3% have 3 citations; 2% have 4 citations; 1% have 5 citations and 1% have 6 or more citations. The average number of citations per year in the sciences is 3.20 and the IPL distribution implies that 96% of all science papers published are below average in the number of times they are cited.

Here the manager that uses the number of citations as a direct measure of the lab's quality of the research is applying a linear world filter to the data and is not being objective about the quality of the lab's research. Most of the research published is in alignment with the 96% [451]. If the number of citations to the lab's papers is average for the discipline then that research falls in the upper 4% category. How a manager ought to use the information on publications, citations and other such criteria, for purposes of evaluation is therefore strongly nonlinear.

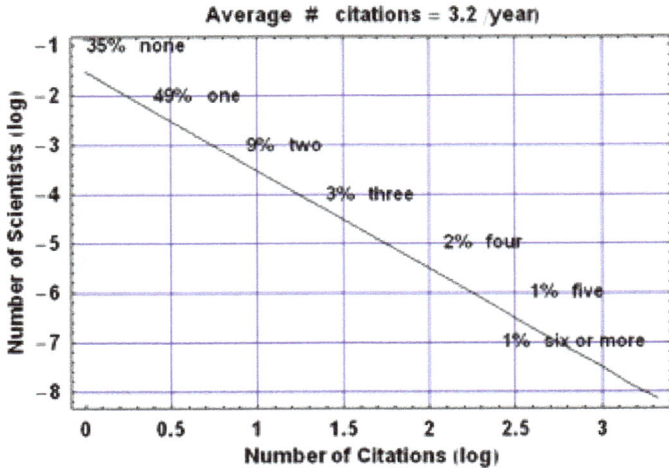

Figure 2.12: The number of scientists publishing papers with a given number of citations within a given year is graphed on log–log graph paper. The data for the IPL curve is taken from de Solla Price [440].

2.4.3   Research aggregates

There is a great deal of hype surrounding the research rank of a university, particularly now that they have become big business, as in the United States. This situation often elicits the question: What fraction of the hype is based on quantifiable information regarding the research carried out by individual faculty or research collectives within the institution and what fraction is generated by innovation on the part of the public relations department? All too often the scientific ranking of a school is viewed in much the same way as the league rankings of their football or basketball team. The latter, overly simple view, was critiqued by van Raan [446], who emphasized the difficulties in making bibliometric assessments of the research performance of institutions such as universities. One constraint is that the complex world of scientific research is compressed into what appears within the categories of output and impact. The output is the number of scientific publications, books, letters, papers, essays, etc. appearing in a given time interval and the impact is the number of citations over a time interval to a published work.

One strategy to characterize scientific institutions using citations was made by Chatterjee *et al.* [139], who were seeking to use citation data to determine universal properties of universities worldwide. We know that the scientific publication rate was determined by Shockley [495] to be dictated by a lognormal distribution, but that the tail of the citation distribution of individual publications, as well as the number of papers, decay as an IPL in the independent variable [440]. The consensus is that both are true. Chatterjee *et al.* use this result to establish, for the first time:

> ...that irrespective of the institution's scientific productivity, ranking and research impact, the probability $P(c)$ that the number of citations c received by a publication is a broad distribution with an universal functional form.

The universality of the form of the distribution resulting from the joining of the lognormal and IPL PDFs depicted in Fig. 2.13 argues for the underlying mechanisms that produce the PDF to be irrelevant and the PDF's functional form to be the result of criticality and the citing of a paper to be considered an event. The particular curve we have elected to show in Fig. 2.13 contains citation data covering a five-year interval, but the same result is obtained on a year by year basis, with suitably adjusted parameters. For each year of collected citation data for an academic institution, the distribution $P(c)$ was observed to be quite broad. Chatterjee *et al.* [139] rescaled the value of citations for each year by the average number of citations per publication $\langle c \rangle$, and plotted this quantity versus $\langle c \rangle P(c)$ in which case all the distributions collapse onto the universal curve shown independently of the variability in output of the different institutions.

Another useful parameter for the research institutions is the k-index, which is interpreted to give the top cited $(1 - k)$-fraction of papers, which have k-fraction of citations. The empirical value obtained in the study is $k = 0.75 \pm 0.04$ suggesting that the top 25% of the articles garner 75% of the total citations for institutions. Recall the discussion of the 80/20 rule for Pareto IPL distributions in which 80% of a network's function is accomplished by 20% of the network's members. This would suggest that for this measure, the tail is wagging the dog.

One indication that these general results are of broad scientific interest is contained in the results obtained by Néda *et al.* [392] depicted in Fig. 2.14. In this latter figure we see the same universal behavior observed in Fig. 2.13, but these latter data are fitted to a hyperbolic universal form and not to a lognormal plus an IPL as was fit to the data in Fig. 2.13. Another difference is that these authors have included Facebook shares along with scientific citations. Here again we emphasize that the universal form

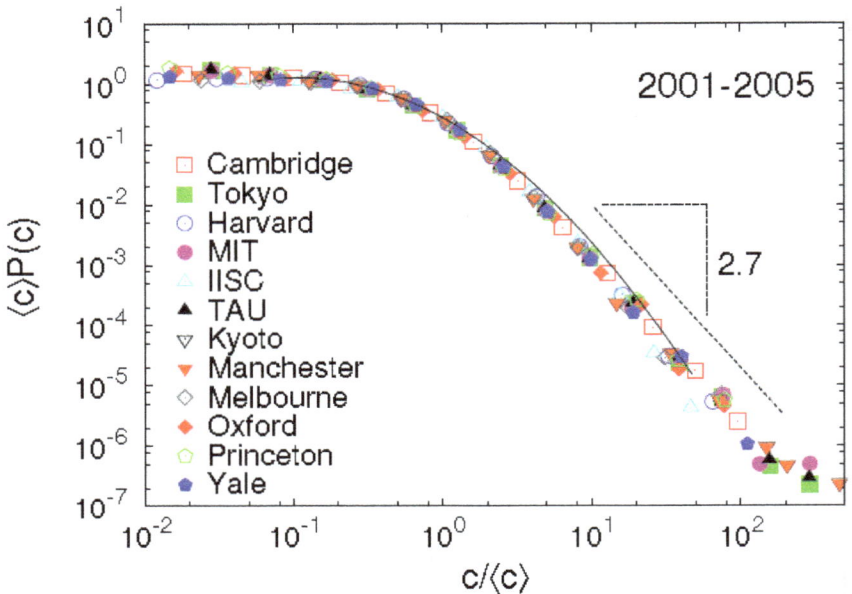

Figure 2.13: **Probability distribution of citations for academic institutions, 2001–2005:** Probability distribution $P(c)$ of citations c rescaled by average number of citations $\langle c \rangle$ to publications from the period 2001–2005 for several academic institutions. Most of the range of the data fit well to a lognormal function with $\mu = -0.60 \pm 0.04$, $\sigma = 1.28 \pm 0.02$, but the highest citations fit to an inverse power law, $c^{-\alpha}$, with index $\alpha = 2.7 \pm 0.3$. Adapted from [139].

of the citation PDF emerges from the underling complexity typical of critical dynamics. Science and Facebook show the same popularity pattern described by a hyperbolic PDF. Chapter 3 shows that the hyperbolic PDF can be understood using complex dynamic network models [604].

The universality form of the hyperbolic PDF suggested to Néda *et al.* that Western society acts responsibly and in a selective manner in retransmitting information. We believe that their results do not justify this interpretation and are rather a consequence of the efficiency of information transfer in dynamic networks undergoing criticality, as we subsequently show herein. We speculate that they reached this interpretation based on the way scientists like to think that information is transferred within a citation network in science. However, given the universal form of the PDF, we

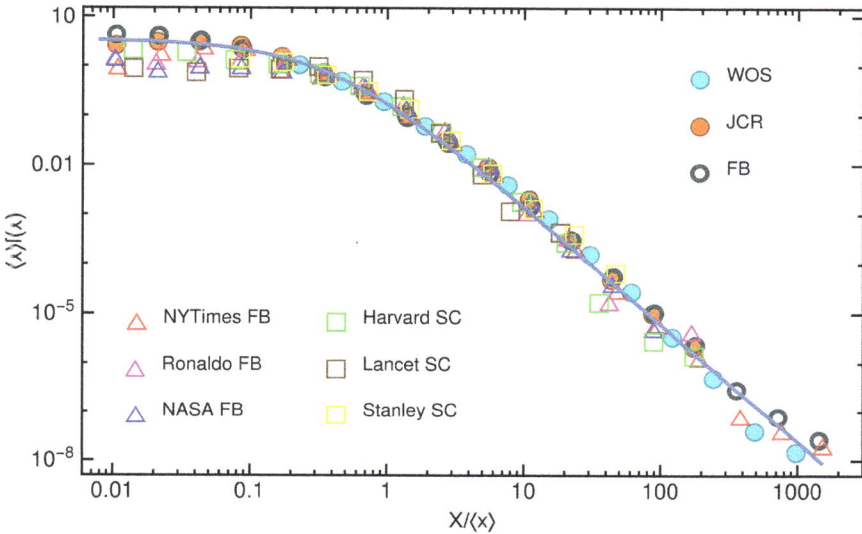

Figure 2.14: **Rescaled distribution of the citation (share) numbers.** $P(c) = f(x)$ is the probability density (PDF) for one paper (post) to have $x = c$ citations (shares). We present the $\langle c \rangle P(c)$ value as a function of $c/\langle c \rangle$ ($\langle c \rangle$ the mean value, or first moment of the PDF). For high citation number a clear inverse power-law trend is visible. Different symbols are for different datasets as illustrated in the legend. The considered datasets are described in [392]. For high $c/\langle c \rangle$ a clear IPL trend is visible. The entire curve can be well-fitted with a hyperbolic distribution with $\mu = 2.4$ and $\langle c \rangle = 1$. Adapted from [392].

conclude that the underlying dynamics are critical. Thus, we know that we cannot reconstruct the microdynamics from the emergent dynamics since the emergent macro behavior cannot be tied to a unique micro behavior. Consequently, the fact that Facebook shares and scientific citations have the same universal form of a hyperbolic PDF does not imply that they share common microdynamics.

Chapter 3

Nonsimplicity, Dynamics & Information

> *Science is said to proceed on two legs, one of theory (or, loosely, of deduction) and the other of observation and experiment (or induction). Its progress, however, is less often a commanding stride than a kind of halting stagger — more like the path of the wandering minstrel than the straight-ruled trajectory of a military marching band. The development of science is influenced by intellectual fashions, is frequently dependent upon the growth of technology, and in any case, seldom can be planned far in advance, since its destination is usually unknown. — Ferris [190]*

The phenomenon of synchronization was first documented by Huygens in a letter to the Royal Society dated February 27, 1665. He elaborated on his findings in a contemporaneous letter to his father in which he described observations made while he was confined to his room by a brief illness. He determined that two pendulum clocks, mounted on the same wall, swung at exactly the same frequency and 180 degrees out of phase. Moreover, after perturbing one pendulum, the synchronization was restored within thirty minutes and remained indefinitely. An excellent brief history of the phenomenon is given by Bennett *et al.* [84], as well as a modern treatment of the associated nonlinear mathematical problem associated with the coupling of the two pendula. As they point out, the onset of synchronization and unique phase relations is a fundamental problem that arises in phenomena ranging from neurobiology and brain function [456] to animal locomotion [513]. A much more extended history of this remarkable discovery is given by Strogatz [514] in his extraordinary book on the general phenomenon of synchronization and its far reaching implications.

In subsequent chapters we shall see that non-ergodic renewal processes (NERP) represent an attractive form of complexity, requiring a revision and extension of the fundamental tools of non-equilibrium statistical physics, such as the FDT of the first kind. In the present chapter, we establish that synchronization is a basic property behind the emergence of this form of complexity. This model is close to neuronal dynamics and may have a direct connection with the complex nature of the brain. Here we focus on the case $\mu < 2$, implying a diverging average time interval between consecutive events, thereby yielding ergodicity breakdown [81] and having all the properties of NERPs.

The human brain is considered by some to be the most complex network in the universe and attempts to construct formal representations of that complexity have met with only limited success. Consequently, investigators have relied on capturing certain identifiable characteristics within models to understand the possible mechanisms within the brain. One such property is that of synchronization and its potential role in connecting neurophysiological processes to consciousness and higher brain function. The relation between synchronization and the graphical analysis of complex networks for brain science has been reviewed [509].

The timing of biological oscillators such as pacemaker cells in the human heart are notoriously poor clocks, which is to say, that the firing period varies from cell to cell and from firing to firing. However, collections of such naturally variable entities, when allowed to interact, can become phase-locked and lose their individual variability. How natural phenomena accomplish this phase synchronization has been gaining increasing attention since Wiener's early work on understanding the alpha rhythm in EEG data [614] and Turing's work on the modeling of collective phenomena [539]. More recent mathematical investigations into the coupling mechanisms of such collectives was made by Kuramoto [305] using a network of coupled nonlinear oscillators to mimic natural synchronization in oscillating chemical reactions. A similar strategy was employed by Winfree [619] in his studies of cardiac oscillations. Glass and Mackey [213] extended our understanding of biorhythms to include modern nonlinear dynamics and chaos theory. But the phenomenon of synchronization became mature with the synthesis done by Strogatz [514] in which he presents a compelling case for synchronization being a fundamental mechanism for organizing confusion into nonsimple collective behavior in natural phenomena, whether it is synchronized clapping by an audience, the coordinated blinking of swarms of fireflies, or the orchestrated firings of collections of neurons within the brain.

Modern mathematical representations of biological clocks are often chaotic attractors, that are periodic, but not harmonic, for example, the Rössler oscillator has been used to represent the output of single neurons [508]. Wood *et al.* [621] have shown that coupled stochastic clocks can manifest cooperative, that is to say, synchronized behavior. Varela *et al.* [545] postulate that brain function, such as cognition, rests on the cooperative behavior of collections of many neurons. The strategy of Bianco and Grigolini [94] is to map brain activity onto networks and consequently study the dynamics of the collections of neurons modeled by such complex networks. Moreover, a collection of BQDs appears to have the same dynamical behavior as brain activity [303, 304, 398, 491]. Bianco and Grigolini [94] find that the changes in the topology of the network describing brain activity are driven by a NERP.

3.1 Nonsimplicity Replaces Complexity

As noted, science is not a smooth progression from hypothesis, to experiment, to theory, which enables the formulation of a more refined hypothesis, initiating the spiral to the next higher level; although it sometimes appears to have that form. As Ferris observed, research resembles more the hesitant steps of a wandering minstrel, than the confident stride of a marching band. In the real world, the systematic study of any complicated phenomenon, by multiple investigators, generates many explanations, from a variety of perspectives, eventually requiring synthesis to achieve understanding. One such synthesis has created the field of non-equilibrium statistical physics, applied to the understanding of complex dynamic physical networks. The keyword here is complex and over the past century the concept of complexity has undergone a remarkable metamorphosis. If you type the word 'complexity' into *Google* on the computer, you are greeted with over a 100 million hits and if you try to restrict that number by typing the more specific term 'complex networks', you actually double the number of hits.

Thus, to avoid continuous semantic digression to qualify the contextual meaning of complexity, it is convenient to replace the word complexity with nonsimplicity. Note that in this replacement we follow the lead of mathematical physicists by defining an extraordinarily difficult concept by what it is not. Complexity may not be defined, but simplicity can be unambiguously defined and therefore we are quite clear about what is not being discussed. This chapter emphasizes the ideas of information, nonsimplicity and how they are interrelated. We begin with a discussion of how nonsimplicity has shaped our understanding and modeling of animate, as opposed to physical and therefore inanimate, phenomena.

Sociophysics was a new science spawned at the dawn of the 19th century, followed soon after by biophysics, both of which became fully outlined in the early 20th century. They were attempts to formalize the social and life sciences, respectively, using techniques from, what was at that time, the more quantitative science of physics. A major difficulty encountered outside the physical sciences was modeling the nonsimplicity of even the most familiar phenomena. Unlike simple things, in which there is a linear progression from cause to effect that enables a tractable mathematical formulation of the process, nonsimple phenomena have large numbers of entangled components that typically cannot be decomposed into fundamental linear cause-effect chains. As pointed out by West and Grigolini [586] nonsimplicity is a delicate balance between regularity and randomness; the stability and adaptability of a nonsimple process can be lost through an imbalance favoring one or the other. Consequently, extending the early modeling of physical phenomena to the behavior of living organisms, either individually or collectively, was almost uniformly disappointing.

The initial successes of physics modeling relied in large part on Newton's concept of mechanical force, which seemed to have no direct correspondent in the social sciences, except in a metaphorical sense, and did only a little better in the life sciences. The lack of a physical force in these latter sciences is not surprising, given that the unique dynamical variables indigenous to their study are not physical in nature. Such restrictive statements must be viewed with caution, however, since disciplines such as physiology are made up of physical objects, which do obey the laws of physics. The stress tolerance of a bone before breaking is certainly a physically measurable property, but how that bone heals after breaking is much more complicated. How one changes the laws of science when transitioning from the world of the inanimate to that of the animate remains mostly a mystery and a challenge. Consequently, the genesis of a non-mechanical force must be traced to its source.

We begin our hunt for the source(s) of non-mechanical forces by briefly discussing the formulation of the underlying principles that culminated in Newton's force law. We do this in order to determine if that procedure might be extended to encompass the social and life sciences. Following this path we discover that nonsimplicity itself might very well be the genesis of at least one class of non-Newtonian forces.

Physics has historically involved a search for the most parsimonious description of nature's behavior and as a science it has been wildly successful when it comes to describing the dynamics of the inanimate. Consider the empirical notion of energy conservation, which states that energy cannot be created or destroyed; it can only change its form, switching back and forth between potential and kinetic. We might comment here that the strict acceptance of/adherence to physical laws has a certain amount of arbitrariness built into it. Historically, when there is empirical evidence of a violation of a law, a new mechanism is introduced to reestablish the law and the new mechanism is then tested. This strategy works for simple processes where the deviations from expected behavior is more systematic and easier to reincorporate into an existing framework through a generalized law, but perhaps fails for more complex processes where the deviations are more difficult to characterize.

Take, for example, the notion of heat, which is generated in nonconservative systems, whose physical properties are described by the phenomenological theory of thermodynamics. With the inclusion of heat, we obtain the first law of thermodynamics, which is the general statement of the conservation of energy, with the introduction of the concept of useful work.

Emmy Noether [400] took the formal difference between kinetic and potential energy and established that a dynamical symmetry in a physical system entails the existence of a conserved quantity. For example, translational invariance implies the conservation of linear momentum, whereas insensitivity to the direction of time entails the conservation of energy. Thus, the elegantly simple notion of symmetry gives rise to much of what we know about the physical world, from string theory and elementary particles to cosmology and black holes [243]; but this view does not include phenomena in the social or biological realms. Although some scientists believe that the formulation of principles in the social and life sciences is just a matter of time and that we have already begun to see how such principles might be related to scale, see [610] for one view from physical sciences and [77] for another.

The world is filled with animate and inanimate objects, but what stands out is one or more qualities of how they smell, taste, feel and sound. In short, we know the world by how we interact with it, exerting forces on various objects and they, in turn, exerting forces back on us. It is useful to distinguish between the forces we encounter through contact with physical objects from those that arise from psychological discomfort and/or social stress. I should say up front that this is to partly answer such psychological questions as: Is the surprise we experience from a nearby lightning bolt any less 'real' than the sound generated by the collapsing air mass, or the light shed from the electrical discharge? Why is the sting of a reprimand by the boss stronger when done in the presence of co-workers?

In one school of thought, the fear response to the sudden lightning strike is not *real*, because it only exists in the mind of the observer. Consequently, each person reacts differently to the same stimulus and therefore the psychological force of the strike is subjective and lacks objective 'reality'. The notion that the force should be dependent on the observer has a place in quantum phenomena, but certainly appears out of place in the macroworld in which we live our lives. On this scale, if it is not objective, it is not scientific and therefore lacks physical reality.

On the other hand, there is another school of thought, which emphasizes that only the subjective is real. Only that part of the world that can be directly experienced by the individual has a reality and that reality is imparted by experience. But such analysis takes us far from the transformation of Natural Philosophy into science made when Newton introduced mathematics into the general discussion of the nature of physical reality. We shall elaborate on this noted difference in subsequent parts of this essay.

Is the 'force of history' merely a metaphor constructed by social scientists to dramatically capture the apparent influence that the past exerts on the social decisions that are made in present time? Or does a social force have a reality more like that of a slap in the face? Getting an unexpected promotion, or receiving a failing grade, can produce as violent a physical reaction as being slapped across the face, but we know that the source of the reaction is very different in the two cases. Or do we?

We have all experienced someone with a dominating personality. The force of personality might result in being overpowered to buy a particular product, or laughing until your sides hurt, or generating an interest in history, or English literature, or mathematics, due to the efforts of that one individual teacher. The person could also have been the bully in high school who embarrassed you in front of your friends. The point is that the force of that person's personality, whoever she was, can be every bit as real as a stone falling on your toe.

But the physiosocial force cannot be expressed in terms of dynes or foot-pounds, because it has not been objectively measured quantitatively. This does not mean that the social force is not real. A more accurate interpretation might be that a social force is different in kind from a physical force, because it involves one or more people, as well as an organization. Consequently, a social force is dependent on both the sender and receiver. In this way, a given action on the part of an organization may exert very different forces on different members of the organization. An unexpected promotion could be welcomed by one person as being long overdue, but terrify a second person, who does not feel up to the task. Conversely, the same activity on the part of different individuals may elicit very different responses on the part of the organization. In the military, one person may be awarded a medal, whereas another may be court marshalled, for essentially the same action.

The question that interests us is whether or not social or psychological forces are merely convenient abstractions we humans have developed to describe how we interact with one another and how we explain ourselves to ourselves and to one another, or whether these forces are real, in the same way a piece of furniture is real. It is not that an organization is held together in the same way physical objects are bound into matter. But it is held together by a different kind of force, and this latter force is as measurable and tangible as the force between two magnets. What is this different kind of force? Can it be understood in the same way that we understand physical forces? This essay attempts to answer these and similar questions in the affirmative.

3.2 Simplicity

Here we are concerned with how to understand nonsimple social and biological systems that typically do not have dynamics determined from first-principles. On the other hand, biological systems should obey the first and second laws of thermodynamics so that when dynamic equations are available for living networks they are typically derived from energy or thermodynamic arguments. Consequently we review many of the reasons why the traditional Newtonian view of nonsimplicity is not sufficient for our investigation of non-physical dynamical systems. We follow this strategy to avoid the trap of using physical analogs to construct models whose underlying assumptions can be shown to be inappropriate for the non-physical system being modeled. For example, we show that the definition of nonsimplicity in a stochastic process has shifted from the conventional exponentially short memory of Poisson processes, towards the action of non-Poisson renewal events, with IPL PDFs for the time intervals between events, which we subsequently gather under the heading of *crucial events*. A great deal more will be said about crucial events after we have laid the groundwork, but for the moment we note that the observation of phenomena generated by crucial events has forced investigators to go beyond the traditional physical models of nonsimplicity. These new models establish closer contact with the nascent field of nonsimple networks, outside the physical domain, as well as with the fractional calculus. The latter calculus has integral and differential operators that are not integer, which may sound strange, but that nonsimplicity forces us to think differently about things we thought we understood. But before we dive into any of this we must first clearly define what we mean by a simple system and to do that we review the meaning of linearity.

3.2.1 Linearity

A simple system is typically one whose defining properties are linear, but not always, as we shall see. The first property of linearity is that of proportionality. Consider your response on meeting a person for the first time: she smiles, you smile; she speaks in a low pleasant voice, you do likewise; he offers a firm handshake, you balance the pressure. This is the typical interaction in the professional world of everyday life and is a social interaction that is operating at a simple or superficial level.

The same social system has many potential levels, most of them being nonsimple. Consider the situation when you nod to a colleague you pass in the hall and he does not respond with a reflexive nod, or bland smile. Instead you receive an angry glare. You can be confident that the simple domain of linear response has been left behind. You must now reevaluate your relationship with this person and estimate how disproportionate his response would be if you asked what was wrong.

A second property of linearity is that the total response to a force of a system, comprised of many (N) components $\{F_1, F_2, ..., F_N\}$, is equal to the sum of the responses of each of the individual components. This is the property of independence of the individual responses. The more individuals I can convince to push my stalled car, the greater the subsequent speed achieved prior to my releasing the clutch and starting the engine. You do remember about standard transmissions and clutches? Well even if you do not remember such quaint vehicles, each individual push adds another independent increment of physical force to move the car, so the total response is the sum of the contributions.

One of the first physical principles discovered, exploited the concept of linearity and was developed through the letter correspondence of John (father) and Daniel (son) Bernoulli, beginning in 1727. They established that a system of N point masses has N independent modes of vibration, that is, eigenfunctions and eigenfrequencies; thereby formulating the *Principle of Superposition*. The principle states: a complicated event can be segmented into a number of simple components in order to understand the components separately and then recombine back into an organized whole that can be understood in terms of the properties of the components. This could be done without making particular assumptions about how to construct the equations of motion as long as the coupling between system components is sufficiently weak.

This principle was the first statement of the use eigenfunctions and eigenvalues in physics. As West [573] explains in *An Essay on the Importance of Being Nonlinear*, a complicated system consisting of N degrees of freedom can be completely described by specifying one function per degree of freedom. If one knows what the eigenfunctions are (for waves these are harmonic functions) and one knows what the eigenvalues are (for waves these are frequencies) then one knows what the motion is of the composite system (a linear wave field). Given that the principle of superposition was the formulation of the first general law pertaining to a system of particles, one might date theoretical physics from the 1755 publication by Daniel Bernoulli [89] in which the present day form of the principle was fully articulated.

3.2.2 Linear resonance

We are interested in the phenomenon of resonance in a variety of settings and therefore we briefly review how the concept arose in the context of linear harmonic oscillations, the most ubiquitous model in theoretical physics. It is the basis for the understanding of every phenomena from water waves, to lattice phonons, to all the -*ons* in quantum mechanics. The stature of the lowly harmonic oscillator derives from the notion that nonsimple phenomena have equations of motion that can be generated by means of an unperturbed Hamiltonian H_0 and a sequence of weak perturbations producing small deviations in the vicinity of that dominant behavior. When the conditions for modeling a phenomenon using a perturbative approach are realized, the harmonic oscillator is a good lowest-order approximation to the dynamics.

To make this discussion explicit, consider the Hamiltonian for a system described by the displacement Q and its conjugate momentum P. The equations of motion for a unit mass linear oscillator are determined by the Hamiltonian:

$$H_0 = \frac{1}{2}P^2 + \frac{1}{2}\omega_0^2 Q^2, \tag{3.1}$$

whose equations of motion have the simple harmonic solution for the oscillator displacement:

$$Q_0(t) = A \cos[\omega_0 t + \phi]. \tag{3.2}$$

Here ω_0 is the natural frequency of the oscillator, with the amplitude A and phase ϕ determined by the oscillator's initial conditions. A single linear oscillator driven by a periodic perturbation of amplitude A_p is described by the time-dependent Hamiltonian:

$$H = \frac{1}{2}P^2 + \frac{1}{2}\omega_0^2 Q^2 - QA_p \cos \omega_p t, \tag{3.3}$$

whose equations of motion have a solution whose amplitude diverges as the frequency of the perturbation ω_p approaches the natural frequency of the oscillator. The solution for the oscillator displacement now becomes:

$$Q_p(t) = \frac{A_p}{\omega_0^2 - \omega_p^2} \cos[\omega_p t + \phi_p]. \tag{3.4}$$

Equation (3.4) is actually the inhomogeneous part of the total solution, with divergence as $\omega_p \to \omega_0$, thereby overwhelming the homogeneous part of the solution. In a real system, this mathematical divergence is eventually quenched by dissipation, which is to say, the amplitude of the solution may become substantial, but it must remain finite even at the resonance condition $\omega_p = \omega_0$. Said differently, the enhanced amplitude eventually triggers a mechanism that violates the weak perturbation assumption and triggers a change in dynamics. The changes in dynamics can be as subtle as a shift in the linear frequency to as dramatic as a complete shutting down of the oscillator.

There are a number of ways to quench the divergence resulting from a linear resonance, for example, as we mentioned above, every thermodynamic network has fluctuations and dissipation, and for a sufficiently large amplitude of the oscillator the influence of the environment can no longer be neglected. Let us consider the total Hamiltonian for a network and its environment, which is discussed more in the next chapter on modeling nonsimple networks with Eq. (5.61). If we introduce a periodic perturbation into the total Hamiltonian, the average homogeneous solution to the unperturbed oscillator (unperturbed by the periodic driving force, but still in contact with the heat bath modeling the influence of the environment) is

$$\langle Q_h(t) \rangle = A e^{-\lambda t} \cos[\omega_1 t + \phi], \tag{3.5}$$

$$\omega_1 = \sqrt{\omega_0^2 - \lambda^2}, \tag{3.6}$$

$$\phi = \tan^{-1}\left(\frac{\omega_1}{\lambda}\right), \tag{3.7}$$

with indicated shifted frequency ω_1 and phase ϕ; the brackets $\langle \cdot \rangle$ denote an average over the thermal fluctuations. In the case of a periodically driven oscillator, we again have two terms in the solution, the homogeneous and inhomogeneous. In the asymptotic limit the average homogeneous term given by Eq. (3.5) vanishes due to dissipation and the inhomogeneous term has a large, but finite maximum amplitude, even at the resonance condition. The average of the particular solution in the driven case is

$$\langle Q_p(t) \rangle = \frac{A_p}{\sqrt{\left(\omega_0^2 - \omega_p^2\right)^2 + \lambda^2 \omega_p^2}} \cos[\omega_p t + \phi_p], \tag{3.8}$$

with the phase

$$\phi_p = \tan^{-1}\left(\frac{\lambda \omega_p}{\omega_0^2 - \omega_p^2}\right). \tag{3.9}$$

After the transient fades away, the average steady-state response of the oscillator to the driver has the same frequency as the excitation, but it is phase-shifted as indicated in Eq. (3.8). This is the under-damped case where the resonance line has a width determined by the magnitude of the dissipation parameter and the average resonance amplitude is a factor $(\lambda \omega_p)^{-1}$ greater than the amplitude of the driver, which can be substantial for weak dissipation.

The importance of resonance: Resonance is a key mechanism underlying some of the most useful measures of physical and physiological phenomena. Magnetic resonance imaging (MRI) used in medical science to study normal anatomy and function of the human body, is one such example and in clinical science it is used to uncover and diagnose pathology [567]. A MRI scanner polarizes the hydrogen nuclei in water molecules in the body by means of a powerful static magnetic field. These polarized nuclei are excited by the spatial gradient of a much weaker time-varying magnetic field for encoding the local spatial structure of the body and finally a weak radio-frequency (RF) field is applied to manipulate the polarized nuclei to produce a measurable signal, which is collected by an RF antenna.

Nuclear magnetic resonance, the phenomenon on which MRI is based, is determined by the quantum mechanical magnetic properties of an atom's nucleus, that being the magnetic moment. It is the magnetic moment (nuclear spin) that is manipulated in MRI. The external magnetic field creates an energy gap between nuclei aligned and anti-aligned with it, so that if the frequency of the RF signal is matched to this energy gap a resonance, or matching, is created and the population of the two states are equalized.

Although traditionally MRI scanners created a two-dimensional image of a thin "slice" of the body, modern MRI instruments are capable of producing images in the form of three-dimensional blocks. An MRI scanner images hydrogen-based objects, so that bone, which is calcium based, is imaged as a void, and will not affect soft tissue views. Consequently, MRI scanners are excellent for imaging joints, the central nervous system and the brain.

A perhaps less familiar matching process is referred to as a limbic resonance. As one might surmise from the name this resonance involves the limbic system of the brain. The physicians Lannon *et al.* [313] maintain that mammals are open systems and the scientific understanding of their behavior must be evaluated taking that openness into account. Thus, the chemistry induced within the limbic system of the brain by sharing deep emotional states is posited to have the evolutionary foundation:

> Within the effulgence of their new brain, mammals developed
> a capacity we call 'limbic resonance' — a symphony of mutual

exchange and internal adaptation whereby two mammals become attuned to each other's inner states.

We mention this here, not because we believe that the scientific validity
of limbic resonance has been established, but because it has attracted a
great deal of attention recently. Moreover, similar concepts have appeared
under different names in different contexts. For example, in a social context
limbic regulation, the process by which our systems synchronize with one
another, can be identified with mood or emotional contagion. Of course,
we would identify limbic regulation with the CME for which there is ample
experimental evidence, as briefly discussed in Chapter 1.

On the importance of linearity: The point of this extended discussion
of linearity is to emphasize the importance of this concept to science in
general. In one sense, all the physics of the nineteenth century and most of
the physics of the 20th century addresses simple phenomena. The physical
models rely on the assumption of simplicity and therefore the models are
linear. The linearity assumption must be abandoned when the phenomenon
being studied becomes nonsimple, but how the loss of linearity becomes
manifest is not always clear. In a dynamic system such as a child's swing,
we say things become nonlinear when we push too hard, that is we increase
the energy input above a given level, but the proper nomenclature when
things are stochastic is not so evident. That is why we replace the word
complex with nonsimple in this essay. In dynamics, to be nonsimple is to
be nonlinear and in statistics, to be nonsimple is to be non-normal, or non-
Poisson and most of all non-ergodic. Not the number of times quantities
are defined in terms of the negation of simpler quantities.

A word of caution. Although a simple process is linear, a linear process
is not necessarily simple. A process being linear does not even guarantee
that it can be easily understood. An immediate example is the basis of
quantum mechanics, the linear Schrödinger equation. Quantum phenomena
described by this equation have challenged the greatest minds in physics
for over a century. Consequently, a simple dynamical description may have
a nonsimple interpretation. Another example is viscoelasticity in which
the dynamics of a non-Newtonian solid, or liquid, depends on the entire
history of the dynamics. For example, pulling on a piece of taffy is very
different from pulling on a spring and jumping into a mud bath is a very
different experience from sprawling in a jacuzzi. The viscoelastic equations
of motion are non-local in time, but still linear, so we again have the nature
of linearity and simplicity diverging. A final word of caution concerns
a mathematical theorem which proves that a low-dimensional nonlinear
dynamic system of equations has an equivalent infinite-dimensional linear

dynamic representation; the Carlman embedding theorem. The price one pays for linearity is infinite dimensionality, just as was done for quantum phenomena.

3.3 Nonsimplicity

Nonsimplicity was first treated systematically in physical systems as a way to account for the variations in individual particle trajectories, with such concepts as mean-free paths and relaxation times in many-body physical systems. Such trajectories were assumed to be in full agreement with Newtonian microdynamics since, in principle, macrodynamics should be derivable from Newton's microdynamic equations. Of course, the transformation of energy is used to generate the dynamics of physical systems in which the total energy is conserved and the dynamics determined by how energy is exchanged between its potential and kinetic forms. Implicit in Newton's equations of motion is the remarkable definition of mechanical force, expressed in terms of a field as the negative gradient of the potential energy.

The main difficulty in deriving macrodynamics from microdynamics was the need to account for the actions of a very large number of degrees-of-freedom. This was accomplished in one context by replacing the single-particle trajectory by an ensemble of such trajectories, followed by the further replacement of the single-particle dynamics, with an average over an ensemble of such trajectories, described by an ensemble distribution function. This first kind of nonsimplicity had to do with distinguishing between single-particle dynamics and many-body dynamics, while recognizing along with Anderson [36] that *more is different*, which is to say that as a system becomes larger and larger there is more opportunity for behavior to emerge that could not exist within smaller (simpler) systems.

However, trajectories only define one kind of dynamics. The existence of events, such as the abrupt transitions from one many-body state to another, could be captured by such formalisms as the CTRW description of nonsimplicity, which was proven to be consistent with the Newtonian view. This is complemented by stochastic equations for the dynamic variables, as well as, phase space equations for the dynamics of the PDFs, e.g. see [327].

Of course, nonsimplicity does not necessarily imply randomness, and statistics may result from other sources of irregularity. Newton's deterministic equations of motion were shown by Poincaré, near the end of the 19th century, to have solutions that could not be treated by classical analysis. Trajectories generated by non-integrable equations, were shown, a half century later, to break up into a spray of disconnected points in phase space. The theory describing this behavior is associated with the mathematicians, Kolmogorov, Arnold and Moser, who, building on the ideas of Poincaré, formulated one kind of nonlinear dynamics that falls under the heading of *chaos theory* and was given the name KAM theory [174]. In this Hamiltonian-based approach, a system with low energy is linear and stable, whereas at high energy, instabilities emerge and the trajectories dissolve into a spray of disconnected points, much like a liquid evaporating into a gas.

The name *chaos* was coined by Li and Yorke [322], during their investigation of nonlinear dynamic systems. Solutions having a sensitive dependence on initial conditions were found to emerge, when the system has three or more degrees-of-freedom. This sensitivity resulted in initially nearby trajectories exponentially separating from one another in time; an instability they called chaos. The nonsimplicity resulting from chaos was determined to arise in systems which conserve energy, as well as, in dissipative system that do not conserve energy, where the dynamics unfolds on a *strange attractor*. The latter is yet another kind of nonsimplicity.

Another fundamental area of science that relies on the concept of linearity is statistics, beginning with the seminal papers of Gauss [209] and Adrian [6] that provided the foundation for the *law of frequency of errors* at the turn of the 19th century. The normal or Gauss PDF was the first PDF systematically used to characterize the nonsimplicity of many-body physical systems. This introduced statistics into physical science as a quantitative way to model nonsimplicity. Poisson did the same in the social sciences, for example, in jurisprudence.

So far we have departed from simplicity in terms of nonlinear dynamics, as well as, in terms of fluctuations and statistics, treating each separately. We now turn to a study of their combined effects through a discussion of stochastic resonance (SR)

3.3.1   Stochastic resonance

Nonlinear dynamical networks with additive fluctuations often have surprising properties. One such unexpected property was identified over a quarter century ago and has to do with the counter-intuitive response to the addition of noise into certain weakly periodically driven nonlinear dynamical networks. Experience would suggest that the signal-to-noise ratio (SNR) in such a network would decrease as the amplitude of the added noise is increased. This is not always the case, however. This became evident in the phenomenon of SR.

The phenomenon of SR was originally hypothesized by Benzi *et al.* [86] in the context of modeling the alternation of the Earth's climate between ice ages and periods of warm weather with a period of about 10^5 years.

The eccentricity of the Earth's orbit varies with that period, but according to existing theories is not sufficiently strong to determine the observed large changes in climate. They conjectured that there may exist a cooperative phenomenon between the weak change of eccentricity and the natural random fluctuations so as to account for the periodic and strong climate changes. Their hypothesis was postulated independently and contemporaneously by other scientists, as well [396]. The theory developed to support this conjecture turned out to be very successful and after 28 years [86] had garnered over a thousand citations. Gammaitoni et al. [206] published a report on stochastic resonance in 1998 and in half that time had collected twice the number of citations of the ground breaking paper.

Here we briefly review the phenomenon of SR, noting that the SNR is, in fact, a non-monotonic function of the fluctuation strength. Consequently, the dynamic response to noise increases as the noise amplitude increases in certain parameter domains. The first experimental verification of the existence of SR in a laboratory was made in 1988 using a bistable ring laser [365].

To demonstrate SR in a specific network we consider a nonlinear oscillator described by a quartic potential, which will come up again in the discussion of nonsimple networks,

$$V(Q) = \frac{A}{4} \left(Q^2 - a^2 \right)^2 . \tag{3.10}$$

This potential has the three extrema, $Q = 0, \pm a$, determined by the vanishing derivative of the potential. The sign of the potential's second derivative reveals that the point $Q = 0$ is unstable $V'' < 0$, and the points $Q = \pm a$ are stable $V'' > 0$, as shown in Fig. 3.1. The two minima are separated by a local maximum in the potential and this energetic barrier is necessary in order for the system to manifest SR.

We require a signal in addition to the potential and therefore impose a periodic forcing function on this isolated nonlinear oscillator in the same way we did on the isolated linear oscillator. The driven system of interest is described by the Hamiltonian for a periodically perturbed unit mass oscillator in a double-well potential:

$$H_N = \frac{1}{2} P^2 + \frac{A}{4} \left(Q^2 - a^2 \right)^2 - QA_p \cos \omega_p t. \tag{3.11}$$

Consequently, by opening the system up to the influence of the environment in terms of a linear dissipation $-\lambda P$ and a random force $\xi(t)$:

$$\frac{dQ}{dt} = \frac{\partial H_N}{\partial P},$$

$$\frac{dP}{dt} = -\frac{\partial H_N}{\partial Q} - \lambda P + \xi(t).$$

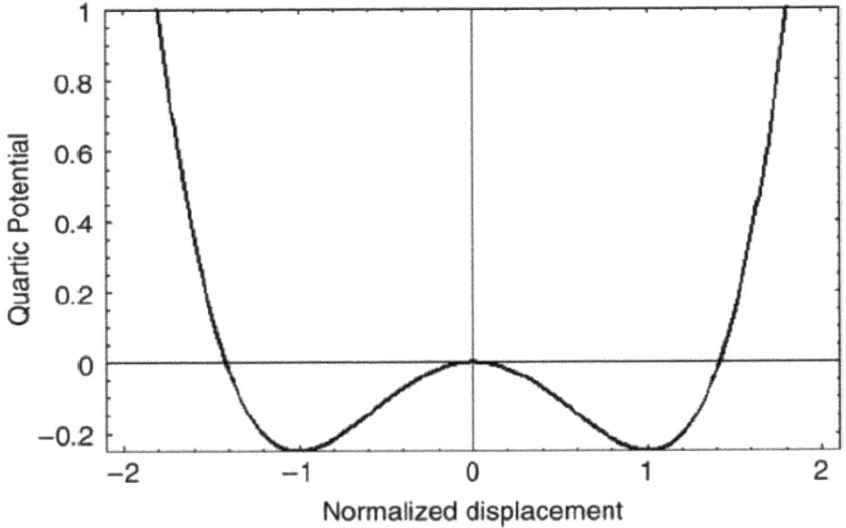

Figure 3.1: The quartic or double-well potential given by Eq. (3.10) with unit amplitude is plotted against the normalized variable, that is, $a = 1$. The height of the potential barrier is $\frac{1}{4}$ for the values of the parameters chosen.

Inserting the designated Hamiltionian into these equations we obtain the equations of motion:

$$\frac{dQ}{dt} = P$$

$$\frac{dP}{dt} = -A(Q^2 - a^2)Q + A_p \cos\omega_p t - \lambda P + \xi(t). \tag{3.12}$$

The effective potential is obtained by grouping terms in Eq. (3.11) to define the effective potential

$$V_{eff}(Q) = \frac{A}{4}(Q^2 - a^2)^2 - QA_p \cos\omega_p t \tag{3.13}$$

in which the Brownian particle moves in the time-dependent potential depicted in Fig. 3.2. In this figure, it is clear that when the periodic forcing has its maximum value, the left-hand minimum of the potential is the absolute minimum and when the periodic forcing has its minimum value, the right-hand minimum plays this role. Using the Smoluchowsky approximation on Eq. (3.12), that is, set the inertial term to zero, we can write the

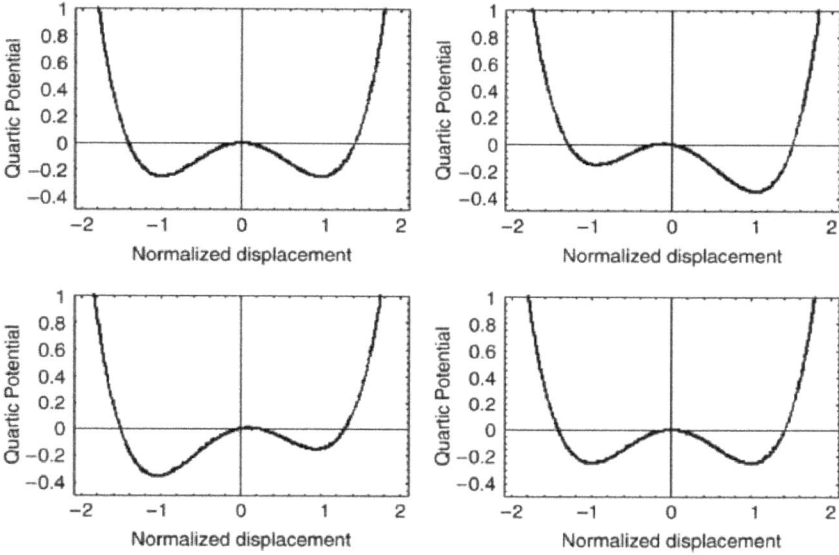

Figure 3.2: Here the effective potential is plotted versus the normalized variable ($a = 1$) and the driver has amplitude $\kappa = 0.1$. Going in a circle, the different graphs are separated by one half cycle of the periodic perturbation.

equation of motion for the Brownian particle in the effective potential as

$$\frac{dQ}{dt} = -\Phi'(Q) + \eta(t) \tag{3.14}$$

where $\kappa = A_p/\lambda$, $\Phi(Q) \equiv V_{eff}(Q)/\lambda$ and $\eta(t) = \xi(t)/\lambda$.

Here we can see that the Brownian particle oscillates in the vicinity of one of the minima and the potential rhythmically changes the relative position of the minima as depicted in Fig. 3.2. If the rate of change in minima matches the fluctuations, a particle can be propelled over the potential barrier. Such a symmetric quartic potential essentially imposes a synchronization of the thermally activated hopping events between the potential minima and the weak periodic forcing [206]. Kramers [300] calculated this rate of transition between such potential minima to be

$$r = \frac{\omega_a \omega_0}{4\pi\lambda} \exp\left[-\frac{\lambda \Delta V}{D}\right], \tag{3.15}$$

where the frequency of the linear oscillations in the neighborhood of the stable minima is

$$\omega_a = |V''(a)|^{1/2};$$

the frequency at the unstable maximum is

$$\omega_0 = |V''(0)|^{1/2}$$

and ΔV is the height of the potential barrier separating the two wells. The mean-square strength of the fluctuations is related to the temperature of the bath by the FDR.

The average long-time response of the nonlinear oscillator does not look very different from that obtained in the case of linear resonance. However here, because the potential is nonlinear, we use LRT to obtain the solution for a small amplitude periodic forcing function [206]:

$$\langle Q(t)\rangle_\eta = \langle Q^2\rangle_0 \frac{2r A_p}{D\sqrt{r^2 + \omega_p^2}} \cos[\omega_p t + \phi_p], \tag{3.16}$$

and phase shift

$$\phi_p = -\tan^{-1}\left[\frac{\omega_p}{2r}\right], \tag{3.17}$$

where Kramers' rate r replaces the frequency difference obtained in the case of linear resonance. Inserting the expression for the rate of transition Eq. (3.15) into the asymptotic solution for the average signal Eq. (3.16) allows us to express the average signal strength as a function of the intensity of the fluctuations. Extracting the D-dependent function from the resulting equation

$$G(D) = \frac{\exp\left[-\frac{\lambda \Delta V}{D}\right]}{D\sqrt{4\exp\left[-\frac{2\lambda \Delta V}{D}\right] + \omega_p^2}} \tag{3.18}$$

and choosing the parameters for the potential to be $A = a = 1$ and the frequency of the driver to be unity, we arrive at the curve shown in Fig. 3.3.

The average signal strength determined by the solution to the weakly periodically driven stochastic nonlinear oscillator is proportional to the curve depicted in Fig. 3.3. It is clear from the figure that the average signal increases with increasing D for small fluctuations levels, which is to say; weak noise facilitates the average signal level. At some intermediate level of fluctuations, determined by the parameters of the dynamical network, the average signal reaches a maximum level. For more intense fluctuations the average signal decreases, asymptotically approaching an IPL response, that is,

$$G(D) \sim 1/D.$$

This non-monotonic response of the network to noise is a defining characteristic of SR.

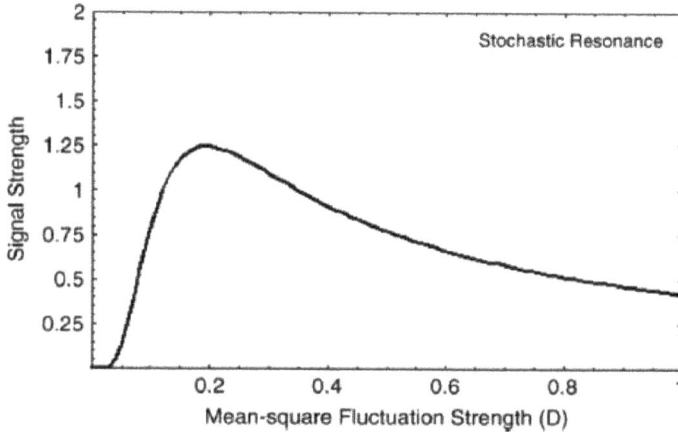

Figure 3.3: The function given by Eq. (3.18) is graphed versus the mean-square level of the fluctuations given by the dissipation parameter D.

3.3.2 SR applications

Stochastic resonance is a statistical mechanism whereby noise influences the transmission of information in both natural and artificial nonsimple networks, such that the SNR is no longer monotonic, but reaches a maximum at an intermediate value of noise intensity. Put it another way, SR is manifest in the performance of a nonlinear system being enhanced by introducing noise and considered by some to be one of those rarest of things, a correct and useful scientific concept [207].

There has been an avalanche of papers developing the SR concept, totalling more than 2,300 publications in the interval (1981, 2007) [363], across a landscape of applications, starting with the original paleoclimatology study [86, 396], all the way to the influence of noise on the information flow in sensory neurons, ion-channel gating and visual perception [389]; see [87, 363] for relatively recent reviews of SR. The majority of theoretical analyses have relied on the one-dimensional bistable double-well potential sketched above, or a reduced version given by a two-state model, both of which are now classified as dynamic SR. Another version of SR, that most commonly found in biological networks, involves the concurrence of a threshold, with a sub-threshold signal and noise. The latter is called threshold SR [389]. Here we briefly comment on how adding noise to sub-threshold signals in nonlinear sensory systems can enhance sensor information processing.

Psychophysics dates back to the middle of the 19th century [183] and is concerned with the accurate detection and characterization of human perception from the physical signals of the human sensory networks. Visual perception is one area in which there have been a great many experiments done to demonstrate the existence of a threshold. The SR mechanism has been found to be operative in the perception of gratings, ambiguous figures and letters, and can be used to improve an observer's sensitivity to weak visual signals [389]. The nonlinear auditory network of human hearing also manifests threshold SR behavior. The absolute threshold for the detection and discrimination of pure tones was shown [628] to be lowered in normal hearing people by the addition of noise. It was also shown that the same effect is manifest for individuals with cochlear or brain-stem implants by the addition of an optimal amount of broadband noise.

The most success of SR, as pointed out by one of the leaders in the field [389], has been associated with modeling and simulation of networks of neurons, where the effects of noise on sensory function match those of experiment. Consequently, the indications are that threshold SR theory explains one of the fundamental neuron operation mechanisms. This does not carry over to the brain, however, since the SR mechanism cannot, in this case, be related to a specific function. One conservative conclusion is that the evidence does not support the interpretation [389]:

> ..that naturally occurring noise actually enhances information transmission and processing, nor is it documented that neuronal systems to optimize the noise intensity for maximum information or efficacy of processing. Yet some indications that SR models may reflect mechanisms that are operative in the CNS seems to exist and justifies research....SR appears to be a ubiquitous and remarkable phenomenon congruent to the available theories on brain function...

3.3.3   Information resonance

In the previous discussion, the measure of the quality of the SR mechanism was the SNR. The ratio of the SNR of the input to that of the output of an SR network provides a measure of the information being transferred through the network and, when optimized, yields the maximum information transfer. However, the SNR measure only provides information about the entropy of the signal versus noise and the degradation of this entropy during information transfer [291]. This measure does not address the channel capacity, which is to say, it is concerned with information in *bits* and not in how the information changes in time in *bits/sec*, as would be appropriate for the information transfer rate. This information transfer approach to modeling SR in neurons has also been discussed as an 'information resonance' [561].

Kish *et al.* [291] directly evaluated Shannon's formula for the channel capacity to estimate the information transfer rate of neurons in the SR region. Shannon's formula is

$$C = B_s \log_2 \left[1 + \frac{S_s}{S_n} \right] ; \qquad (3.19)$$

with B_s the maximal bandwidth of the signal, S_s is the maximum mean-square signal amplitude and S_n the maximum mean-square noise amplitude. Shannon interprets the two factors of Eq. (3.19) to be the rate at which information is refreshed given by B_s and the potential amount of information available at each refreshing event given by the logarithm. Without going into the details of the calculation we write their result for the channel capacity [291]:

$$C = \frac{B_{n,in}}{2\sqrt{3}} \exp\left[-\frac{U_t^2}{2B_{n,in}S_{n,in}} \right] \log_2 \left[1 + \left(\frac{2AU_t}{B_{n,in}S_{n,in}} \right)^2 \right] \qquad (3.20)$$

where $B_{n,in}$ is the bandwidth of the input noise, U_t is the excitation threshold potential of the neuron, $S_{n,in}$ is the power spectral density of the input noise and A is the root-mean-square amplitude of the input signal. To get a sense of the dependence of Eq. (3.20) on the mean-square level of the input noise we set all other parameters to one and graph the resulting function in Fig. 3.4, much like we did in Fig. 3.3, thereby obtaining the SNR by another measure.

In Fig. 3.4 we can see that the measure of the information transfer rate is non-monotonic, reaching the characteristic SR peak for an intermediate level of input noise. Consequently, the network capacity is a viable measure of the ability of a complex network to transfer information. Here again the maximum rate of information transfer is achieved when the network capacity is matched to a given level of fluctuations.

3.4 Optimal Information Exchange

Nonsimple networks, which are referred to in the literature as complex networks, are a challenging area of modern research, overarching and drawing together the common characteristics of the traditional scientific disciplines. The transportation networks of planes, highways and railroads; the economic networks of wages, stock markets and global finance; the social networks of nation states, theocracies and terrorist groups; the communication

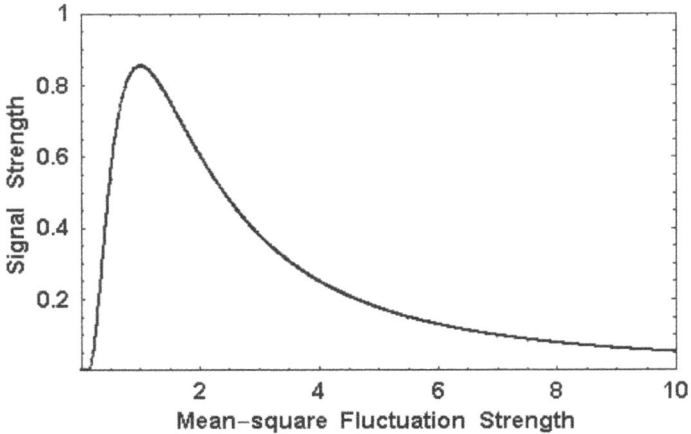

Figure 3.4: The function given by Eq. (3.20) is graphed versus the mean-square level of the fluctuations when the other parameters of the network have been set to unity.

networks of iphones, twitter, and the Internet; the physical networks of the WWW and global warming; the biological networks of the brain, physiology, and food webs, all share a number of apparently universal properties. As networks become increasingly nonsimple, they have fewer and fewer mechanisms in common, and yet their dynamics, in apparent contradiction, converge onto a common set of emergent properties, a number of which we discuss.

Ubiquitous aspects of nonsimple networks emerge that are in conflict with traditional assumptions made about the nature of network dynamics, some dating as far back as Isaac Newton's *Principia*. The most egregious transgressors are the appearance of non-stationary and non-ergodic statistical processes, often manifest as IPL PDFs. Herein we examine traditional dynamic and phase space methods for modeling networks as their nonsimplicity increases and focus on the limitations of traditional procedures used to explain them.

The appearance of non-ergodic behavior ought to have been expected, but it was not, in part, because we have developed our intuition about uncertainty using simple statistical models of reality. Recall that an ergodic process has averages that can be obtained either by using one realization of a long time series generated by the dynamic process, or by using a distribution of an ensemble of realizations of the dynamic process. If the process is ergodic, the two ways of averaging yield the same result. Reasonable arguments can be made to justify ergodicity for simple dynamics, those being linear additive stochastic processes. However, as the processes being studied become more complex (nonsimple) the central limit theorem (CLT) breaks down and the two ways of calculating averages yield very different results.

Immediate examples of weak ergodicity breaking have been found in systems whose dynamics is captured by IPL waiting-time PDFs, whose first moment diverges. Bel and Barkai [78] explain that the relation between ergodicity breaking and diverging waiting times can be understood by noting that one condition for ergodicity to exist is that the ratio of the observation time to the characteristic time scale of the problem is large. However, this condition is never satisfied if the microtime scale, that is, the average trapping time, is infinite.

3.4.1   Nonsimple networks

Of course, we are not able to review the entire field of network science given its explosive expansion over the past decade, so we limit ourselves to an examination of how certain nonsimplicity-induced barriers to understanding have been surmounted, using newly applied theoretical concepts, among which are aging, renewal, non-ergodic statistics, the fractional calculus, but most importantly, crucial events. One phenomenon that emerges from the technical discussion is that of information exchange between nonsimple networks. We posit the management of information by means of the POIE, which is based on establishing the criteria necessary to maximize, much less to determine the information exchange between nonsimple networks. The POIE concept requires a fundamental change in viewpoint that we facilitate by examining how the familiar stochastic resonance idea is changed using the new concept of matching and/or management of nonsimplicity.

A long-standing problem in communications engineering is matching the flow of information to the physical structure of the channel supporting that flow in such a way as to enhance efficiency. To optimize the amount of information transmitted over a channel per unit time, requires a rigorous definition of information, but in addition we also need a measure of how much information transfer a channel can support under ideal conditions (channel capacity). The mathematician Wiener [611] and engineer Shannon [488], among others, narrowly defined these things over half a century ago to make the nascent field of information science manageable. Given the multiple forms of information used in the cyber world, science is today rapidly extending the ideas of these pioneers to the broader context of nonsimple networks. Shannon's notion of a channel has been stretched to include networks from the social and life sciences including the properties of the sender and receiver. Consequently, we find it convenient to mainly avoid the use of the term channel altogether and rely on the more encompassing term networks.

Over the last century, many books have been published that address the nonsimplicity of networks: books on biological networks, focusing on the macrolevel [331], on the scaling of organisms [129, 340], the broad study of the phenomenon of synchronization [514]; tomes on social networks, focusing on the interaction among groups of people [482], and on the citations to scientific papers forming an invisible college [160]; monographs on networks of natural phenomena, including Mandelbrot's now classic first book on fractals [354], which introduced the concept of fractals, the growth of river basins [455] and the morphogenesis of physiologic organs [565]; tracts on information networks, exploring the interface between man and machine [611], initiating the modern theory of communication [489], and quantifying nonsimplicity, using information measures [324].

Each nonsimple network is generated by a mechanism specific to the particular discipline of the phenomenon being considered; the psychology of the individual is modulated by group interactions in the social domain; the across-scale connections from the social domain appears again in the scaling within ecological dynamics; the explosive growth of information in the critical behavior of nonsimple dynamical networks. On the other hand, there are properties common to a myriad of nonsimple phenomena that enables us to identify them as networks. To understand the dynamics of these nonsimple networks, we concentrate on the analysis of information traffic, within and between networks. Traditionally, the flow of information, was assumed to follow an exponential PDF of the time duration of messages, which was consequently characterized by Poisson statistics of traffic volume. However, with the increasing nonsimplicity associated with computer traffic, the traditional methods were abandoned and replaced with less familiar statistical techniques. These techniques had been developed in the diaspore of the physical sciences literature on nonsimple networks and are discussed herein in due course.

There are a number of excellent review articles on nonsimple networks, each with its own philosophical slant. The fundamental aspects of nonsimple networks can be discussed from the perspective of non-equilibrium statistical physics [10], or organized around applications, including social, information, technological and biological networks and the mathematical rendition of the common properties that are observed among them [395].

The approach taken herein is a hybrid, using non-equilibrium statistical physics as a starting point, including stochastic differential equations and the probability calculus. Recent experiments done on nonsimple networks are then used to highlight the inadequacies of traditional approaches. More importantly, we use these data to motivate the development of new mathematical modeling techniques necessary to understand the nonsimple nature of the emergent properties entailed by the underlying network structure.

An intriguing aspect of the dynamics of nonsimple networks is their almost complete invalidation of historical assumptions, including the appearance of non-differentiable, heterogeneous deterministic processes, as well as, non-stationary, non-ergodic, non-Poisson, renewal, statistical processes, which, as mentioned, we label *crucial events*. These properties are manifest through IPL PDFs that not only challenge traditional understanding of nonsimplicity in physical networks, but demand new strategies for understanding how information flows within such networks and is exchanged between such networks. Herein we discuss how the traditional methods of statistical physics, used to characterize the dynamics of nonsimple phenomena, have been extended in the analysis of such living phenomena as networks of neurons, biofeedback techniques and the brain's response to music; and in such geophysical phenomena as global warming. Moreover, we review how these extensions apply to the problem of information exchange between nonsimple networks, independently of the specific interactions among the elements within a network.

One of the most useful mechanisms for extracting weak signals from noisy backgrounds has historically made use of resonances. A linear dynamic process responds strongly to a harmonic perturbation with matching frequency and this linear resonance has been used in a variety of detectors, such as in magnetic resonance imaging (MRI) [566]. Effective two-state stochastic phenomena respond strongly, when the period of a harmonic perturbation matches the transition rate of the stochastic process. Such SR [86], is one strategy for modeling the transmission of information through random (Poisson statistics) media that can enhance the SNR [206], as we discuss in Chapter 5.

If a perturbed network generates a signal described by a Poisson process, using ordinary LRT [301], it is straightforward to go beyond the aperiodic SR [144, 145, 146], which is confined to the case of slow perturbation, and to derive the phenomenon of rate matching, which has been overlooked in the SR literature. Allegrini *et al.* [23] and Aquino *et al.* [42] address the challenging situation where the network being perturbed is neither Poisson, nor ergodic, so as to meet the conditions emerging from nonsimple networks. The case where the network to be perturbed is not Poisson, but is still ergodic, has been studied [223]. The non-ergodic condition [79, 80, 357, 358] violates the conditions for the traditional form of LRT [301], and has been studied by Barbi *et al.* [62], who conclude that a nonsimple network described by intermittent fluctuations, with non-Poisson statistics, does not respond to external periodic perturbations in the long-time limit [503]. We show that the non-ergodic property, which violates LRT, can be explained by adopting a proper definition of nonsimplicity. But, even more importantly, this non-ergodicity leads us to a method for information transport, which bypasses limitations pointed out by various investigators [62, 503].

3.4.2   Synchronization and NERPs

Synchronization is not a simple concept [519], but has a variety of subtleties arising in physical, chemical, social and neural networks whose mathematical analyses have provided distinctions between complete synchronization [135], phase synchronization [459, 461], lag synchronization [454, 462, 486, 629], generalized synchronization [101, 119, 293, 464, 487], intermittent lag synchronization [101, 462], imperfect phase synchronization [625], almost synchronization [187] and finally anticipating synchronization [558]. Consequently, we restrict our discussion to a very small piece of this literature, that piece which most closely relates to the non-ergodic statistics discussed in subsequent chapters, including the influence of age.

Two-state stochastic clocks: In this section, we follow [520] and adopt the general arguments of the decision making model (DMM) made by Turalska *et al.* [535, 536, 597]. In the DMM the two-state clocks are placed at the nodes of a two-dimensional lattice network, where either only the nearest neighbors are coupled, or each element is coupled to every other element in the network (all-to-all, ATA). Here we examine the ATA case in terms of the mean field variable defined by:

$$\xi(K,t) \equiv \frac{N_1(K,t) - N_2(K,T)}{N}, \tag{3.21}$$

where $N_j(K,t)$ is the number of elements in the state $j = 1,2$, at time t given a coupling parameter K. The transition rate from state 2 to state 1 is of the exponential form:

$$g_{12}(\xi;K) = \frac{1}{2}ge^{-K\xi} \tag{3.22}$$

and the transition rate from state 1 to state 2 is:

$$g_{21}(\xi;K) = \frac{1}{2}ge^{K\xi}. \tag{3.23}$$

West *et al.* [597] explored the critical behavior of a network of N clocks in the ATA configuration using the mean field variable in a two-state master equation for each individual. In the mean field case, when $N \to \infty$, the single network becomes a statistical ensemble of identical clocks. When the mean field limit is not taken, there is a finite number of clocks and the dynamical picture stemming from the above master equation is changed. Bianco *et al.* [93] pointed out that in the finite number case, the two-state master equation is formally the same as that for the infinite number situation, except the transition rates fluctuate. The source of the fluctuations in social networks is therefore the finite network size and not the thermal agitation found in physical phenomena.

When substituting these exponential rates from Eqs. (3.22) and (3.23) into the two-state master equation [520], the Langevin equation for the mean field variable $\xi(K, t)$ becomes:

$$\frac{d\xi}{dt} = g\sinh(K\xi) - g\xi\cosh(K\xi)$$
$$+ \sqrt{\frac{2g}{N}}\sqrt{\cosh(K\xi) - \xi\sinh(K\xi)}\eta(K; t). \qquad (3.24)$$

This interacting system is known as the DMM [96, 535, 536, 597]. Here the fluctuations introduced into the rates appear as a delta correlated Gaussian process denoted by $\eta(K; t)$.

Following the expansion procedure outlined in [520], we determine the coefficients to be given by:

$$a_n(K) = g\frac{K^n}{n!} \qquad (3.25)$$

along with the difference coefficients:

$$\lambda_n(K) = g\left(\frac{K^n}{n!} - \frac{K^{n-1}}{(n-1)!}\right) = g(K - n)\frac{K^{n-1}}{n!}, \qquad (3.26)$$

with the K-independent term being the transition rate of the isolated individual:

$$\lambda_0 = a_0 = g. \qquad (3.27)$$

The all-to-all DMM is characterized by a second-order phase transition at the critical point $K_c = 1$ and falls within the (mean field) Ising universality class. This becomes evident upon studying the expanded version of the Langevin equation Eq. (3.24):

$$\frac{1}{g}\frac{d\xi}{dt} = (K - 1)\xi + \frac{1}{6}K^2(K - 3)\xi^3 + \cdots$$
$$+ \sqrt{\frac{2}{gN}}\sqrt{1 + \frac{1}{2}K(K - 2)\xi^2 + \cdots}\eta(K, t). \qquad (3.28)$$

In particular, the coefficient of the linear term is:

$$\lambda_1(K) = g(K-1), \tag{3.29}$$

which vanishes identically when the control parameter is at the critical point:

$$K_c = 1. \tag{3.30}$$

This is a pitchfork bifurcation associated with a second-order phase transition [428, 460]. When $\lambda_1 < 0$, there is one stable equilibrium point at $\xi = 0$ about which the mean field fluctuates. When $\lambda_1 > 0$, there are two stable equilibrium points:

$$\xi_{eq} = \pm\sqrt{\frac{\lambda_1}{|\lambda_3|}} \propto |K - K_c|^{1/2}, \tag{3.31}$$

symmetric about the now unstable equilibrium at $\xi = 0$. Therefore the population is in a disordered phase when $K < K_c$ and an ordered phase when $K > K_c$. The mean field near the critical point is proportional to the square root of the reduced control parameter in Eq. (3.31), which is indicative of the Ising universality class.

The dynamical behavior near the critical point is nonlinear $K - 1 = \varepsilon \ll 1$:

$$\frac{d\xi}{dt} \approx \varepsilon\xi - \frac{1}{3}g\xi^3 + \sqrt{\frac{2g}{N}}\sqrt{1 - \frac{1}{2}\xi^2}\,\eta(t), \tag{3.32}$$

and has been associated with non-ergodic fluctuations and enhanced information transport properties [338, 536, 543]. The term temporal complexity (nonsimplicity) was introduced to describe the special nature of the stochastic dynamics near criticality [536, 597]. These features are relegated to the critical region which, due to finite-size fluctuations, is of size N^{-1} about the theoretical value $\varepsilon = 0$. The critical region becomes more extended for smaller populations, implying that some benefits might be gained by limiting the population size, but signatures of the bifurcation can be lost if the size becomes too small [460]. The population size was found to strongly affect the ergodicity of the mean field dynamics near criticality [535]; the network can be moved from a non-ergodic condition to an ergodic one upon decreasing its size. Note that once the control parameter K has been fixed to a near-critical value, there are no free parameters left in the DMM. This fixes the phase transition and prevents the possibility of generating the extended critical behavior discussed in [520].

Collective behavior: The random fluctuations induce transitions between the two states of the potential well. Thus, for a network with a finite number of clocks the phase synchronization of Eq. (3.28), obtained when $\eta(t) = 0$, is not stable. The mean field variable fluctuates between the two minima for the coupling parameter greater than the critical value as depicted in Fig. 3.5. The single clock follows the fluctuations of the global clock variable ξ, switching back and forth from the condition where the state 1 is preferred statistically to that where the state 2 is preferred statistically.

The PDFs of the sojourn times in the state $\xi > 0$ or $\xi < 0$ as shown in Fig. 3.5 are identical since the potential is symmetric. Thus, we denote them by $\psi(t)$. Let us consider a condition where the coupling constant K is close to the critical value $|K - K_c| \ll 1$. In this case, the height of the barrier, separating the wells of the potential, is smaller than, or comparable to, the intensity of the fluctuations in Eq. (3.32). Under these conditions, we expect an IPL of the form:

$$\psi(t) \propto t^{-3/2}$$

for an extended interval of sojourn times [93, 357]. This is exactly what

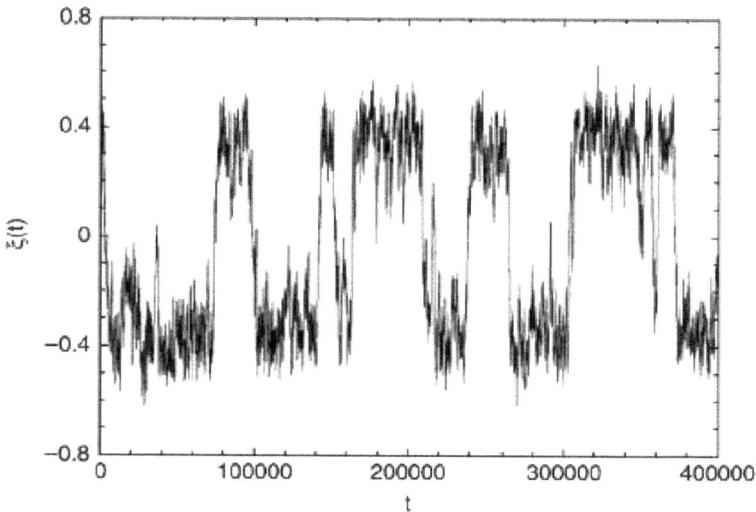

Figure 3.5: The global variable $\xi(t)$, as a function of time, is graphed for the parameter values $K = 1.05, g = 0.01$ for a network with 103 clocks. Taken from [597] with permission.

is observed in Fig. 3.6, where the full line denotes the survival probability $\Psi(t)$, namely, the probability of observing a sojourn time larger than t, which in this case would be $t^{-1/2}$. For any fixed value of the unperturbed rate g, the height of the potential barrier increases as the value of the coupling parameter increases. Eventually, the barrier height becomes much larger than the intensity of the fluctuations, so consequently, the IPL behavior depicted in Fig. 3.6 vanishes, the theoretical arguments given above [93, 357] lose validity and an exponential behavior emerges, as predicted by Kramers theory [300, 597].

The influence of age: The influence of age is established by showing that the transition between the states in the potential well is a renewal process. We evaluate the survival probability $\Psi(t_a, t)$ of age t_a by observing a sojourn time larger than t, when the observation starts at a time t_a after a crossing from $\xi > 0$ to $\xi < 0$, or vice versa. Note that with this definition of an aged survival probability, we have $\Psi(t_a = 0, t) = \Psi(t)$. Now we compare the aged survival probability with the expected survival probability evaluated according to renewal theory $\Psi_{ren}(t_a, t)$ as discussed in a subsequent chapter.

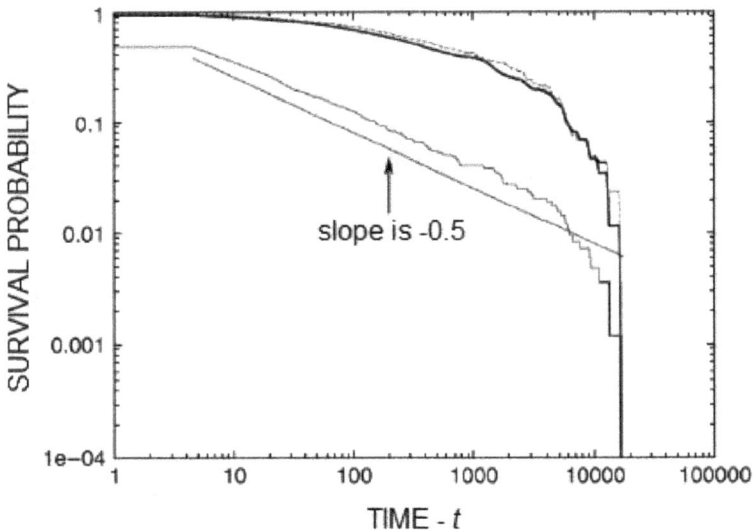

Figure 3.6: Three survival probabilities are depicted. One for the original time series, the second delayed for a time t_a which superposes on the first. The third is shuffled and has an IPL index of 0.5 for over three decades of scale. Taken from [597] with permission.

If the two survival probabilities are equal:

$$\Psi(t_a, t) = \Psi_{ren}(t_a, t)$$

for all ages t_a, the process described by the survival probability $\Psi(t)$ is a renewal process. If $\Psi(t)$ is not an exponential function, $\Psi(t_a, t)$ yields a slower decay than $\Psi(t)$, a condition that is denoted as 'aging' [93]. Figure 3.6 shows that the transition between the minima of the potential well is a renewal process, that is, the above equality is established numerically and consequently there is aging.

On the other hand, as the value of the coupling parameter K is increased, the aging property is lost, because the survival probability $\Psi(t)$ becomes an exponential function [94]. Even with this loss of aging the renewal property persists, as is also evident in Fig. 3.6. The survival probability in this figure is terminated by an exponential resulting from the finite size of the network.

In summary, for an ensemble of networks, each with an infinite number of two-state clocks, a phase transition occurs at the critical value of the coupling parameter $K_c = 1$. The phase transition mirrors a statistical 'preference' for a single clock and for the ensemble a global phase synchronization. If the number of clocks in the network is finite the phase synchronization is not stable. In this case, the variable describing the collective motion of the network is characterized by non-Poisson intermittent behavior, see Fig. 3.6. At the onset of the phase transition ($|K - K_c| \ll 1$), the zero crossings of the global variable define a series of events, with the properties: (1) The PDF of inter-event intervals (full line in Fig. 3.6) has a non-Poisson non-ergodic character ($\psi(t) \propto t^{-3/2} \Rightarrow$ infinite mean inter-event interval). (2) The sequence of inter-event intervals satisfy the renewal aging condition, cf. Fig. 3.6. These two properties are also observed properties of the statistics of blinking quantum dots [303, 304, 398, 491] and brain activity, suggesting that a network consisting of a finite number of coupled two-state clocks may be a useful model for exploring the dynamics of both processes, as well as others that share certain generic features.

3.5   The Dunbar Effect

The social brain hypothesis of Dunbar [176, 178] has generated substantial interest due to its bridging the gap between cognition and sociology. The number of people with whom a single individual can establish stable social relations is the "magic" [48] number 150 and is related by Dunbar to the connectivity of neurons within the brain. Physicists have historically introduced such "magic numbers" to highlight important patterns in complex datasets for which no underlying theory had yet been established, but for which the need was evident.

A quarter century ago, Dunbar [175, 177] in considering the social organization consisting of primates was able to relate measures of the neocortical ratio C to the mean group size N of a sample of 36 primate genera and constructed a version of the social brain allometry relation (SBAR):

$$N = aC^b, \tag{3.33}$$

where a and b are the empirical constants:

$$a = 1.239, \quad b = 3.389. \tag{3.34}$$

Here C is the ratio of neocortical volume to that of the total volume of the brain minus that of the neocortex. The neocortex ratio for humans is $C = 4.1$ [510], which when inserted into the SBAR model yields an average group size of $N = 147.8$. This value rounded off to 150 is now called the Dunbar number.

G. West, no relation to the author, in a recent popularization of his two decades of collaborative research on allometry relations [610], illustrated the attempts at explaining the Dunbar hypothesis including the conjecture that the number 150 may reflect the desire of individuals within a group to maximize their assets while realizing the maximal filling of social space. The still more recent book of Bahcall [48] relates the Dunbar effect to the occurrence of phase transitions in nonsimple phenomena. Bahcall conjectures that, as far as nonsimplicity management issues are concerned, the long-term survivability of a group, company, or organization, depends on its adopting the form of organization that supports innovative ideas favoring societal progress. The recent work of Mahmoodi *et al.* [342] proposes a theory by which a nonsimple network can realize self-organization, and provides a theoretical foundation supporting the speculations of both G. West and Bahcall just cited.

According to the self-organized temporal criticality (SOTC) model [342] the process of self-organization is determined by the actions of single individuals to receive the largest individual payoffs, and they accomplish this by simultaneously reaching maximal agreement between their opinion and the opinions of their nearest neighbors. The direct calculation of the transmission of information from one SOTC system to another was found to be non-monotonic with increasing network size. The maximum information transfer occurred when the number of units N of the two self-organizing systems is in the range [100, 200].

The SOTC proposed by Mahmoodi *et al.* [342] shifts the focus from the nonsimplicity of amplitude avalanches [51] to the nonsimplicity of time intervals between consecutive events [605]. The spontaneous transition of nonsimple dynamics to criticality generates crucial events that are responsible for the sensitivity of the network dynamics to the environment. In addition crucial events entail the maximization of information transport from one to another SOTC system. The time interval τ between consecutive crucial events is given by waiting-time PDFs sharing in the intermediate asymptotic regime [64] the same IPL structure as:

$$\psi(\tau) = (\mu - 1)\frac{T^{\mu-1}}{(T+\tau)^{\mu-1}}, \tag{3.35}$$

with $1 < \mu < 3$. This condition generates $1/f$-noise with the IPL spectrum:

$$S_p(f) \propto \frac{1}{f^\beta}, \tag{3.36}$$

where

$$\beta = \begin{cases} \beta = 3 - \mu, & \text{if} \quad \mu < 2 \\ \beta = \mu - 2, & \text{if} \quad 2 < \mu < 3 \end{cases}. \tag{3.37}$$

Note that the IPL structure holds only in the intermediate asymptotic regime being exponentially truncated in the long-time regime. This truncation limits the efficiency of information transmission and it is important to explain the Dunbar number $N = 150$. In fact, decreasing N has the effect of not only increasing the intensity of temporal fluctuations, but also that of decreasing the extension of the nonsimple intermediate asymptotics.

In the last few years, the conjecture has been made that biological systems function best when their dynamics are close to criticality [386]. Thus, it is reasonable to implement the associations made among nonsimplicity, criticality and collective behavior to address the issue of cognition using the concept of a collective mind [154]. Long-range correlations are amplified at the onset of a phase transition and are often studied by means of the Ising model [131]. On the other hand, the Ising model at criticality generates intermittent [149] and crucial events [481], which according to Paradisi *et al.* [411] is a manifestation of consciousness.

In this section we address the issue of consciousness and nonsimplicity by exploring the connection between criticality and the Dunbar hypothesis. We do this by relating the consciousness/nonsimplicity issue to the use of scaling theory in the search for the origin of anomalous diffusion series $\xi(t)$ and using a mobile window to transform the fluctuations characterized by $\xi(t)$ into an ensemble of diffusional trajectories $Q(t)$. The purpose of the procedure is to establish that the departure of $\xi(t)$ from a completely random function could be detected through the departure of the scaling of $Q(t)$ from ordinary diffusion by means of a scaling index being different from $\delta = 0.5$.

We herein extend this technique by interpreting $Q(t)$ as the carrier of crucial events that may be different from those hosted by $\xi(t)$. In other words, we interpret $Q(t)$ as a time series in its own right, which can be analyzed by studying its diffusive properties. If the subsequent diffusion is anomalous, we denote that behavior with the nomenclature from the sociology literature of the subsequent dynamics being intelligent.

We statistically analyze the time series generated by two models of criticality-induced intelligence, with a method recently proposed to detect crucial events by Culbreth *et al.* [157]. This method shares the same purpose as that of an earlier paper [472], based on converting the time series data into a diffusion process and is called DEA to determine the scaling of a diffusive process. When criticality-induced intelligence becomes active the constructed process is expected to depart from ordinary diffusion which is signified by having a scaling index different from $\delta = 0.5$. The modified DEA (MDEA) illustrated in [157] overcomes the limits of the original DEA technique [472] that were pointed out by Scafetta *et al.* [475]. As noticed in the latter reference the original version of DEA cannot assess if the deviation from $\delta = 0.5$ is due to the action of crucial events, or to the infinite memory contained in fractional Brownian motion (FBM) [353]. The present analysis, using MDEA, filters out the scaling behavior of infinite stationary memory, when it exists, and the remaining departure of the scaling index from $\delta = 0.5$ is solely a consequence of crucial events.

The MDEA applied to the signal $\xi(t)$ generated by the criticality-induced intelligence implements the original DEA in conjunction with the Method of Stripes (MoS), which is discussed in some detail in Chapter 7. In the MoS the ξ-axis is divided into many bins of equal size and an event, either crucial or not, is detected when $\xi(t)$ moves from a given stripe to an adjacent stripe. A random walker (RW) step is triggered by such an event and the RW takes a forward step of constant length each time an event occurs, thereby generating a diffusional trajectory $Q(t)$ affording information on the persistence of opinion. To detect this information we apply the MDEA method to $Q(t)$.

We have selected two network models that generate criticality-induced intelligence. The first is the DMM briefly reviewed in the preceding section where N individuals have to make a choice between two conflicting decisions. They do that under the influence of their nearest neighbors. This model falls into the Ising universality class, thereby making it possible for us to compare our results to the predictions of Tagliazucchi *et al.* [523]. The second model considered is that of swarm intelligence proposed by Vicsek *et al.* [548]. We evaluate the intelligence of the DMM network using both the signal $\xi(t)$ and its time derivative $\eta \equiv d\xi/dt$.

Figure 3.7 illustrates the results of those analyses. The qualitative

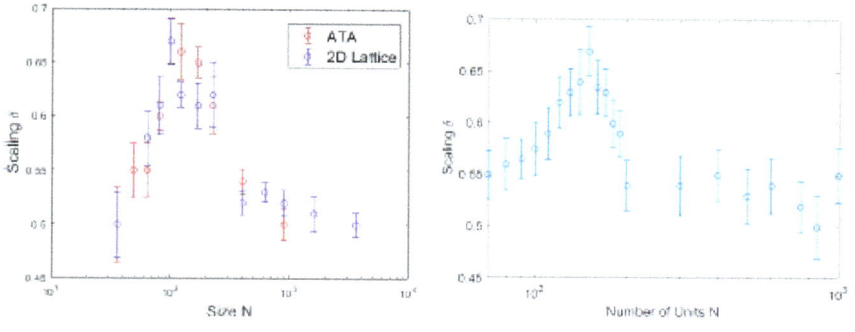

Figure 3.7: Scaling detection of the Dunbar number is obtained by calculating the non-montonic dependence of the scaling index δ on a network of size N. On the left, two calculations are depicted using a DMM [605]; the red points with an all-to-all interaction, the blue points with a nearest neighbor interaction on a two-dimensional lattice. On the right, the same calculation is carried out using a model of swarm intelligence proposed by Vicsek *et al.* [548].

agreement between DMM, in the left panel, and swarm intelligence, in the right panel, is remarkable. The maximum scaling index δ is very close to the value $\delta = 0.67$ when N is in the vicinity of the Dunbar number 150 and falls quickly to $\delta = 0.5$ on the left, for $N < 150$ and more slowly to the same value on the right, for $N > 150$. One is tempted to interpret this result as a sign that intelligence emerges in the vicinity of the Dunbar number $N = 150$.

The left panel of Fig. 3.7 contains the global fields of DMM with the ATA condition realized using:

$$\eta \equiv \dot{\xi} = g \sinh K(\xi + f(t)) - g\xi \cosh K(\xi + f(t)), \qquad (3.38)$$

where $1/g$ defines the time scale of the single units, when they work in isolation. The random noise $f(t)$ serves the purpose of mimicking the small size nature of the social system insofar as its intensity is inversely proportional to the small number of units N. In the ATA case the process is realized by the observation of η. It is beyond the scope of our purpose here to discuss the technical issue of whether the observation of η may reveal crucial events without using MDEA.

The fact that optimal information exchange between networks occurs at the size of the Dunbar number is supported by direct numerical calculations. Let us consider two nonsimple networks, e.g., two flocks of birds, A and B

at criticality. They are identical, each having the same number of units N and they interact with one another for a time L. The global field is denoted by $\xi(t)$ for system A and the global field is denoted $\zeta(t)$ for system B. We evaluate the cross-correlation function:

$$C(\tau) = \frac{\int_0^{L-\tau} dt(\xi(t) - \bar{\xi})(\zeta(t+\tau) - \bar{\zeta})}{\sqrt{\int_0^L dt(\xi(t) - \bar{\xi})^2 \int_0^L dt(\zeta(t) - \bar{\zeta})^2}}, \tag{3.39}$$

where $\bar{\xi}$ and $\bar{\zeta}$ denote the time averages of the field $\xi(t)$ and $\zeta(t)$, respectively.

This cross-correlation numerical experiment is done in two ways and the results shown in Fig. 3.8. In the first case a small percentage, 5% of units of A, randomly chosen, make their choices on the basis of the choices made by their nearest neighbors and one randomly chosen unit of the system B. The system B is influenced by the system A through the same interaction process. As a result of this reciprocal interaction, the cross-correlation time is expected to be symmetric around $\tau = 0$. In the second case, we expect that the cross-correlation function shifts to the right as a consequence of the fact that the information about A transmitted by 5% of B units perceiving the motion of A does not have the immediate effect of making all the other units adopt the motion of A.

Cavagna *et al.* [137] made the conjecture that the transmission of information from the lookout birds to all the other birds of the system occurs through a diffusion process. Lukovic *et al.* [338] argued that the change of direction of the flock requires a sufficiently large number of origin re-crossings, namely, they assigned to the visible crucial events an important role in information transmission. Of course, theoretically predicting the Dunbar number does not establish that the size of the network influences the transmission of information, much less that the Dunbar size of a network optimizes the exchange of information between networks. We can, however, determine network efficiency from the cross-correlation of the time series for the perturbing network A and the perturbed network B.

Figure 3.8 shows that the time delay between the driven and driving networks is extremely small when $N = 150$ (the delay time is $\tau \approx 0$), whereas larger networks require a finite non-zero time to reorganize and maximize their correlation. The larger the deviation in network size from the Dunbar number the greater the delay in transmitting the information throughout the perturbed network. This is an evident sign that the Dunbar effect facilitates the transport of information from the lookout birds (the 5%) to all the other birds of the flock. More generally, this is an evidence that the Dunbar effect facilitates the transport of information from the individual who first acquires the information to all the other individuals in the network.

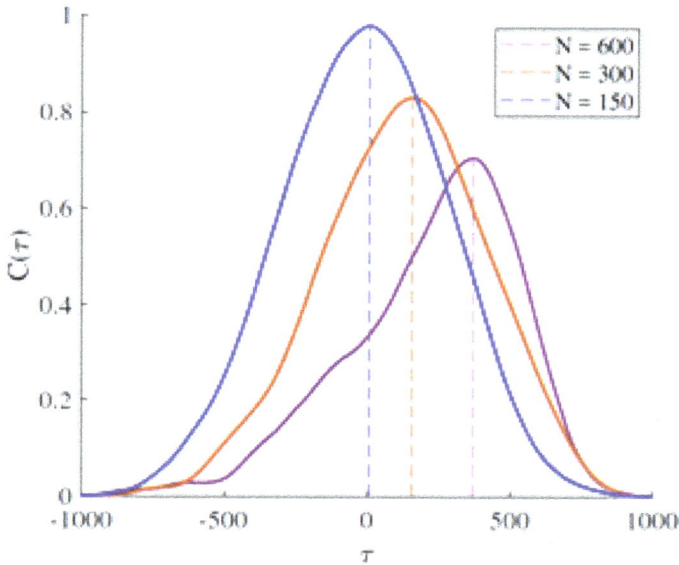

Figure 3.8: The cross-correlation function of two interacting networks A and B, at criticality are depicted for three sizes of networks. Here 5% of A units determine their behavior by selecting the average behavior of their six nearest neighbors and one unit of B. The unit of B is influenced by the 5% of A, but otherwise interacts normally with the other members of its group. The cross-correlation function is calculated using a Z-variable obtained by centering each time series on its time-averaged value and normalizing the difference variable to the standard deviation to obtain: $C(\tau) = \langle Z_A(t)Z_B(t + \tau)\rangle$, where the brackets denote a time average over the interval L.

3.6 Information and Nonsimplicity

Maxwell's Demon: James Clerk Maxwell, considered by many to be the leading scientist of the 19th century, recognized that the kinetic theory of gases, which he invented and refined along with Ludwig Boltzmann, who others believed held this distinction, could in principle, lead to a violation of the second law of thermodynamics. His argument used information about the motion of molecules to extract kinetic energy to do useful work. In his 1871 book, *Theory of Heat*, Maxwell introduces his now infamous demon as depicted in Fig. 3.9:

..a being whose faculties are so sharpened that he can follow every molecule in his course, and would be able to do what is presently impossible to us....Let us suppose that a vessel is divided into two portions. A and B by a division in which there is a small hole, and that a being who can see the individual molecules opens and closes this hole, so as to allow only the swifter molecules to pass from A to B, and only the slower ones to pass from B to A. He will, thus, without expenditure of work raise the temperature of B and lower that of A, in contradiction to the second law of thermodynamics.

In his remarkable book, *Science and Information Theory*, Brillouin [116] revealed that the paradox of Maxwell's demon is at the nexus of information theory and physics. He reviewed the many 'resolutions' to the paradox that have been proposed over the nearly hundred years between the publications of the two books. He pointed out that Szilard [522] was the first to explain that the demon acts using information on the detailed motion of the molecules, actually changing information into negentropy. Information from the environment is used to reduce the entropy of the system. Brillouin,

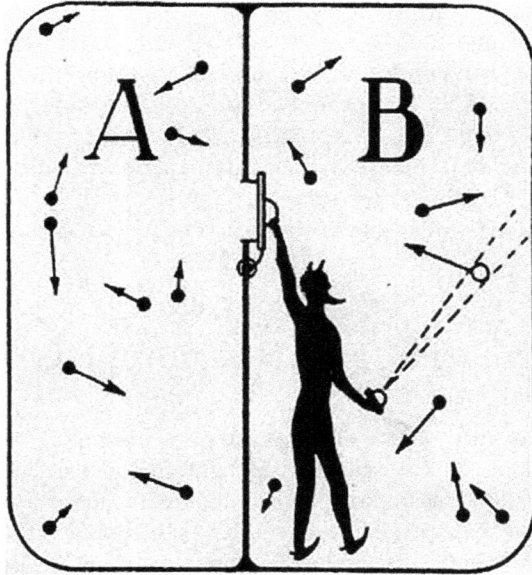

Figure 3.9: Maxwell's demon is depicted controlling the passage of particles between compartments by means of a hole that he can open and close.

himself, resolved the paradox using photons, against a background of black-body radiation, that the demon must absorb to see a molecule of gas. We do not present Brillouin's discussion here, but he subsequently concludes that the average entropy increase is always larger than the average amount of information obtained in any measurement:

> ..every physical measurement requires a corresponding entropy increase.

Norbert Weiner, in his book that started the field of *Cybernetics* [611], suggested that the demon must have information about the molecules in order to know which one to pass through the hole and at what times. He acknowledges that this information is lost once the particle passes through the hole and puts information and entropy on an equal footing by observing that the demon's acquisition of information opens up the network. The demon-gas network has an increasing total entropy, consisting as it does of the sum of the entropy of the gas, which is decreasing, and the negentropy (information), which is increasing. Ball [54] notes that Szilard had previously concluded that:

> ..the second law would not be violated by the demon if a suitable value of entropy were attributed to the information which the demon used in order to reduce the entropy of the rest of the system.

The key feature in the resolution of Maxwell's demon paradox is the existence of dissipation, that is, the erasure of memory, in the information cycle. This occurs in Brillouin's argument through the requirement that the photon energy exceeds that of the blackbody radiation and in Wiener's discussion, with the observation that the particle forgets its origin once it passes through the hole. Landauer [309] indicates that these early arguments can be summarized in the statement:

> The erasure of the actual measurement information incurs enough dissipation to save the second law.

Landauer's *erasure principle* [310, 311] interpreted information to be a physical phenomenon. He argued that rather than being the abstract quantity that forms the basis of intense mathematical discussion in texts on information theory, which it certainly does, information itself is always tied to a physical representation. Whether it is a spin, a charge, a pattern of human origin, or a configuration of nature's design, information is always tied to a physical process of one kind or another. Consequently, information is physical in the same sense that entropy and energy are physical.

Bennett [85] was able to pull the threads of all the various arguments together, in the context of reversible computation, and in so doing he obtained what is considered to be the final resolution of the Maxwell's demon paradox. This world view suggests that we adopt a perspective that focuses on the occurrence of events rather than on the dynamics of network variables. We adopt this point of view and briefly review the changes that arise in the physical descriptions of nonsimple networks when this is done.

3.6.1 Information/nonsimplicity interchangeability

It is possible to identify information with nonsimplicity in unique and useful ways. Bialek *et al.* [92] do this by noting that information only has value when it results in a prediction and that most empirical information does not satisfy this constraint. In their analysis, the mutual information between the past and future behaviors of a time series of length L is used to define the predictive information $\mathcal{I}_p(L)$. Consequently, in the limit of long observation times $\mathcal{I}_p(L)$ is either: a constant, independent of L; increases as log L; or increases as L^α, with $\alpha > 0$. The links between predictive information and measures of nonsimplicity have been established by analyzing physical systems, using nonlinear dynamics and statistical physics. They argue that the ways in which the different contributions to the total information diverge gives rise to a unique measure for the nonsimplicity of the dynamics generating the time series.

They avoid the pitfall that many have identified while attempting to use entropy as a direct measure of nonsimplicity, since for a totally random process, the entropy is maximum and gives the appearance of having maximum nonsimplicity, as well as, maximum information. As a pedagogical example, of how to circumvent this problem of predictive information being overwhelmed, they [92] consider a long chain of two-state clocks (spins) and define the number of distinct sequences of digits W_k, $k = 0, 1, ..., 2^N - 1$, which for a long sequence provides an estimate of the PDF $P_N(W_k)$ of having different words W_k of length N. The entropy of words of length N, using a binary alphabet, drawn from a very long sequence of two-state elements, is

$$S(N) = - \sum_{k=0}^{2^N-1} P_N(W_k) \log_2 P_N(W_k). \qquad (3.40)$$

As they observe, the entropy is an extensive property of N, so consequently, $S(N)$ is asymptotically proportional to N for any long chain of two-state clocks:

$$S(N) \approx S_0 N. \qquad (3.41)$$

The goal of physical applications, such as the magnetization of a piece of material, is to provide a model for S_0. But mechanistic modeling of parameters is not our purpose here.

The entropy $S(N)$ for a chain of one billion two-state clocks was calculated for three types of interaction, as depicted in Fig. 3.10 and the asymptotic extensive nature of the resulting entropy is clear. The extensive nature of the entropy is independent of whether the interaction strength is constant, short-range or long-range and is therefore truly universal. It is also clear that if these were constructed from empirical datasets they would be indistinguishable.

They go on to remove the extensive part of the entropy from the total entropy as follows:

$$S(N) = S_0 N + S_1(N) \tag{3.42}$$

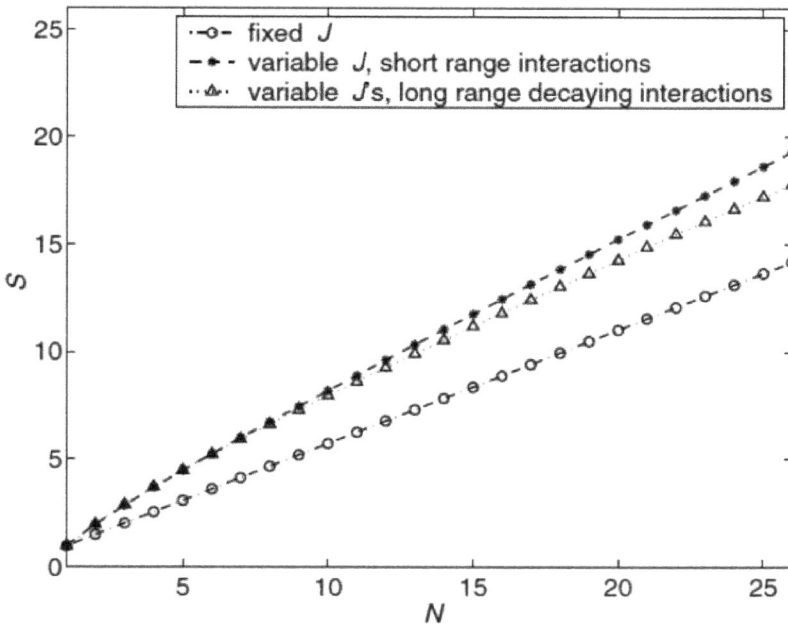

Figure 3.10: Entropy as a function of the word length for two-state clock chains with different interactions as indicated in Appendix A. All lines start from $S(N) = \log_2 2 = 1$, since at the values of the coupling investigated, the correlation length is much smaller than the chain length (10^9 two-state clocks). From [92] with permission.

and obtain the dramatically different result depicted in Fig. 3.11, where only the sub-extensive entropies $S_1(N)$ remain. Here the difference in the interaction strengths is evident: the constant interaction strength produces a constant sub-extensive entropy; a short-range interaction produces a logarithm in N; a long-range interaction induces a power-law in N. The sub-extensive part of the entropy is postulated [92] to contain predictive information about the underlying dynamics.

3.6.2 Diffusion entropy analysis

These results are now shown to be useful in the analysis of time series generated by nonsimple phenomena. It is proven to be particularly successful

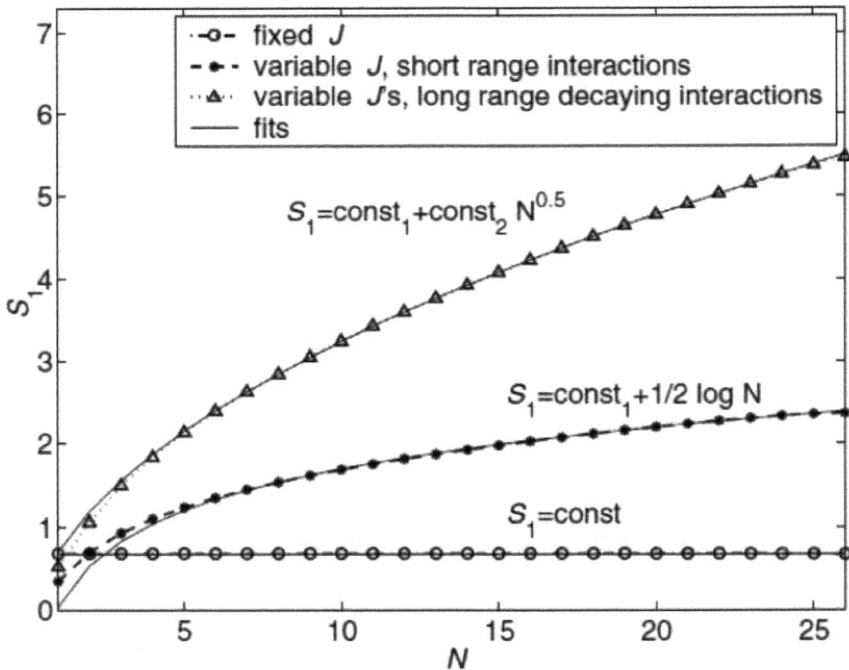

Figure 3.11: Entropy as a function of the word length for two-state clock chains with different interactions as indicated in Appendix A. Each line starts from its own value of $S(N) - S_0 N$, since at the values of the coupling investigated, the correlation length is much smaller than the chain length (10^9 two-state clocks). From [92] with permission.

in the determination of the scaling properties of the empirical time series generated by nonsimple networks. Consider, for example, the time intervals between beats of the heart, known as heart rate variability (HRV). It has been determined that such HRV data are a mixture of crucial events and ordinary Poisson events [29]. A method involving the use of entropy enables the detection of crucial events in the presence of Poisson events by measuring the difference in the ways the two statistical processes scale [102]. The desired difference in scaling is detected in the following way.

Grigolini *et al.* [232] use the detected events to generate a diffusion process $Q(t)$ by means of the rule that the random walker jumps ahead when an event, either crucial or Poisson, occurs. The scaling generated by the Poisson events has a power-law index $\delta = 0.5$, whereas the scaling power-law index δ of the crucial events is given by

$$\delta = \frac{1}{\mu - 1}, \tag{3.43}$$

when the IPL index of the crucial events is given by μ. Note that the latter scaling dominates asymptotically in the time due to Eq. (3.43) resulting in $\delta > 0.5$ when the condition $2 < \mu < 3$ applies [232]. To be explicit, we generate a fluctuation $\xi(t)$ with the value 1 when an event from the empirical time series, either crucial or Poisson, occurs, and a value zero when no event occurs. The diffusion variable $Q(t)$ is obtained by integrating the following equation:

$$\frac{dQ(t)}{dt} = \xi(t). \tag{3.44}$$

We follow the diffusion entropy analysis (DEA) procedure outlined in [103], who used the diffusion entropy to establish a connection between rhythmic behavior and fluctuations in brainwave data. They introduced a moving window of size l to generate a PDF $P(x, l)$ by sweeping the window through the time series and constructing a histogram from the resulting ensemble of random walk trajectories. The empirical PDF is then used to define the Wiener/Shannon information entropy for the ensemble of random walks generated by the empirical dataset:

$$S(l) = - \int_{-\infty}^{+\infty} dq P(q, l) \log_2 P(q, l). \tag{3.45}$$

The empirical PDF constructed from the diffusion process frequently has the scaling form

$$P(q, l) = \frac{1}{l^{\delta}} F\left(\frac{q}{l^{\delta}}\right), \tag{3.46}$$

where $F(\cdot)$ is an undetermined analytic function of its argument. Then inserting Eq. (3.46) into Eq. (3.45), after some algebra, yields the diffusion entropy

$$S(t) = A + \delta \log_2 l, \tag{3.47}$$

where A is the entropy constant:

$$A \equiv - \int_{-\infty}^{+\infty} dy F(y) \log_2 F(y). \tag{3.48}$$

Note that Eq. (3.47) yields the result for the non-extensive entropy in Fig. 3.11 when $\delta = 1/2$, which it would be for a dataset having Gauss statistics.

It is important to stress that a significant advance of the theoretical dynamic justification of Eq. (3.47) based on an extension of the theory of *self-organized criticality* (SOC), incorporating nonsimplicity in the time domain, is called *self-organized temporal criticality* (SOTC) [297, 298, 299]. This new theory provides a rationale for the crucial IPL index μ, which we subsequently explain more fully. In fact, according to SOTC, the processes of spontaneous self-organization, in general, and especially those behind the statistical analysis used here, namely physiological processes, naturally evolves to a critical state generating crucial events. As a consequence of temporal nonsimplicity emerging in the intermediate time scale, the diffusion entropy $S(l)$ is not a straight line when expressed as a function of $\log_2 l$, however, the diffusion entropy is a straight line in the intermediate time region. In the intermediate time regime its slope is used to define the statistics of crucial events occurring within that region through the IPL index μ, established by Eq. (3.43).

In Fig. 3.12 the diffusion entropy is graphed versus the natural logarithm of the time as generated by an EEG dataset for a healthy young adult [103]. The slope of the curve yields $\delta = 0.81$ from Eq. (3.47), which when inserted into Eq. (3.43) yields the IPL index $\mu = 2.23$ for the crucial events. The intermediate time regime in this figure for which the IPL holds is greater than three decades of time. The reasons for the deviation of the DEA from the straight line at each end is an artifact of the data processing technique and is discussed at length in Bohara *et al.* [103].

3.7 Information (Entropy) a 5th Force of Nature?

There has been an ever increasing number of investigators arguing that entropy is actually a 5th force of nature; one separate and distinct from the

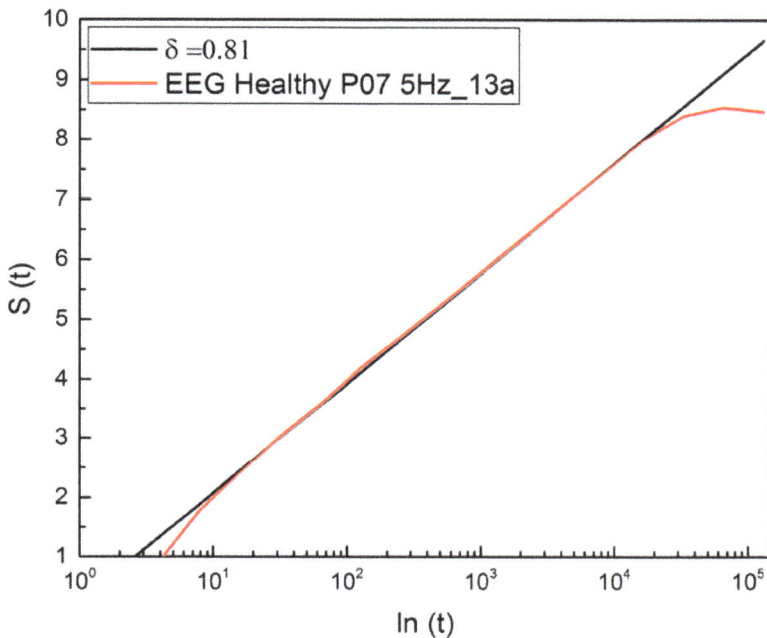

Figure 3.12: Diffusion entropy, as a function of the natural logarithm of the time, was generated by an EEG time series. The slope of the entropy curve yields $\delta = 0.81$ clearly different from a Poisson process that would have $\delta = 0.50$. From [102] with permission.

gravitational force, the electromagnetic force, the strong force and the weak force. The argument for a 5th force goes as follows: the entropic force is a force generated by entropy, see for example, [132, 257, 458]. A tautology, right? Not quite. An entropic force will give rise to an entropic flux, but through the Onsager relations of thermodynamics it may also give rise to a dazzling variety of other fluxes, as well, including the cross effects of the thermoelectric effect and thermodiffusion [203].

3.7.1   Entropic forces

Consider, for example, the phenomenon of osmosis, which occurs across a selectively permeable membrane separating two fluids in a vessel. A selectively permeable membrane is one that allows for the free passage of one kind of molecule, but not of another. In the simplest case, on one side of the membrane the fluid is pure water, on the other side the fluid has a concentration of a solute dissolved in water, say sugar. The semipermeable membrane allows the free passage of water molecules from one side of the vessel to the other, but obstructs the passage of the larger sugar molecule across the membrane, thus restricting the sugar molecules to one side of the vessel. The thermal agitation of the water molecules, the same mechanism that results in diffusion, traps the water molecule on the high concentration side of the barrier once it has been traversed. This is a consequence of the water molecule moving more slowly on the high concentration side with the larger sugar molecules. The osmosis phenomenon is depicted in Fig. 3.13 where the lowering of the water level on the low concentration side of the membrane, due to what is termed osmotic pressure, is really quite dramatic.

From the above description, one might conclude that osmosis is nothing more than the simple diffusion of water through a membrane. It is not.

Experiments have been done that indicate that the diffusion rate of water is much less than the flux of water, observed using tracer particles, due

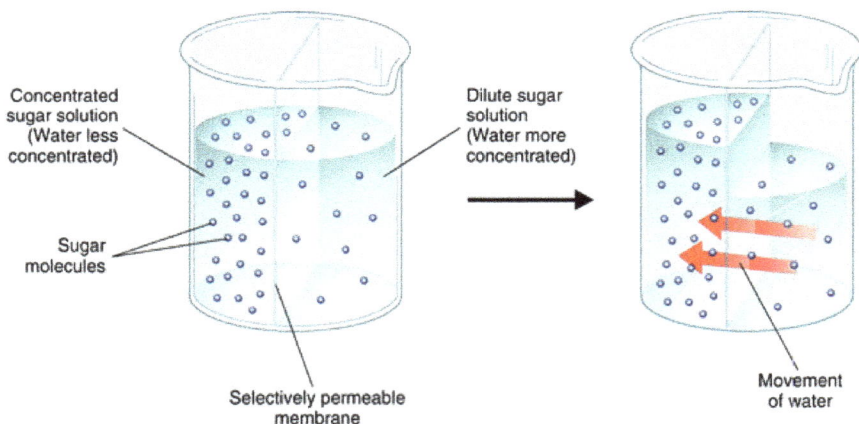

Figure 3.13: On the left is the initial state of the separation of two fluids consisting of water and sugar. A high sugar concentration to the left of the membrane; a low sugar concentration on the right of the membrane. Osmotic pressure moves the water from the low to the high concentration side of the membrane, thereby lowering the height of the fluid on the low concentration side, since there is no back-flow to replace the water. The water surface stops falling when the hydrostatic pressure, which produces a reverse osmotic pressure, just balances the osmotic pressure across the membrane. The equilibrium occurs at a specific height that is dependent on the initial concentration gradient across the membrane.

to osmotic gradients. Thus, even though the energy source for both osmosis and diffusion is the same, that being thermal motion of the molecules, the enhanced flux of water due to osmosis is a consequence of a non-energetic mechanism; one resulting from the entropy gradient and not an energy gradient. This is extremely important because the osmotic force is very common in living matter, facilitating all manner of physiologic functions, from the transport of nutrients across cell membranes, to the determination of water movement between the intestinal lumen and the blood, which is produced by osmotic forces determined by solute gradients.

On the other hand, Sokolov [506] argues that entropic forces in a physical system at equilibrium can be traced back to conditional averages of constraint forces. In his pedagogic paper, he discusses several examples where such constraints are modeled as limiting values of mechanical interactions, such as the constraint of a wall being given by the infinitely steep limit of a harmonic potential. He begins with the Boltzmann equation for the conditional entropy of a polymer chain:

$$S(x) = k_B \ln W(x), \qquad (3.49)$$

where x can be the fixed end-to-end distance of the chain and $W(x)$ is the number of possible configurations. Neglecting all interactions except the mechanical constraints holding the chain together and assuming the internal energy U to be independent of the chain configuration, the free energy is $F(x) = U - TS(x)$ and allows the entropy force to be written as:

$$f(x) = -\frac{dF(x)}{dx} = T\frac{dS(x)}{dx}. \qquad (3.50)$$

Consequently, using the total number of configurations by W the PDF can be expressed by the relative frequency $p(x) = W(x)/W$, thereby reducing Eq. (3.50) to the form

$$f(x) = k_B T \frac{d}{dx} \ln p(x). \qquad (3.51)$$

Thus, the entropic force is determined once the PDF is established. But the question remains: Where does this force come from?

Sokolov's arguments to determine its source, invariably demonstrate that the mysterious entropic forces appear to always have a mechanical origin. Furthermore, the entropic force can be explained using statistical thermodynamics and may well be restricted to equilibrium systems in which one has equipartition of kinetic energy. However, to be fair to Sokolov, his stated intent was to demystify the concept of entropic forces in polymer physics. He accomplished this by showing that entropic forces can be understood in terms of average constraint forces corresponding to non-holonomic constraints.

3.7.2 Information forces

Information entropy was defined in terms of the Wiener/Shannon entropy in Sec. 3.6 and now we get to the central point of this chapter, that being, whether or not there are information forces that can be defined that directly parallel entropic forces. But perhaps even more strongly. Are there information forces that depend solely on the properties of information and which are as tangible as entropic forces, but not reducible to other thermodynamic-like quantities such as diffusion or average constraint forces?

Let us first assume that each microstate in the information entropy is equally likely, such that $p_k = 1/W$. Inserting this value for the probability of each microstate into the Wiener/Shannon information entropy yields Boltzmann's equation for the thermodynamic conditional entropy being interpreted as k_B times the information entropy. In information theoretic terms, this amounts to the information entropy of a system being the quantity of "missing" information needed to determine a microstate, given a known macrostate. This provides an information theory interpretation of the statistical mechanical foundation of thermodynamics in terms of entropy maximization as developed over a half century ago by Jaynes [269]. One lesson that may be drawn from Jaynes' approach is that information or entropy maximization is a consequence of the global statistical properties of a system being made consistent with average constraints. This is related to the average constraint forces that Sokolov argued in the previous subsection were the origin of entropic forces.

Of course, entropy maximization is not the only option for defining an information force. In his book Ben-Naim [83] goes beyond entropy maximization to predict a PDF and abandons the concept of entropy altogether, replacing it with information, or rather with missing information. He argues that entropy was a poor choice of words by Clausius to capture the physical concept and the confusion stemming from that initial error has only become magnified over the intervening years. A more appropriate term is either *missing information* or *uncertainty*, either of which would be more compatible with the concept of nonsimplicity. Also for those that may believe that Shannon had some deeply held association of information with entropy, I offer the following story in refutation from the preface to Ben-Naim's book [83]:

> What's in a name? In the case of Shannon's measure the naming was not accidental. In 1961 one of us (Tribus) asked Shannon what he had thought about when he had finally confirmed his famous measure. Shannon replied: "My greatest concern was what to call it. I thought of calling it 'information' but

the word was overly used, so I decided to call it 'uncertainty'."
When I discussed it with John von Neumann, he had a better
idea. Von Neumann told me, "You should call it entropy, for two
reasons. In the first place your uncertainty function has been
used in statistical mechanics under that name. In the second
place, and more important, no one knows what entropy really
is, so in a debate you will always have the advantage."

Information is typically measured in *bits* and the foundation of the
physical world, as tersely described by Harte [239], is measured in terms
of interacting things moving about in space and time, that are tagged *its*.
The theoretical physicist J.A. Wheeler's view of a mechanism-free universe
is *its from bits*, which is interpreted by Harte in its strong form to mean that
' *its*" are an illusion and only "*bits*" exist. However, in its more reasonable
weak form, our knowledge of *its* derives from *bits*. It is with this latter
interpretation of *its from bits* that Harte, in his book *Maximum Entropy
and Ecology: A Theory of Abundance, Distribution and Energetics* [239],
developed his METE theory, following the lead of Jaynes, but applying
the entropy (information) maximization formalism to ecosystems and not
thermodynamics. His interpretation would constitute an extreme form of
information forces, since in this case every force would fundamentally be
derivable from information.

The attitude regarding theory we adopt herein is less revolutionary than
that of Harte and more in keeping with the perspective of Ben-Naim. As
the latter author points out [83]:

If one accepts the statistical interpretation of entropy, and agrees
on the Wiener/Shannon definition of information, then the definition of *thermodynamic entropy* as *thermodynamic information*
is inevitable.

Consequently, we associate an information force with an information
gradient, such as would be obtained in a thermodynamics in which the
word entropy had never been introduced, but more importantly the use
of entropy in sociology and biology could be discontinued and replaced
with the more natural use of the terms uncertainty or missing information.
Moreover the information gradient could be the result of an information
imbalance such as in the interaction of the two nonsimple networks we
discussed in Chapter 1. It could also be a consequence of the heterogeneity
of criticality arising in a nonsimple dynamic network, as we subsequently
discuss using an Ising-like dynamic model with interacting dyadic elements.

Chapter 4

Fractal Living Networks

I have always believed that scientific research is another domain where a form of optimism is essential to success: I have yet to meet a successful scientist who lacks the ability to exaggerate the importance of what he or she is doing, and I believe that someone who lacks a delusional sense of significance will wilt in the face of repeated experiences of multiple small failures and rare successes, the fate of most researchers. — Kahneman [279]

This chapter adopts an orientation based on what has been perceived to be a weakness of the modeling strategy historically applied within the social and life sciences. Said strategy was initially adopted in the 19th century with the development of statistics in physics to model the nonsimplicity of many-body effects and its almost immediate acceptance by the Natural Philosophers of the social and life sciences. The problem with this strategy of adopting models from a separate and distinct discipline is that the assumptions on which the models were based, although perhaps appropriate in the domain of the physics discipline in which they were originally developed, were often inappropriate in the more complex social or life science disciplines to which they are applied. All too often, the models of phenomena in these latter disciplines were tailored to have the characteristics of the statistical physics models, rather than tailoring the statistical model to have the observed properties of the phenomena being studied. It should be emphasized that this criticism does not apply to the life or social sciences alone, but rather the remarkable successes of physics in support of the booming of industry that was just coming of age invited imitation by all the other science disciplines, as well.

The critique of living networks presented in this chapter draws inspiration for its rationale, in part, from Daniel Kahneman's book, *Thinking, fast and slow* [279]. Kahneman is a psychologist who was awarded the 2002 Noble Prize in Economics, suggesting that disciplinary borders, between economics and psychology in that case, are self-imposed barriers not supported by the phenomena being studied. One consequence of the psychology experiments done in collaboration with his long time friend and colleague Amos Tversky was that the historical assumptions concerning how humans make decisions, and in particular, how economic decisions on which microeconomics was based, had to be reexamined and some needed to be abandoned altogether. The central assumption brought into question was that regarding strict rationality in humans. The foundational tenet of modern economic theory, that of the rational economic man, turned out to be at odds with the empirical findings of psychology experiments.

The purpose of the present chapter is to review the evidence that although physics models have historically guided the development of quantitative physiologic models, with some extraordinary successes, we are now entering a new era in science that argues against reliance on borrowing as a strategy for model building. It is not much of a stretch to say that typical phenomena in the social and life sciences are significantly more nonsimple than those in the physical sciences. It is also the case that the standard training of the life scientist relies less on the formalism that had enabled the physical scientist to succeed in the construction and application of mathematical models to experimental data within their discipline. However, scientists from both camps saw, almost immediately, the benefit of the fractal concepts for their domains of interest when it was introduced by Mandelbrot [354], given that the exemplars he analyzed and discussed were drawn from a broad swath of disciplines.

Mandelbrot's choices for application were based to a large extent on his intuition, since he was using a kind of mathematics that no one other than himself had seen within the physical, social or life sciences, much less implemented in the modeling of phenomena. Consequently, investigators in each distinct area of science began to develop fractal models based on what was needed to understand the processes and phenomena in their respective disciplines. It was equally clear that the fractal behavior of phenomena in living systems was the norm, and not the exception it seemed to be, at first, in the physical sciences.

The fractal strategy was based on quantifying the nonsimplicity of the phenomenon being investigated, a realization that was being made independently in more and more disciplines. It became increasingly apparent that nonsimplicity itself needed to be understood and that was the origin of a number of new computational techniques, numerical and otherwise. From a certain perspective we expect these new techniques for describing biodynamics, data processing of time series, as well as, information exchange between nonsimple networks in physiology to be disconnected from the reductionistic methods of the past, just as in the case of the phenomena studied by Kahneman. The psychological tasks studied became so nonsimple that the experimental results only made sense when the historic simplifying assumptions regarding how we humans make decisions, were either strongly modified, or abandoned altogether.

A singularity in the development of the science seems to have been reached. At this singular point the foundational models that had been sufficient to explain the behavior of isolated nonsimple networks in the past are no longer adequate to describe the observed emergent behavior arising from the interaction of two or more living networks. The venerable arguments are not able to resolve the contradictions these arcane models pose in explaining the exchange of information between two or more nonsimple systems but we will get to that.

In Sec. 4.1, we briefly review the reasons behind the initial decision to formulate a new view of physiology that by construction was dependent on the mathematical notions of fractal geometry, fractal statistics and fractal dynamics. A notion that was used is to develop the intellectual tools necessary to enable a modeler to think differently about such things as how the body operates as a whole, as is necessary in self-regulation, as well as, in health and disease.

Section 4.2 provides a summary of some of the successes of this new perspective, over the past decade, demonstrating how it might even be argued that fractal physiology has itself become a new science. A new science in the same sense that mechanics became a new branch of physics, separating itself from the study of light and heat by being inherently different from the phenomena studied in the other branches of physics. The unifying concept that holds all these different branches apart, yet supports the theoretical models as being the content of physics, is energy and how it manifests itself in these different branches. The physics-based models must now be left behind because energy is not always the glue that holds the physiologic concepts together; the new adhesive is information.

Finally, we enter the realm of forecasting and speculation in Sec. 4.4 and extrapolate across the knowledge gaps that we can see, into the mists of the potentially knowable unknown. This is done by extrapolating the trajectory of medicine as it has incorporated the nonsimplicity of the modeling techniques being developed into overcoming the limitations of previous methods.

4.1   Origins

Looking back over the past two centuries, it is clear that the perception as to how the human body operates paralleled our growth in understanding of how the technological society in which we are immersed came into being and how it functions. The perspective was developed, in large part, using physical analogs. For example, the *Le Chatelier's principle* (1844), as explained by Norwich [401], asserts that a disturbance to a state in equilibrium drives it away from its resting state, but it also invokes a countervailing influence to suppress the effect of the disturbance. Although we cannot pinpoint the exact time when the mechanical notion of how the body works took root and began to permeate society, in medicine the concept was summarized and articulated in clearest form by the nineteenth century scientist Claude Bernard (1813–1878). He developed the notion underlying homeostasis in his study of stability of the interacting networks within the human body and that idea was popularized half a century later by Walter Cannon (1871–1945) [130]. Homeostasis is considered by many physicians to be the guiding principle of medicine, whereby every human body has multiple automatic inhibition mechanisms that suppress disquieting influences, whether those disruptions are internally generated by malfunctioning internal systems, or externally by the environment.

Healthy living networks give rise to time series that display erratic fluctuations that contain as much, if not more information, than does the regular behavior, if we only knew how to extract it: The statistical properties of physiological fluctuations, such as those that determine the spatial properties of the tree-like structures of the human lung, arterial and venous systems, and other ramified structures [576]; statistical fractals [354] that determine the properties of the distribution of intervals in the time series of the beating of the human heart [420], respiration [40], as well as, in human locomotion [228, 241, 579]. Finally in dynamical fractals [581], as in the firing of certain neurons and found in the time series of posture control [143], which determine the dynamical properties of physiologic networks having a large number of characteristic time scales. These and other fractal physiology processes have been the focus of interdisciplinary research on living networks for more than three decades [70, 111, 597].

The rationale for this persistent interest is related in part to the idea that unlike the thermal fluctuations found in physical networks, which perturb the system but do not contain useful information, physiologic fluctuations are often the result of internal control and therefore often do contain useful information. The goal has been to better understand self-regulatory control systems for nonsimple living phenomena that produce such fluctuations and to describe the dynamics of such phenomena with tools capable of capturing their nonlinear and often exotic statistical character [70].

Bassingthwaighte *et al.* [70] coined the term *Fractal Physiology* to de-lineate the use of fractal patterns and metrics, usually related to fractal dimensions of multiscaled anatomical structures, or the bizarre behavior of time series generated by a physiologic process. The ubiquitous nature of fractal physiologic phenomena was supported by examples of the properties of fractals in space and time. Starting from the definitions, the mathematical logic was presented to provide detailed practical methods for assessing the fractal characteristics of the observed variability in a variety of living phenomena. This was done more in the manner of a handbook than to the rigor of a mathematics article or text.

The scaling of measures of variability were used by them to define internal spatial or temporal correlations within a live fractal system or object, where as simple, recursive rules gave rise to nonsimple physiological structures by various methods. More nonsimple dynamic rules, implied by interactions at the biochemical, cellular, and tissue levels, govern ontogenetic development and therefore play a major role in the growth of an organism. Multiple examples were provided of structural and behavioral fractal phenomena in nerve and muscle, in the cardiovascular and respiratory systems, in cognition and in growth processes. Why microdynamics and nonsimplicity give rise to fractals was explored and how the macroproperties emerged in networks operating at high levels of nonsimplicity was also examined.

In short, Bassingthwaighte and collaborative friends laid out the beginnings of a research program in physiology, with the intent of capitalizing on what was then a new direction for the life sciences. This new direction was stimulated by what was implied regarding the behavior of a living system being fractal over what that behavior would be if the system were mere complicated. Since no one individual, or group, could pursue all the avenues of research suggested by this new paradigm of physiology, the next best thing was to start a journal where such studies would not only be welcome but would be encouraged. This formed the background for the initiation in 2010 of the *Fractal Physiology* section of *Frontiers in Physiology*, for which the author had the honor of being the founding editor.

4.1.1 Nonsimplicity entails criticality

Why are fractals ubiquitous in living systems? The short answer to the question is that time-dependent fractals are entailed by criticality, which in turn are a consequence of nonsimple dynamics and the fact that physiological networks are generically nonsimple. Historically, linear dynamics and normal statistics were the working hypotheses made to model living systems, which eventually became foundational to the understanding of the operation of physiologic networks. The relatively recent replacement of these now arcane assumptions, with nonlinear dynamics and chaos theory, along with non-normal and non-Poisson statistics, have been shown to be related to the IPL PDFs of nonsimple dynamic networks [597].

To better understand the physiologic exemplars discussed in this chapter, it is useful to have a general idea as to how criticality is generated by nonsimple dynamics. The ubiquity of nonsimple dynamic networks in physical, social and life sciences naturally derives from the fact that the structures are composites of many simpler, loosely connected and dynamically interacting elements. However, despite the relative simplicity of the basic building blocks, nonsimple networks are characterized by rich self-emergent behavior, which cannot be understood on the basis of aggregating the single component dynamics. Since in most cases solving a system of coupled nonlinear equations that trace the dynamics of a network composed of N units is not possible, investigations into nonsimple networks has primarily focused on their global behavior.

The macrovariables describing the global behavior of nonsimple networks, such as the mean field in physical systems, or the population density in ecological systems, display emergent properties of spatial and/or temporal scale-invariance, manifest in IPLs of connectivity of individual units and waiting-time PDFs for independent time intervals between events. Typically, these IPLs cannot be simply inferred from the equations describing the dynamics of the individual elements of the underlying network. The self-organized criticality theories have shown how scale-free phenomena emerge at critical points, as described subsequently.

However, it is known that "more is different" [36], which is to say, as the size of a naturally occurring network grows, so does its nonsimplicity and its fundamental character changes. Macroscopic properties emerge that cannot be traced back to the microdynamics, or at least could not have been anticipated prior to their emergence. An emergent macroproperty cannot be described in terms of the micromodels that determine the microdynamics, but requires an entirely new view of how that part of the world works, a view that runs counter to reductionistic thinking.

The limitations of reductionistic thinking in medicine is a consequence of the undeniable nonsimplicity of living networks when the individual is considered as a whole [226], rather than as an aggregation of healthy or diseased parts (organs, tissues, cells, or molecules). The biosocial understanding of disease advocated by Greene and Loscalzo [226] relies on the mathematical infrastructure provided by network science in the formulation of network medicine [330, 598].

A new mechanism introduced is based on the nonsimplicity of living systems and that is criticality. The phenomenon of criticality has long been known in physics, but it has recently gained currency under the label tipping point, in which nonsimple systems can unexpectedly switch from one critical state to another [547]. The theory of critical states has suggested that the rate of relaxation of a perturbation approaches zero as a tipping point is approached. This also goes by the name of critical slowing down in the physics literature and is associated with the divergence of the observed relaxation time of a perturbation. For example, the critical slowing down of mood dynamics has been used to predict future transitions to a state of major depression. As van de Leemput *et al.* [316] observe, a person's mood system may have crucial points wherein positive feedbacks within a network of symptoms can propel a person into a state of major depression.

Type I criticality: It needs to be emphasized at the outset that there is more than one type of criticality and we begin with the most familiar. We see this kind of criticality, call it Type I, in the radiating lines of frost on the window when the outside temperature dips below freezing. We feel it in the cooling effect of sweat evaporating from our skin after exercise, and in the condensation of the water vapor visible on the outside of a beer bottle on a hot summer day. Type I criticality is a physical mechanism the body uses in the homeostatic regulation of its temperature, along with blood flow near the surface of the skin, among others. The physical mechanisms controlling the transformation from one phase of water to another was not understood in depth until the middle of the 20th century. It was, of course, clear that phase transitions control the temperature of the surroundings and the transfer of heat, but the secret of how differing macroproperties of the gas, fluid and solid phases emerge from the microdynamics required a new way of understanding the phenomenon of criticality-induced phase transitions.

Subsequently, phase transitions were recognized to give rise to a kind of universal behavior that is independent of the underlying dynamics of the elements making up the substance. For example, when a liquid is boiled, it becomes a gas and the corresponding volume increases discontinuously as a manifestation of critical behavior. This universal behavior is manifest in the scaling of certain system parameters called critical exponents, on which there is now a vast literature. One critical exponent of water is the correlation length between water molecules, which diverges as the temperature approaches its critical value from above. It is the divergence of the macroscopic correlation length that unambiguously describes the phase transition we experience with our coarse five senses.

The models of the changes in macroscopic behavior during a physical phase transition eventually found their way into the social and life sciences. Type I criticality initiated new ways of thinking about the behavior of collectives, whether they were collections of neurons in the brain, cells in a membrane, birds in a flock, humans in a demonstration, or warriors

on a battlefield. Scientists began to ask questions such as: How do the
large number of pacemaker cells, each of which is a relatively unreliable
time-piece, synchronize their firing to generate normal sinus rhythm? Dc
living systems manifest universal behavior and if they do, what are the
implications for medicine?

Type II criticality: A second kind of criticality, call it Type II, is manifest in the frequency with which earthquakes of a given magnitude occur
as well as, solar flares of a given intensity, brainquakes engaging so many
neurons, species extinction, stock market disruptions involving large numbers of stocks, traffic jams and a myriad of other collective behaviors [563]
that operate without being influenced by an external control parameter
In these Type II fluctuating phenomena, criticality emerges spontaneously
by means of internal dynamics, along with scaling behavior, with no external parameter being involved as is required for Type I criticality. There is
nothing analogous of the temperature with which to control Type II criticality. In this latter type of criticality, the system is spontaneously driven
to a critical state by its own internal dynamics and has been dubbed SOC
and has been used to describe the above applications [563]. Here again
the universality of criticality makes the macroproperties of the system independent of the detailed microdynamics even though the sources of the
criticality are different. Bak [51] observed in his book, tracing the history
of the 'science' of SOC, that the nonsimplicity of a system is a consequence
of criticality.

The biological nonsimplicity of the brain has been argued to be a consequence of the same mechanism as for Type II criticality [523]. The experimental evidence for this is the existence of phase transitions as indicated by
the divergence of correlation lengths in fMRI data [197]; another is brain
activity consisting of neuronal avalanches in neocortical circuits characterized by scale-free IPLs [74, 334]. In addition to the brain, other living
systems that provide direct evidence of Type II criticality through self-
regulation include spiral waves in astrocyte syncytium [277], global gene
expression [532], and intracellular calcium signaling in cardiac myocytes
[399], to name a few.

Type III criticality: There is in addition to the above two types of
criticality, a new kind of self-organizing criticality [342], which we refer
to as Type III. Instead of focusing on the magnitude of the event being
observed, Type III records the time interval since a self-organized event of a
given kind last occurred and determines the PDF of these time intervals and
consequently called SOTC [604]. The time series for the inter-beat intervals
of the heart, inter-breath intervals and inter-stride intervals have all been
shown to be fractal and/or multifractal statistical phenomena having IPL

PDFs. Consequently, the fractal dimension, which is related to the IPL index, turns out to be a significantly more reliable indicator of the quality of an organism's functions in health and disease than the traditional average measures, such as average heart rate, average breathing rate, or average stride interval [589].

4.1.2 Scaling time series

Living system time series are invariably erratic and have scaling properties, often being a member of the class of fractal statistics, as we subsequently explain. Consider a physiologic observable $X(t)$ representing a time series, whose behavior is determined by the homogeneous scaling relation:

$$X(\lambda t) = \lambda^h X(t) \tag{4.1}$$

where λ is a constant and h is a scaling parameter. If the scaling is interpreted in terms of the PDF then the second moment of such time series scale algebraically in time:

$$\langle X(\lambda t)^2 \rangle = \lambda^{2h} \langle X(t)^2 \rangle, \tag{4.2}$$

from which it is determined, using the scaling parameter $\lambda = 1/t$ that the second moment is:

$$\langle X(t)^2 \rangle \propto t^{2h}, \tag{4.3}$$

where the scaling parameter is in the interval $0 < h \leq 1/2$.

The two-time autocorrelation function of $X(t)$ is found to have the form 70]:

$$C(t_1, t_2) \propto |t_1 - t_2|^\beta, \tag{4.4}$$

and the power-law index is given by $\beta = 2h - 2$ in agreement with Eq. (4.3). Note further that when the two-point autocorrelation function is determined to depend only on the time difference, the underlying process is stationary. The autocorrelation function is an IPL in time because the scaling index is in the interval $0 \leq 2h \leq 1$ so that $\beta < 0$. Finally, the power spectrum $S_p(f)$ is an IPL in frequency f:

$$S_p(f) \propto \frac{1}{f^{\beta+1}} \tag{4.5}$$

see [590] for a more complete discussion in a physiologic context.

The IPL nature of these second-order measures is the signature of non-simplicity in general and fractal random processes in particular, for which Mandelbrot and Van Ness [353] coined the phrase fractal Brownian motion (FBM). From this we surmise that heart rate variability (HRV) as measured

by its time series is a fractal random point process [212, 517], as is stride
rate variability (SRV) [228, 276] and breath rate variability (BRV) [585],
among many other nonsimple physiologic phenomena [590]. Consequently,
the dynamics of traditional stochastic processes described by differential
equations for the dynamic variables, or in phase space for the PDFs, are
not sufficient to describe the properties of nonsimple physiologic networks.
The convergence of fractals and statistics requires the use of a new kind
of calculus, the fractional calculus, to model the evolution of the PDF describing the behavior of such systems.

Alan Mutch, recognizing the fractal nature of physiological time series,
reasoned that post-operation ventilators, being operated harmonically, were
not taking full advantage of the natural variability of the respiratory system. He and his colleagues developed a program in which to compare the
relative efficacy of biologically variable ventilation (mechanical ventilation
that emulates healthy variation) over conventional control mode ventilation
(monotonously harmonic). They did the initial comparison in an animal
model of bronchospasm to determine which approach yields better gas exchange and respiratory mechanics. The biologically variable ventilation
was superior to control mode ventilation in terms of gas exchange and respiratory mechanics during severe bronchospasm [364] and in all subsequent
investigations.

Time series are often treated as discrete intervals between events. Here
we consider the $N + 1$ times at which a given event occurs as the set $\{t_j\}$;
$j = 0, 1, \ldots, N$ and the time intervals between consecutive events are given
by $\tau_j = t_j - t_{j-1}$. We consider these time differences to be the time intervals
between crucial events. One way to think about crucial events is to adopt
the engineering language of Cox [155] and define the age-dependent failure
rate of the process to be $g(t)$ through:

$$g(t) = \frac{\psi(t)}{\Psi(t)}, \tag{4.6}$$

where $\Psi(t)$ is the survival probability, namely the probability that a machine keeps working for a time interval t from the time at which it was
created, at which point in time it fails and $\psi(t)$ is the corresponding PDF.

Imagine that a team of engineers takes action the moment a machine
fails, identifying and correcting the machine's malfunction, thereby making
it brand new. In this ideal world, the repair activity takes no time and has
the effect of extending the working life of the machine to the next failure,
at which time it will again require the instantaneous action of the idealized
team of engineers distribution given by the waiting-time PDF:

$$\psi(t) = -\frac{d\Psi(t)}{dt}. \tag{4.7}$$

Using Eq. (4.7) to integrate Eq. (4.6) yields the survival probability:

$$\Psi(t) = \exp\left[-\int_0^t g(t')dt'\right]. \tag{4.8}$$

If $g(t) = r_0$, a time-independent constant, the survival probability would be an exponential and the statistics of the repair process would be Poisson. This is not the case for the time interval between beats of the heart, strides in walking, or a myriad of other pseudo-regular processes in living networks, even though Poisson statistics had been almost universally assumed to be the case in the past [576].

Making the assumption that $g(t)$ decays asymptotically in time as $1/t$, or more precisely, according to the algebraic form:

$$g(t) = \frac{r_0}{1 + r_1 t}, \tag{4.9}$$

when inserted into Eq. (4.8) and integrating, yields the hyperbolic form for the survival probability where here $T = 1/r_1$ and $\mu = 1 + r_0/r_1$. The hyperbolic form of the survival probability is readily differentiated in time to yield the waiting-time PDF given by Eq. (5.10) which is asymptotically an IPL.

We define the occurrence of these 'failures' as crucial events when the IPL index satisfies the condition $1 < \mu < 3$. We note that the average waiting time is given by:

$$\langle \tau \rangle = \int_0^\infty t\psi(t)dt = \frac{T}{\mu - 2}, \tag{4.10}$$

when $\mu > 2$ and when $\mu < 2$ this average waiting time diverges. The crucial events are defined by the time interval separating the occurrence of events being statistically independent of one another, thereby making the crucial event process renewal, with the condition $1 < \mu < 3$ [155].

It bears emphasizing that controlling living networks in order to ensure their proper operation is the presumed function of homeostatic control systems, which are both local and relatively fast. Allometric control systems, on the other hand, take into account correlations that are IPL in time, as well as long-range interactions in nonsimple phenomena as manifest by IPL PDFs. An allometric control network achieves its purpose through scaling, enabling a nonsimple living network such as the one performing physiologic regulation to be adaptive and accomplish concinnity of its many interacting subnetworks. Allometric control is a generalization of the idea of explicit feedback regulation in homeostasis [588].

Multifractals: Different physiologic processes generate different fractal time series because the long-time memory of the underlying dynamical processes can be quite different. Physiological signals, such as cerebral blood flow (CBF), are typically generated by nonsimple self-regulatory networks that handle inputs with a broad range of characteristics. Ivanov *et al.* [265] established that healthy human heartbeat intervals, rather than being fractal, exhibit multifractal properties and uncovered that the loss of multifractality is the cause of congestive heart failure. Note that multifractality is defined by a time series having a spectrum of scaling exponents (fractal dimensions), which is to say the scaling index changes over time [184], resulting in no single fractal dimension, or scaling parameter, characterizing the process. West *et al.* [583] similarly determined that CBF in healthy humans is also multifractal. Figure 4.1 shows the CBF time series to be multifractal and that this multifractality is greatly restricted for people who suffer from "severe" migraines.

As pointed out in [596] the etiology and pathomechanism of migraine headaches have not been explained up to this point in time. However, CBF time series differs substantially between that of normal healthy individuals and migraineurs. High-resolution measurement of middle cerebral artery blood flow velocity has been recorded using Transcranial Doppler ultrasonography. In Fig. 4.1 the multifractal spectra of middle cerebral arterial flow velocity time series for healthy and migraineur groups [583] are displayed. The width of the multifractal spectrum centered on the local scaling exponent for the CBF velocity is greatly constricted, by a factor of three, suggesting that the underlying process has lost a great deal of flexibility. The biological advantage of multifractal processes is that they are highly adaptive, so that the brain of the healthy individual adapts to the multifractality of the interbeat interval time series.

4.2   Rise to Center Stage

The modern world is an interconnected mesh of networks satisfying a myriad of functions: transportation, electrical power, food distribution, finance, and healthcare to name a few. The interoperability of these networks is developed as part of urban evolution over the past two centuries such that these and other webs interconnect to national and/or global networks [402]. This is the engineered webbing of humanity, but there are comparable natural structures in the spheres of biology, ecology, sociology, and physiology. It is not only our external world that is cluttered with networks, but our internal world as well. The neuronal network carries signals from the brain to the physiological networks of the body and back again. This network-

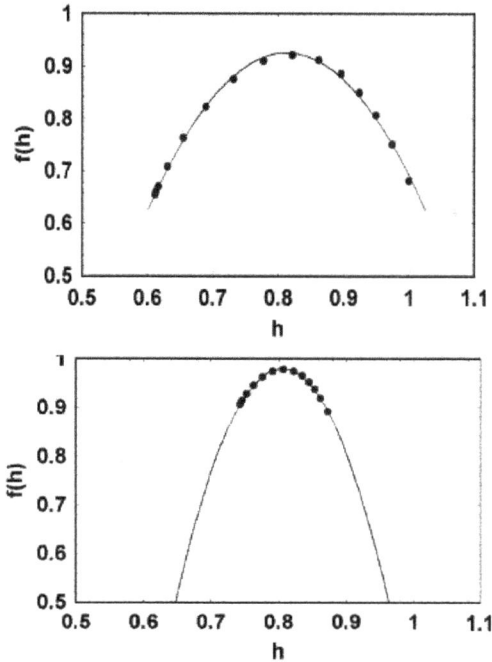

Figure 4.1: The average multifractal spectrum for middle CBF time series is depicted by $f(h)$. (a) The spectrum is the average of 10 time series measurements from 5 healthy subjects (dots). The solid curve is the best least-squares fit of the parameters to the predicted spectrum. (b) The spectrum is the average of 14 time series measurements of eight migraineurs (dots). The solid curve is the best least-squares fit to the predicted spectrum. From 654] with permission.

of-living-networks is even more nonsimple than the modern city, or a typical ecology network. Thus, the basic research into network science spans outward to encompass a multitude of disciplines; understanding each one individually sheds light collectively on all the others. Our scientific understanding of these nonsimple networks and their mutual interactions form the context in which we understand physiology [585].

Fractal physiology is devoted to the many aspects of experimental design, model building, data analysis and theoretical interpretation of nonsimple living networks. Fractals in anatomical structure, in the dynamics of control networks and in the statistics of physiological time series have

proven to be of fundamental importance for understanding the nonsimplicity of living networks. This perspective provides a formal context in which
to understand physiological mechanisms and how these mechanisms might
be altered in disease states. This is achieved by placing particular emphasis
on identifying and interpreting fractality by means of scaling of nonsimple
physiologic phenomena as manifest in structure, time series and dynamics.
Key questions addressing the structure, functions and activities of organs,
tissue and cells, which reveal the coupling across space/time scales, as well
as the communication within and between functional and regulatory networks are answered by fractal scaling. Such behavior has been described by
mathematical models at all levels of sophistication, from the psychophysical, motivated by direct simulation or emulation of phenomena, to the
rigorous application of the mathematics of nonsimple networks.

4.2.1   Fractal brain dynamics

Fractal measures of the neurodynamics of cognitive phenomena has significantly modified our understanding of the operation of the brain over
the past decade. Consequently, one often finds the neuroscience discussions
conducted in two distinct ways. In a neuroscience mode, authors emphasize
physiology in which fractals are introduced as a necessary infrastructure to
systematically capture the observed behavior of the system. In a mathematical mode, scientists extend existing mathematical formalism in order
to capture an anomalous feature of the data that the usual analysis cannot
handle. They use physiology as an application area for nonsimple phenomena to show the scientific relevance of the generalization.

   Consider the often-made assumption that fluctuations in brain dynamics
rapidly relax to the steady state, resulting in only short-time memory, or
no memory at all. Aburn *et al.* [4] point out that studies of human EEG
time series reveal long-time memory, that is, significant memory within the
time series on the scale of minutes. Such memory indicates the influence
of nonlinear dynamical processes. General computational approaches lead
to increased correlation lengths, as well as, IPL scaling resulting from Hopf
bifurcations of the dynamics with tuning of the background input.

   Type II criticality (SOC) is one of the techniques that has been used
to model the apparent need for a fundamentally new way to describe
neuronal dynamics based on the fractal properties of nonsimple adaptive
networks. SOC theory shows how scaling structures emerge from local
interactions [511] and typically generalizes the fractal concept to multifractals. In Sec. 4.1.3 we showed that unlike ordinary fractals, multifractals
use multiplicative rather than additive interactions. This theory describes
the behavior of brain dynamics in terms of non-stationary, non-ergodic

statistics manifest in IPL PDFs [569, 570]. If the probability that the physiologic observable X falls in the interval $(x, x + dx)$ is $P(x)dx$ the PDF has the IPL form:

$$P(x) \propto \frac{1}{x^\mu}, \tag{4.11}$$

where the interval for the IPL index is typically $1 \leq \mu < 3$. The equilibrium neural structures are dominated by local interactions, and the transition to criticality produces IPL PDFs in the number of neurons contributing to a brain quake, also called a neurological avalanche [74].

However, even the spontaneous fluctuations in brain noise have been associated with critical dynamics [197]. In the latter case, the variate in the IPL is not the number of neurons contributing to a brain quake, but the time interval between neural events of a given magnitude, the magnitude being the number of neurons involved in the quake. This is the well-documented $1/f$-noise behavior of the brain at rest. Physiologic time series typically consist of electrical impulses marking the occurrence of an event, such as a beating of the heart, the ring of a neuron, or the taking of a breath. These events can be characterized by the time interval from one event to the next. The PDF of the time intervals between successive events, also called the waiting-time PDF $\psi(\tau)$ has the IPL form:

$$\psi(\tau) \propto \frac{1}{\tau^\mu}, \tag{4.12}$$

where the IPL index is again in the interval $1 \leq \mu < 3$. The theoretical basis for this PDF is Type III criticality (SOTC), recently devised by Mahmoodi et al. [342] and put into a broad context in [604].

However, the criticality hypothesis remains controversial as elegantly presented using the format of a Socratic dialogue [80]. This controversy motivated some investigators to generalize the notion of criticality, as used in physical phase transition, into a more general form for living processes [329, 334]. Dehghani et al. [163] find issue with the presence of critical states in a healthy awake brain and conclude that there is a lack of IPL scaling in the size of avalanches in the cat, monkey and human cerebral cortex. In short, their results do not show clear evidence of IPL scaling or SOC states in the awake or sleeping brains of mammals, from cat to man. In their *Research Topic in Fractal Physiology* Boonstra et al. [111] argue against criticality as a general principle for different mammalian brains, even though they observe that the findings of de Arcangelis et al. [162] do not contradict criticality in neuronal slices and anesthetized states.

On the other hand, Timme et al. [529] found evidence that the nonsimplicity of a neural system is optimized at or near a critical state and this appears to be where neural systems prefer to operate. Perhaps more importantly, this nonsimplicity was found to be independent of precise inter-neuronal spiking relations, but instead to be sensitive to the IPL index of the avalanche PDF, as well as, the average neuron firing rate.

Functional magnetic resonance imaging (fMRI) has also significantly contributed to the understanding of brain function [523]. The fMRI data was used to define a control parameter. This parameter was derived from the blood oxygen level dependent (BOLD) signal, and shown to correspond to a system near a critical point. The resting brain was shown to spend most of its time in the vicinity of criticality, exhibiting avalanches of activity entailed by the same dynamics that drive neuronal events on a smaller scale. These avalanches are characterized by IPL PDFs for the size, duration [162] and power spectra of the electrical signal reproducing the IPL behavior found in human EEG spectra. Mikkelsen and Lund [375] determined that the scaling index, using BOLD fMRI data, is sensitive to the sampling rate and must be treated with caution.

Another mechanism leading to the scaling observed in collective neuronal behavior is produced by nested oscillations leading to hierarchical control of inter-areal synchrony. These nested rhythms could provide a more comprehensive view of the dynamical structure of oscillatory interdependencies in the human brain [380] in which the inter-areal phase synchrony are modulated by the phase of a slower neuronal rhythm using magnetoencephalography (MEG). Botcharova et al. [112] use the phase difference between MEG signals from the left and right motor cortices during rest as well as, during a finger-tapping task. They determine that fluctuations in the phase differences at rest are at or near a critical state, similar to that of a system of Kuromoto oscillators. The finger-tapping task disrupts this collective behavior and drives the system from a strongly correlated precritical state to an uncorrelated random state.

Norbert Wiener, who was the initial mathematical force behind the discipline of *Cybernetics* [612], believed that the brain wave variability observed in EEG signals would ultimately yield to interpretation using *Generalized Harmonic Analysis* [615]. He introduced the idea of synchronized oscillators to describe brain wave dynamics and six decades later crucial events were used to describe $1/f$-noise in the brain and made compatible with the wave-like nature of brain processes [103]. These later authors reviewed the literature to establish that brain dynamics host crucial events generated by criticality, which along with waves are a consequence of spontaneous self-organization.

Differences in nonsimple, random, non-stationary behavior of neural processes in different regions of the brain is significant during task performance as measured with EEG signals. Under a naming task, it was observed that these complexity differences in brain dynamics of patients with schizophrenia was markedly different from a matched cohort of controls. This mismatch was interpreted as indicative of schizophrenia-related failures to adapt brain functioning to the naming task [262].

The nascent science developed to understand nonsimplicity using network dynamics has, as one of its many areas of application, been adopted to explain cognition [500]. These investigators used network-based modeling to demystify cognitive science and establish conditions under which the cognitive state self-organizes in such a way that the subsequent behavior patterns are used to identify and explain how biological networks may be controlled. So let us now turn to the more general perspective of using network science to understand the workings of the large number of mutually interdependent nonsimple living networks.

4.2.2   Nonsimple networking

Network science provides a mathematical infrastructure for living systems that is different in a number of ways from systems engineering, but these differences need not concern us here. Suffice it to say that the network viewpoint focuses on the behavior of an organism as a whole, recognizing that the macrobehavior cannot be associated in a reductionistic way to the dynamics of the microvariables. For example, the time-dependent behavior of the beating heart (macro) is a consequence of the collective dynamics of the pacemaker cells (micro) making up the sinoatrial node, being driven by the nonsimple signals generated by sympathetic and parasympathetic neural networks. The consequent emergent behavior in such time series as HRV has been called the *Network Effect* [598].

One measure of the nonsimple nature of interacting physiologic networks is the width of the multifractal spectra, as we observed previously. The RR-interval fluctuations were described by broad multifractal spectra for healthy individuals and a narrowing of this spectrum was detected in patients with congestive heart failure [265]. More recently, the chronic pain of fibromyalgia (FM) and other symptoms, as pointed out by Reyes-Manzano et al. [453], affect multiple networks providing strong evidence that the autonomous nervous system produces multifaceted alterations of this disease. Specifically, there was no narrowing of the multifractal spectrum for the magnitude of the RR-intervals, but there was a shift in the spectral peak to a higher fractal dimension. Overall they found a decrease in multifractality as well as in nonlinearity and an increased anticorrelation in FM patients compared with a matched cohort of healthy subjects.

Dynamic networking has suggested new ways to transfer information between fractal physiologic networks. The generic information transfer mechanism was suggested by Wiener in 1948, resulting in the CME [586], but the CME was not mathematically proven until the 21st century [43]. The empirical support for the CME in the transfer of information between physiologic networks includes the relief of chronic intractable pain for which transcutaneous electrical nerve stimulation is beneficial [390], and body movements in synchronization with external events [166, 278], see Chapter 2.

A *Frontier Research Topic* helped clarify how the global intelligence of a nonsimple network emerges from the local cooperation of units, whether these units were neurons or people and emphasizes the role played by critical phase transitions in the observed persistence of this cooperation. Turalska *et al.* [537] implemented a social network model, the DMM, exploiting the observation made by an increasing number of researchers, that intelligence emerges from criticality as a consequence of locality breakdown and the onset of long-range correlation, well-known properties of critical phenomena. They focused on the critical condition that had been shown to maximize information transport within a nonsimple (fractal) network and show that their results are compatible with fault tolerance and Hebbian learning.

Mahmoodi *et al.* [344] introduced a two-level model of social interaction and established that criticality is not confined to the structure of the DMM at one level, through the study of the emergence of altruism using the altruism-selfishness model (ASM) at a second level. Both models generate criticality, one by imitation of opinion (DMM) and the other by imitation of behavior (ASM). The dynamic competition between personal and social benefits in this two-level interacting network results in a global evolution towards criticality. They show that the two-level theory of this article is compatible with recent discoveries in the burgeoning field of social neuroscience.

The analysis of Gallos *et al.* [205] identified functional brain modules of fractal structure that were interconnected in a small-world network topology, in which a relatively few long-range links are sufficient to have a locally connected network manifest long-range connectivity properties. They indicated the utility of using percolation theory to highlight the modular character of the functional brain network, presenting a fractal, self-similar topology, identified through fractal network methods. When weaker ties were introduced through lowering the threshold of correlations, the network as a whole assumed a small-world character.

4.2.3 Managing nonsimplicity

As previously mentioned, the transfer of information between nonsimple networks is determined by the level of the nonsimplicity, or information, gradient between the two networks. For example, in certain psychophysical experiments a person is asked to synchronize a tapping finger in response to a chaotic auditory stimulus, and the CME is interpreted as the transfer of scaling of the fractal statistical behavior of the stimulus, to the fractal statistical response of the stimulated subject's brain. This response of the brain, in such motor control tasks, when the stimulus is a multifractal metronome, has been established [510].

Experimental results [166] confirm this interpretation, based on the transfer of global properties from one nonsimple network to another. In these latter experiments the multifractal metronome generates a spectrum of fractal dimensions $f(h)$ as a function of the average singularity strength of the excitatory signal and it is this dimensional spectrum that is captured by the brain response simulation, as depicted in Fig. 2.3. The multifractal behavior manifested by the unimodal distribution provides a unique measure of nonsimplicity of the underlying network. It is worth noting that the same displacement of the metronome spectrum, from the body response spectrum, is observed for walking in response to a multifractal metronome.

The experimental evidence supports the interpretation that the greater the nonsimplicity of the physiologic time series, as measured by the width of the multifractal spectrum, the healthier the individual. In addition, theory suggests that the information transfer between two coupled networks is from the network with the wider spectrum (greater nonsimplicity) to that with the narrower spectrum (lesser nonsimplicity). As pointed out by Almurad et al. [34] interacting systems tend to attune their nonsimplicities in order to enhance their coordination. They further comment that this effect has been observed in a number of synchronization experiments, and interpreted as a transfer of multifractality between systems. In their experiment, older participants were involved in a prolonged training program of synchronized walking, with a young experimenter and synchronization between the duos was dominated by the CME. A restoration of nonsimplicity in the older participants was observed after 3 weeks, and this effect was persistent 2 weeks after the end of the training session. They further remark that this presents the first demonstration of a restoration of nonsimplicity in deficient systems. Thereby suggesting a new strategy for rehabilitation.

4.3   A Calculus for Medicine

Scientists have believed, since the mid-19th century, that certain continuous functions with diverging first-order derivatives do not lend themselves to the description of the dynamics of natural phenomena. However, in the early part of the last century, it became evident that certain of these mathematical curiosities can describe the dynamics of the flow of non-Newtonian fluids, such as honey, blood, tar, as well as, a variety of other viscoelastic materials [580, 590]. The dynamics of these materials that are neither fluids nor solids in the Newtonian sense could only be described by dynamic equations that incorporate memory. Such hereditary dynamics required a new kind of calculus, one involving non-integer derivatives that do not diverge when applied to a continuous function that has a fractal dimension and has come to be called the Fractional Calculus (FC) [600].

FC has been used to describe more than one class of nonsimple phenomena for which other, more traditional methods have been found to be inadequate, which is to say that the familiar approaches lead to predictions that are inconsistent with experimental results. The FC has been used to model the dynamic interlinking of elements and harmony of nonsimple phenomena ranging from the electrical impedance of biological tissue to the biomechanical behavior of physiologic organs; see, for example, Magin [349] for an excellent review of such applications. For an orientation more toward physics, but with physiologic applications as well, see, for example West [589, 600].

As discussed in Sec. 4.2, the empirical evidence supports the interpretation that physiologic time series are described by fractal stochastic processes. Furthermore, the fractal nature of these time series is not constant but may change with the vagaries of the interaction of the network with its environment and internal dynamics; therefore, physiologic phenomena are often multifractal, as manifest in the fractal dimension changing slowly over time. The scaling index, the fractal dimension, or in the case of multifractals, the spectrum of fractal dimensions, suggests new ways to quantify a physiologic network's response to disruption and can be used as an indicator of its state of health. The width of the fractal distribution, along with the average fractal dimension, become two independent measures of the state of health of a living network.

This is where we begin forecasting, as opposed to predicting, how the fractal concept may guide the development of future dynamic models of living networks. At the very least, the future will see the development and application of techniques that go well beyond the straightforward analysis of physiological time series to determine the why, as well as the how, of fractal properties. These new methods will come to terms with ways to mathematically model the dynamic nonsimplicity of physiological systems that entail fractal phenomena. In addition, historical strategies such as using standard rate equations to model the relaxation of a perturbed system back to its dynamic equilibrium will be replaced with fractional rate equations. These latter equations will allow for the history of the physiologic process to modulate and therefore guide its response to any disruption.

But rather than just creatively extracting such arguments from the air, let us examine how these new dynamic models can be systematically developed and organize what we can hopefully learn from them. This is where the transdisciplinarity of network and nonsimplicity science may not only provide the tools necessary to address our problems, but may in addition provide a language with which scientists from various disciplines will be able to communicate with one another. Without a fundamental theory, however, we are forced to rely on data processing methods to infer the science from those techniques that most consistently give the more robust results. In the relatively near future we will be able to develop theory consistent with the successful methods of data analysis.

4.3.1 Allometry relations

G. West in his remarkable book SCALE [610] explains how the fractal paradigm entails the existence of empirical laws of allometry that relate the functionality of a network Y to the network's size X raised to a non-integer power β:

$$Y = \alpha X^{\beta}, \tag{4.13}$$

where α and β are taken to be independent empirically determined constants. In physiology the measure of a body size is taken to be the total body mass (TBM) and functionality may be the metabolic rate, the physiologic time, or temperature regulation, to name a few functions. Literally hundreds of other interspecies and intraspecies allometry relations (ARs) have been experimentally determined, see for example, the classic works [129, 478]. However, what was believed about the underlying causes for allometry was disrupted by the introduction of fractals to describe the way in which living networks scale in order to transport nutrients [565, 607, 608].

The collective insight of the trio of West, Brown and Enquest [608] lead to an extraordinary vision of why living systems function as they do. The title of one of their seminal papers captures the exceptional nature of their insight: *The fourth dimension of life: fractal geometry and allometric scaling of organisms.* However, their original arguments rely heavily on theoretical physics for the generality of their proof, which is transliterated into ordinary speech by one of the trio [610]. In this latter work, G. West also extends his presentation to encompass all manner of nonsimple systems, recognizing that ARs appear in every discipline from Architecture to Zoology, with living networks being front and center.

Returning to physiology, we note that this is not the end of that story. Interpreting the empirical variables in Eq. (4.13) in terms of average values, as is always done in the literature, presents a fundamental problem for theory. The problem is a consequence of the fact that the average of a random variable raised to the power does not equal the average value of a random variable raised to the β power, except in the trivial case $\beta = 1$:

$$\langle X^{\beta} \rangle \neq \langle X \rangle^{\beta} . \qquad (4.14)$$

The resolution of this problematic inequality led to a determination of the PDF entailed by the statistics of a number of allometry datasets. The lowest moments of this PDF produces the empirical allometry relation, see [602] for a narrative that stitches these studies into a coherent story involving the fractional calculus applied to probability theory.

These two general theories are not necessarily inconsistent with one another, but they each present their own challenges in the life sciences. One determines the most general scaling properties of a deterministic fractal and explores the implications for the robustness of branching nutrient transport systems within an organism. The other determines the most general scaling properties of a statistically fractal network and examines what is entailed for the maintenance of stability within such a physiological network. The investigation into the overlap between these two ways of modeling physiologic networks has barely begun, but what has been done promises a bright future for whichever approach turns out to be the more fundamental.

4.3.2   Allometric control

Control of physiologic nonsimplicity is the goal of medicine, which is another way of saying that medicine seeks to understand what is the proper operation of living networks through their sensing and control. We mentioned earlier how we control physiological networks by means of homeostasis and contrasted that with allometric control. The direct modeling of the nonsimplicity of physiologic networks, such as the autoregulation of heartbeat variation, human gait variability, and cognition, have more intricate feedback arrangements than assumed in simple homeostatic feedback models. Such nonsimple physiologic networks have more intricate feedback arrangements because scaling requires each sensor to respond to its own characteristic set of frequencies, the feedback control must carry signals appropriate to each of the interacting subnetworks. The coordination of the individual responses of the separate subnetworks is manifest in the scaling of the time series in the output and the separate subnetworks select that aspect of the feedback to which they are the most sensitive. In this way, an allometric control network not only regulates, but also adapts to changing environmental and biophysical conditions [597].

We can relate the allometry approach to a recently developed branch of control theory involving the fractional calculus. The generalization of control theory to include fractional operators enables the designer to take into account, memory and hereditary properties; properties that are traditionally neglected in integer-order control theory, such as in the traditional picture of homeostasis. Podlubny [433] proved that if a 'real-world' process has the dynamics determined by a fractional-differential equation, then attempting to control it with an integer-order feedback, leads to extremely slow convergence, if not divergence, of the system output. On the other hand, a fractional-order feedback, with the indices appropriately chosen, lead to rapid convergence of output to the desired signal. Thus, one might anticipate that dynamic physiologic systems with scaling properties, since they can be described by fractional dynamics [583], ought to have fractional-differential control systems.

From this perspective, the loss of nonsimplicity is the loss of the body as a cohesive whole; the body can be reduced to a disconnected set of organ systems. One of the traditional views of disease is what Tim Buchman calls the "fix-the-number" imperative [120], which asserts that if the bicarbonate level is low, then give bicarbonate; if the urine output is low, then administer a diuretic; if the bleeding patient has a sinking blood pressure, then make the blood pressure normal. Buchman critiques this imperative by observing that such interventions are commonly ineffective and even harmful. Consequently, one's first choice of options, based on an assumed simple linear homeostatic relationship between input and output, is probably wrong and a more circumspect intervention based on a fractal perspective is warranted.

4.3.3 Dynamic disease

As Leon Glass [212] pointed out, dynamical disease refers to illnesses that are associated with striking changes in the dynamics of some bodily function. It is not merely a new kind of control system that is suggested by the modified scaling of physiologic time series, however. Scaling also implies that the historical notion of disease, which has the loss of regularity at its core, is inadequate for the treatment of dynamical diseases. Instead of loss of regularity, the loss of variability is identified with disease, so that a disease not only changes average measures of health, such as heart rate or breathing rate, but is manifest in changes in variability at very early stages. Loss of variability not only implies a loss of physiologic control, but the loss of the flexibility to adapt to a changing environment as well. This loss of variability is reflected in the change of fractal dimension, that is, in the increase in the scaling index of the corresponding time series. The change in fractal dimension with age and with disease suggested the new definition of disease is interpreted as a loss of nonsimplicity, rather than a loss of regularity [219, 220, 598]. However, this new definition has not been universally embraced.

The well-being of a body's network-of-networks is measured by the fractal scaling properties of the dynamic subnetworks, and such scaling determines how well the overall harmony is maintained. Once the perspective that disease is the loss of nonsimplicity has been adopted, the strategies primarily used today in combating disease must be critically examined. Life-support equipment is one such strategy, but the tradition of such life-support is to supply blood at the average rate of the beating heart, to ventilate the lungs at their average rate, and so on. So how does the new perspective regarding disease influence the traditional approaches to assisting the healing of the body?

Alan Mutch applied the lessons of having fractals in physiology to point out that blood flow and ventilation are delivered in a fractal manner in both space and time in a healthy body. However, he argues, during critical illness, conventional life-support devices deliver respiratory gases by mechanical ventilation, or blood by cardiopulmonary bypass pump, in a monotonously periodic fashion. This periodic driving overrides the natural aperiodic operation of the body. It was not so long ago that a cardiac pacemaker delivered such a monochromatic stimulus to the heart. Mutch speculated that these devices result in the loss of normal fractal transmission and, consequently, life support winds up doing more damage the longer it is applied and becomes more problematic the sicker the patient [391].

Similar reasoning can be applied to the origin of pathophysiology. We know that two nonsimple networks have optimal information exchange when their levels of nonsimplicity are matched, which is to say they have the same value of the fractal measure of nonsimplicity [586]. It is therefore not much of a stretch to speculate that nutrients are supplied optimally when the nonsimplicity of the system delivering the nutrient is matched in nonsimplicity to the system receiving it. This could, for example, explain the pathology of traumatic brain injury (TBI). In its original design the network supplying CBF is matched to the nonsimplicity of the brain mass to which the blood supplies oxygen. When trauma is imposed on the brain either the nonsimplicity of the CBF, or that of the brainscape, or both, is disrupted, as is the CME. The severity of the TBI is proportional to the degree of mismatch imposed and the amount of oxygen supplied to the brain is proportionately less than required for the brain to function normally.

4.4   Future Research Directions

Predicting how a future technology will entail new ways of thinking, such as may be provided by an innovative data processing tool is almost as difficult as anticipating the new thinking entailed by a scientific breakthrough. And

It is probably foolish to say which is the more difficult to achieve, or which may have the larger impact on an existing discipline. While it is true that if only a single tool is available, you may become inclined to treat all problems as either unsolvable, or amenable to the existing tool. The way the tool is applied depends on a theory of the problem, typically one devised to optimize, or at least to take advantage of the tool's utility. Predicting the future of research in a particular problem area, given a single tool, is straightforward and entails refining the tool until it either solves every version of the problem, or it leads to a paradox when it cannot penetrate the problem's core.

In medicine when a patient has a malignant tumor, there are a number of tools or options to deal with the problem: cut it out, often the first choice of surgeons; poison it, with medicine delivered by an oncologist; or starve it, by strict diet, using homeopathy. However, because the malignancy is part of a living system, each of these options entails one or more negative responses on the part of the organism. It is the nonsimplicity of the organism through its networking structure that makes correctly predicting the response to pathology so elusive and so necessary. One difficulty lies in the fact that nonsimplicity entails contradiction and modeling nonsimple phenomena inevitably leads to paradox [605].

A transformational understanding of the nonsimple way the human body adapts to external changes and combats illness occurred in medicine when Jenner determined that the way to protect against a life-threatening disease was to expose a patient to that very disease. What could be more paradoxical than a vaccine? This was the beginning of a science revolution in medicine, requiring the synthesis of multiple disciplines to understand what was demonstrably true about vaccinations. New theory entailed by the resolution of empirical paradox invariably leads to new kinds of knowledge that are incompatible with prior understanding and this was certainly the case in medicine [604]. It has been argued that the source of nonsimplicity is criticality [51], which is the emergent collective behavior of natural nonsimple systems. Moreover, the cooperative behavior of critical dynamics has been shown to ultimately resolve empirical paradox [343, 605] and to be intimately related to fractal dynamic processes.

The future of fractals in physiology has been anticipated in a number of workshops hosted by *Frontiers in Physiology*. Some of the contributions to these workshops focus on what can be learned by adopting the techniques of network science, others concentrate on the emergent collective properties determined by criticality, some relate data to system properties determined by their underlying nonlinear dynamics, a few even exploit the factional calculus to relate a system's dynamics to its fractal scaling. These were considered at one time to be esoteric mathematical concepts and can be brought under the broad headings of *Network Science* and the *Fractional Calculus*. They are strong candidates for being the foundational sciences necessary to understand living networks in the 21st century.

It occurred to me that the success of any research activity is determined by how well it is communicated to the science community as a whole, resulting in its ultimate integration into the body of scientific knowledge. This does not occur across the disciplinary boundaries we have constructed without conscious intent. For example, the construction of "interdisciplinary laboratory without walls" was a phrase I first encountered in a paper by my friend and colleague Ary Goldberger *et al.* [221] in regard to the then future utility of *Physionet*, which was intended to foster friendly competition and collaboration. Its utility was to be accomplished by having massive datasets for medical pathologies available online along with modern data processing techniques, as well as, handbook-like instructions on how to use them together. From my perspective, this site has been remarkably successful, but along with that success, it has also highlighted a pedagogic limitation regarding the implementation of new data processing techniques

This is one of the future directions where fractals and network science can play a leading role. As the data processing methods capture and quantify more subtle properties of physiologic time series, the underlying mathematics becomes more obtuse to those not having mathematics as a first, or at least a second language. A prime example of which is the fractional calculus. The use of this calculus is necessitated, for example, by the increasing importance of multifractal time series in describing pathophysiology, along with the importance of memory and non-locality in space. The changing fractal behavior of a process entails new ways to connect the dynamics of physiologic processes with the processing of time series data.

This is similar to the situation we found ourselves in, during the mid-1980s, when we were working to convince colleagues that much could be learned about medicine in general and physiology in particular by applying concepts from nonlinear dynamics, chaos theory and yes, even fractals, to the study of living systems. That battle is still being fought, only now is it being done somewhat more subtly under the banner of Network and/or Complexity Science. We have learned a great deal from the now arcane intellectual skirmishes encountered, not the least of which is the importance of translating mathematical ideas into medical concepts.

Actually, it was unreasonable to expect such translating to be initiated by the members of the medical community. As Wiener observed in 1948 [611]:

> If a physiologist who knows no mathematics works together
> with a mathematician who knows no physiology, the one will be

unable to state his problems in terms that the other can manipulate, and the second will be unable to put the answers in any form that the first can understand.

Consequently, each member of any ongoing transdisciplinary collaboration must work to become conversant in the other's language, or as Wiener went on to say so clearly:

Dr. Rosenblueth has always insisted that a proper exploration of these blank spaces on the map of science could only be made by a team of scientists, each a specialist in his own field but each possessing a thoroughly sound and trained acquaintance with the fields of his neighbors; all in the habit of working together, of knowing one another's intellectual customs and of recognizing the significance of a colleague's new suggestion before it has taken on a full formal expression.

Let me close this chapter with an apology to all of those that I have not cited, but whose research has contributed to making fractal physiology a reality. My only explanation for this lack of acknowledgment is the overwhelming success this idea has enjoyed over the last decade to the point that no one person could read, absorb and synthesize it all. With history being prologue, I anticipate that fractal physiology will continue to enjoy success, with its cleaving to network science and the FC, into the future.

Chapter 5

Modeling Uncertainty

I know that you believe you understand what you think I said,
but I am not sure you realize that what you heard is not what I
meant — Anonymous

The ordinary differential calculus along with the analytic functions that
solve the differential equations resulting from Newton's force laws have been
seen by many scientists as not only necessary, but sufficient, to provide a
proper and complete description of our nonsimple physical world. On the
other hand, experiments indicate that a broad range of physical, biological
and social phenomena cannot be understood using the analytic functions
we have come to rely on in physics. These functions do not capture the uncertain dynamics of common physical phenomena, such as earthquakes and
hurricanes [507]; everyday social phenomena, including opposing groups
reaching consensus [535], transitions from peaceful demonstrations into
riots [227], economic unpredictably, as in stock market crashes [356], high
frequency finance [161] and healthcare networks [515]; or the familiar psychological activity of cognition and habituation [591]; in computer and communication networks such as ethernet traffic [317], internet topology [182]
and in the understanding of why networks fail [426]. As Perrow [426] notes,
catastrophic failures may emerge because of unanticipated interactions in
nonsimple systems; redundancy alone is not an effective strategy for preventing catastrophic failure. The inherent nonsimplicity of these phenomena and many others is beyond the scope of familiar 19th century analysis,
which, to a large extent, forms the mathematical foundation of present-
day physics and engineering. Understanding nonsimplicity, as an extended
class of phenomena, with common structural and mathematical properties,
requires a new way of modeling and consequently more innovative thinking.

5.1   Statistics and Nonsimplicity

Nonsimple phenomena and aging in the context of spin glasses and poly-
mers have attracted significant attention. The reason for the interest in
nonsimplicity in these phenomena has to do with the breakdown of certain
fundamental assumptions made in equilibrium statistical physics, when applied to strongly disordered systems. One example is the Onsager Principle
(OP), which is the relaxation of a perturbed network back to its equilibrium state, described in terms of an unperturbed autocorrelation function,
being violated by anomalous diffusion and relaxation. Papers on this phenomenon are devoted to studying aging in diffusion processes occurring on
low-dimensional lattices [385], in low-dimensional environments [306], and
in the quantum dynamics of dissipative free particles [436]. There has also
been interest in the manifestation of aging in processes described by means
of the CTRW formalism [18, 66].

The relaxation of perturbations captured by the OP is one of the basic
tenets of statistical mechanics. The OP establishes a connection between
a property of the equilibrium state, the unperturbed autocorrelation function of a given physical observable A, and the regression to equilibrium of
a macrosignal resulting from a perturbation of that observable. For this
reason, the OP has been judged to be a fundamental measure, establishing
the connection between dynamics and thermodynamics. It is important to
stress, as clearly stated by Onsager in his original publication [405], that
what would ultimately become the OP holds true for aged networks, namely
networks in contact with heat reservoirs that are in thermal equilibrium. If
the regression to equilibrium is very fast, it is not necessary that the bath
be in equilibrium at the moment when measurements of the regression to
equilibrium of the perturbed network are begun. In fact, in the specific
case where the bath is responsible for fluctuations that can be assumed
to be white, that is, to have a flat frequency spectrum, the regression to
equilibrium of the bath is essentially instantaneous. The OP refers to a
variable of interest whose dynamics are made stochastic by the interaction
with a bath. Thus, when we discuss the process of regression to equilibrium, we need to specify if we are referring to the network, or to its bath
(environment).

If we adopt the white noise approximation to describe the fluctuations
that are responsible for the erratic motion of the network variable, then
the regression to equilibrium of the reservoir is virtually instantaneous and
we can easily satisfy the condition for the validity of the OP. The network's dynamic observable $a(t)$, put in standardized dimensionless form as

in Eq. (2.3) is:

$$A(t) = \frac{\Delta a(t)}{\sqrt{\langle \Delta a(t)^2 \rangle}}, \tag{5.1}$$

where the deviation from the average value of the dynamic variable, denoted by brackets, is given by:

$$\Delta a(t) \equiv a(t) - \langle a(t) \rangle. \tag{5.2}$$

The normal form of the dynamic variable is used to write the stationary autocorrelation function:

$$\Phi_A(t, t') = \langle A(t) A(t') \rangle = \langle A(0) A(t - t') \rangle, \tag{5.3}$$

which only depends on the time difference $t - t'$. The stochastic behavior of the network variable is caused by the interaction between the network and bath, and this kind of process is often studied by means of the master equation method discussed in the previous chapter. However, these conditions need not be met in nonsimple networks and the autocorrelation function, in general, need not be stationary.

To create a master equation compatible with the OP we consider the case when the bath relaxation is not infinitely fast and we need criteria for entanglement between the network and bath. This is not a trivial problem, since the departure from the Poisson condition can generate an infinitely extended memory, and the network-bath entanglement is the result of a rearrangement process with an indefinite time scale. In fact, the traditional fluctuation–dissipation relation needs modification in such networks.

Recall that the CTRW has also been used in foundational discussions of statistical physics and the connection between the master equation formalism and the CTRW has been debated continuously starting from the pioneering work of Bedeaux *et al.* [73] published a half century ago. The focus of the discussion is often based on the waiting-time PDF $\psi(t)$ and it was proven [73] that the Markov master equation is compatible with the CTRW, if the stepping process is Poisson and the waiting-time PDF is exponential. This result raises the related issue of the connection between a CTRW, a non-Poisson stepping process, and a non-Markov master equation. This problem was solved using the GME [286] and which we demonstrate in this chapter can be made compatible with the OP. The GME was demonstrated to unify the fractional calculus and CTRW [372] and is discussed, subsequently. But first we focus attention on how to make the GME stationary and thus compatible with the OP.

The OP makes it possible for us to derive the autocorrelation function from the GME. If we require the GME and CTRW to be equivalent, we

find the result that the two are compatible only when the statistics are
Poisson. The reason for this restriction is that a departure from Poisson
statistics entails memory, which makes the GME incompatible with the
Markov condition. This incompatibility implies that the structure of the
GME is dictated by initial conditions. If the process is not stationary,
the resulting GME is not a *bona fide* transport equation [196]. On the
other hand, if the waiting-time PDF $\psi(t)$ is not exponential, there are
aging effects. These observations require that we allow the network to age
until it reaches a state where the OP is valid. In this aged situation, it is
possible to establish a GME that is a *bona fide* transport equation and to
establish this last relation we also establish a complete equivalence between
the CTRW and the GME [18].

The dynamics of the physical variables to which the OP apply can be
described using two different kinds of equations: (1) the Langevin equation
a stochastic differential equation for the dynamical variable, considered in
this chapter and (2) the phase space equation for the PDF, discussed in
Sec. 5.5.

In the present chapter, we show that the emergence of memory in the
GLE can be thought of as a form of higher-order nonsimplicity than is
contained in the fluctuations of the ordinary Langevin equation. Moreover
a number of formalisms that establish a connection between the fluctuating
force driving the process and the dissipative memory kernel recording the
network's dynamic history are considered.

5.2   Intermittent Stochastic Processes

A number of nonsimple phenomena have been shown to have non-Poisson
statistical properties, including collections of neurons in the human brain
[94]. Bianco *et al.* [95] have shown, using a network of coupled two-state
stochastic clocks that with the onset of phase synchronization, at a critical
value of the coupling coefficient, the dynamics of the network becomes
that of a non-Poisson renewal process, operating in the non-ergodic regime.
The breakdown of the collective phase structures occur, with no memory of
previous state changes, consequently yielding a non-Poisson renewal process
(NPRP), which under well-defined conditions, we refer to as crucial events.

5.2.1   Nonsimplicity and IPLs

NPRP are characterized by a waiting-time PDF between events, indicated
by $\psi(\tau)$. The regression to equilibrium of an ensemble of crucial events is

given by the survival probability:

$$\Psi(t) = \int_t^\infty \psi(\tau)d\tau; \tag{5.4}$$

which is the probability that no event occurs in the interval $(0, t)$. The nomenclature *survival probability* is a consequence of these ideas having been developed within the statistical theory of failure. Herein we focus most of our attention on nonsimple hyperbolic statistics, yielding the asymptotic IPL PDF for the survival probability:

$$\Psi(t) = \left(\frac{T}{T+t}\right)^{\mu-1} \propto t^{1-\mu}, \tag{5.5}$$

with index $1 < \mu < 3$. This PDF corresponds to processes that can violate the finite time-scale assumption, so often made in statistical physics and engineering. Said differently, the mean time between events is given by Eq. (4.10), where T is a parameter characteristic of the waiting-time PDF, which is defined by the time derivative of Eq. (5.4) and diverges for $\mu \leq 2$. More generally, for $1 < \mu < 3$ these events are defined as being crucial.

One way a non-exponential decay, such as the IPL, can be expressed is as the superposition of infinitely many Poisson components. But if these components are independent of one another, as are the single dynamic elements in the absence of cooperation, there are no NPR events, and consequently no crucial events are generated [22]. The production of crucial events is a sign of close cooperation among distinct dynamical elements, thereby offering a rationale as to why the NPRPs do not respond to harmonic perturbations [62, 503], as would single independent Poisson dynamical elements. NPR processes reflect a condition shared by the phenomenological models of glassy dynamics [114], laser cooling [63] and models of atomic transport in optical lattices [339]. Other NPR processes have been found at the core of the correlations in DNA sequences [379], heart rate variability (HRV) [17, 585] and earthquake statistics [368].

In Chapter 6 we review a relatively new theory of information transport, that is, the determination of how one nonsimple network responds to an excitation by a second nonsimple network as a function of the mismatch of the measures of nonsimplicity of the two networks. The nonsimple network considered is NPR and the measure of nonsimplicity is taken to be the IPL index. More precisely, we consider a nonsimple network, whose waiting-time PDF $\psi(\tau)$ has the IPL index $\mu < 2$ and we review, in analogy with SR theory [206], the case where the rate of production of jumps, a kind of renewal event, is modulated by an external excitation. This leads to the new concept of nonsimplicity management, where the exchange of information between two nonsimple networks is maximized when their non-simplicity is matched as discussed in Chapter 1. The nonsimplicity matching phenomenon is a special property of non-ergodic renewal (NER) processes, which are a special class of NPR processes.

5.2.2   Manneville mapping

When a physical network is outside the domain where the linear approximation is valid, say, beyond the elastic limit in a piece of metal, then nonlinear interactions dominate and the Hamiltonian, or any other generator of the dynamics, must be modified to include terms beyond the quadratic. As noted earlier, another way to model the increased nonsimplicity of a network is to put it in contact with the environment, which in the simplest physical case provides interrelated fluctuations and dissipation. In classical diffusion the statistics of the fluctuations are empirically determined to be Gaussian, delta correlated in time, with the response of the diffusing particle at equilibrium to these fluctuations determined to be a canonical distribution. However, the statistics of the heat bath need not be Gaussian, nor delta correlated in time, so another way nonsimplicity can enter network models is through generalizing the statistical properties of the fluctuating force in the Langevin equation. In this section we present a model for the generation of intermittent fluctuations, as a prototype of crucial event generators. Crucial events are the main property of the form of nonsimplicity that we discuss in detail in the next two chapters.

We adopt a dynamical model, introduced by Manneville [355] to study turbulence, based on a particle moving in the positive direction along the q-axis and confined to the unit interval $I = [0, 1]$, [8, 19]. The equation of motion is chosen to be the nonlinear rate equation:

$$\frac{dq(t)}{dt} = aq(t)^z \qquad (5.6)$$

where the coefficient is small, that is, $0 < a \ll 1$. Statistics are introduced into the solution to the deterministic equation Eq. (5.6) through a boundary condition. Whenever the particle reaches the border $q = 1$, it is injected back into the unit interval I to a random position, having uniform probability on the interval I. Throughout this discussion we refer to this back-injection as an event; an event that disconnects what happens in one sojourn on the interval to any other.

We now construct a PDF for the particle's sojourn-time in the unit interval. One can construct such an argument for $z = 1$, with the result that the PDF for the number of events (reinjections) is Poisson and the sojourn-time PDF is exponential. For $z \neq 1$ the situation is entirely different and one obtains a non-Poisson PDF. The time τ it takes for the particle to reach the boundary $q(\tau) = 1$, starting from the initial point $0 < q(0) < 1$, is determined by quadrature:

$$\int_{q(0)}^{1} \frac{dq}{q^z} = a\tau, \tag{5.7}$$

by integrating Eq. (5.6). The result is a relation between the random initial condition $q(0)$ and the random sojourn-time τ, obtained after some algebra:

$$q(0) = \frac{1}{[1 - (1 - z)a\tau]^{\frac{1}{1-z}}}. \tag{5.8}$$

The sojourn-time, or waiting-time, PDF $\psi(\tau)$ is consequently determined by the identity between PDFs:

$$\psi(\tau)d\tau = p(q(0))dq(0). \tag{5.9}$$

We assume the reinjection PDF is uniform on the unit interval, so that $p(q(0)) = 1$, and obtain, after some algebra using Eq. (5.8), the hyperbolic PDF:

$$\psi(\tau) = \left| \frac{dq(0)}{d\tau} \right| = \frac{(\mu - 1)T^{\mu-1}}{(T + \tau)^{\mu}}. \tag{5.10}$$

The parameters μ and T are determined to be related to those from the dynamics determined by Eq. (5.8):

$$\mu = \frac{z}{z - 1} \quad \text{and} \quad T = \frac{1}{(z - 1)a}. \tag{5.11}$$

The hyperbolic PDF given by Eq. (5.10) is asymptotically IPL:

$$\lim_{\tau \to \infty} \psi(\tau) = (\mu - 1)\frac{T^{\mu-1}}{\tau^{\mu}}, \tag{5.12}$$

and is a model we find useful, as a prototype, for any dichotomous renewal network. The IPL depends on two parameters, μ and T, which have very different interpretations. The parameter T is the time lapse necessary for the expression on the RHS of Eq. (5.10) to become identical to a strict IPL. In the asymptotic region, T is also a measure of the weight of the fat tail of the PDF. The second parameter μ, on the other hand, is a fractional IPL index that marks the presence of nonsimplicity. It can, for instance, be due to a hierarchical scaling superposition of exponentials, indicating an invisible (inaccessible) exploration of structures within structures, or in a physical chemistry context, it can be the effect of an Arrhenius-activated process with a fluctuating rate, subject to another local temperature. It may also be the expression of many other mechanisms, with many or few degrees-of-freedom (returns in RWs in exotic topologies or energy landscapes). When we record a relatively rare event, such as a threshold passage, or a structure change, all these networks yield a point process in time that may or may not be renewal, in which the time intervals between successive events are independent of one another. In many cases, it is also possible to assign a sign to these rare events. It is evident that, in the renewal case, a dichotomous model can be adopted.

The sojourn-time hyperbolic PDF given by Eq. (5.10) is properly normalized for $\mu > 1$. As mentioned, the mean sojourn-time is finite for $\mu > 2$, but diverges if $\mu < 2$. Consequently, Eq. (5.10) is normalizable, but it has a diverging mean sojourn-time for $1 < \mu < 2$. The probability that no event occurs up to time t defines the survival probability given by Eq. (5.4), which was found to be very useful and continues to be of value in subsequent discussions.

5.2.3 Poisson distribution

It is a simple matter to substitute these expressions for the parameters Eq. (5.11) in terms of z back into Eq. (5.10) to obtain:

$$\psi(\tau) = a[1 + (z-1)a\tau]^{-\frac{z}{z-1}}, \tag{5.13}$$

which in the limit $z \to 1$ yields the exponential distribution in the limiting form:

$$\psi(\tau) = ae^{-a\tau}. \tag{5.14}$$

The waiting-time PDF for n events occurring in the time interval t is given by the convolution:

$$\psi_n(\tau) = \int_0^t \psi_{n-1}(t-t')\psi_1(t')dt', \tag{5.15}$$

and associating $\psi_1(t)$ with the exponential Eq. (5.14), this convolution equation can be integrated to yield:

$$\psi_n(t) = \frac{a(at)^{n-1}}{\Gamma(n)}e^{-at}. \tag{5.16}$$

In the renewal theory [155] and queueing theory [237] literature Eq. (5.16) goes by the name of the Erlang distribution, which for the probability of n events occurring in a given time interval gives the Poisson distribution:

$$P_n(t) = \frac{(at)^n}{n!}e^{-at}. \tag{5.17}$$

Thus, we see that an exponential distribution of waiting times implies a Poisson distribution for the number of events occurring in a given time interval, indicating the statistics of the time intervals and the statistics of the number of time intervals are not the same, but they are related.

5.2.4 Hyperbolic PDFs

Hyperbolic PDFs asymptotically become IPLs and consequently provide one of the most familiar manifestations of statistical nonsimplicity. They are discussed in the context of avalanches [341]; identified in biological speciation [617, 624]; connected to linguistics [630]; seen in the tails of the distribution of income [414]; dominate the number of scientific citations [160]; uncovered in the number of scientific publications [331, 332]; revealed for fetal lamb breathing [521]; associated with the branching of bronchial trees [550, 576]; proven in small-world phenomena, involving connections on the Internet and the WWW [59, 564], and the list goes on and on. We give more extensive discussions of experimental datasets at the appropriate points in our presentation of the underlying theory.

Nonsimplicity, as we have seen, can be addressed starting from dynamic equations, by introducing more and more degrees-of-freedom that are allowed to interact nonlinearly with one another. With each new variable the network becomes increasingly nonsimple as previously discussed. Alternatively, we can start with the evolution of the PDF in a homogeneous isotropic phase space. Things become increasingly nonsimple as additional structure is introduced into the space, structure in the form of spatial heterogeneities or temporal memory. Historically, a number of approaches have been developed to generate IPLs [37, 333, 353, 384, 551] and in Chapter 2 we introduced some of the empirical evidence for $1/f$-phenomena leading to such distributions, but for the moment we focus on how various nonsimplicity mechanisms fit into a random walk perspective. One strategy to systematically include these effects is with the CTRW, where after each step the random walker pauses for a time interval of random duration before taking her next step.

Let us consider the hyperbolic waiting-time PDF, introduced using a dynamic argument, resulting in Eq. (5.10). This PDF can be obtained by means of the transformation [123]:

$$\tau = T \left(\frac{1}{y^{\mu-1}} - 1 \right), \tag{5.18}$$

which changes the variable y, defined to be uniformly distributed over the interval $I = [0, 1]$, into the numbers τ distributed according to the hyperbolic PDF given by Eq. (5.10). Note that the numbers y coincide with the initial conditions $q(0)$ used in Eq. (5.7). In other words, the algorithm Eq. (5.18) has a dynamical origin, corresponding to a slow and regular motion towards a condition of abrupt change, that is, to a quake, or catastrophe. The time interval between consecutive quakes is generated by regular dynamics that is not linear.

In summary, renewal events are defined as those events whose occurrence has the effect of erasing the memory of earlier events. It is evident that the transformation Eq. (5.18) accomplishes this erasure. In the dynamic case given here by the convolutions in Eq. (5.15), the lack of correlation among the waiting times τ_i, and the consequent renewal character of the process, is the result of deterministic chaos.

We refer to the work in the following section, based on [24, 102], as an example of a network having event statistics described by Eq. (5.10), but chosen to violate the renewal condition. We emphasize that the adoption of the transformation defined by Eq. (5.18) makes it easy to realize the renewal condition through the random selection of y. Adopting other methods, for instance choosing fluctuating Poisson dynamics, can produce renewal events that do not have anything to do with the dynamics leading to the hyperbolic PDF given by Eq. (5.10). The method of fluctuating Poisson dynamics produces renewal events that are embedded in a cloud of Poisson events, thereby generating a significant departure from the genuinely renewal condition, and presents a challenge for the detection of renewal properties as well [24]. If the fluctuations of the Poisson parameter are made infinitely slowly, the renewal events are completely annihilated, and the physics of such networks, for instance, the response to external excitations, may depart from the physics of the renewal events, in spite of sharing the same hyperbolic waiting-time PDF given by Eq. (5.10).

Grigolini *et al.* [234] explained that there are two main complexity categories, the Renewal Approach to Complexity (RAC) and Complexity Without Renewal (CWR). The term complexity is used here when it is necessary to make contact with this earlier discussion, otherwise, we continue with the use of its replacement term, nonsimplicity. We adopt that perspective here for the following reasons.

It is well known that $1/f$-noise is considered to be a manifestation of nonsimplicity, as described in the excellent review on $1/f$-noise in physical networks [179]. However, it is not so well known that the most popular approaches to explaining $1/f$-noise are different forms of CWR. We refer the reader to the excellent book of Jensen [271], who clearly shows the connection of nonsimplicity with self-organized criticality (SOC) [50, 51] and superstatistics [71], with an approach to nonsimplicity based on the slow fluctuation of a Poisson parameter, in accordance with more recent observations [52]. Thus, we conclude that SOC, as originally formulated, is another form of CWR. One may use Jensen's arguments that SOC is not a satisfactory approach to $1/f$-noise, to decide the issue of whether $1/f$-noise rests on the CWR, or on the RAC.

Here we focus on the approach to $1/f$-noise proposed by Voss and Clarke [554] in their study of music discussed in Chapter 2. This makes it possible to show that in the case where the ergodic condition $\mu > 2$ applies, the RAC yields results equivalent to the CWR, thus creating the mistaken impression that the two categories are physically equivalent. As discussed in the next section, it is possible to express the signal $\xi(t)$ as a sum of harmonic oscillations in such a way as to obtain for its normalized autocorrelation function [127]:

$$\Phi_\xi(t) = \frac{\langle \xi(0)\xi(t) \rangle}{\langle \xi^2 \rangle} = \left(\frac{T}{T+t} \right)^\beta, \tag{5.19}$$

with $\beta < 1$. This hyperbolic autocorrelation function entails the power spectral density $S_p(\omega)$ to have the IPL form:

$$S_p(\omega) \propto \frac{1}{\omega^\alpha}, \tag{5.20}$$

where $\alpha = 1 - \beta$. This very simple argument yields the same conclusion as that of Voss and Clarke, concerning the origin of $1/f$-noise, with α slightly smaller than 1.

On the basis of these remarks, one might be tempted to adopt the CWR philosophy to model the dynamic effects of nonsimplicity. In fact, the superposition of many oscillations, such as the Fourier transform of a random time series, does not generate renewal events. However, let us use the hyperbolic PDF Eq. (5.10) to create a renewal time series $\{\tau_i\}$ from which the time sequence t_i can be generated according to the prescription $t_{i+1} = \tau_i + t_i$, where the subscript denotes the order in which the time interval occurs in the succession and the initial time is $t_0 = 0$. Let us imagine that the event occurring at time t_i is a coin tossing, used to define the sign of the entire time region between t_i and t_{i+1}, such that, the fluctuations between successive intervals at most vacillate between $+1$ and -1. This prescription creates a fluctuating function $\xi(t)$, whose autocorrelation function is determined by renewal theory to be [211]:

$$\Phi_\xi(t) = \frac{1}{\langle \tau \rangle} \int_t^\infty d\tau (\tau - t)\psi(\tau), \tag{5.21}$$

where again $\psi(\tau)$ is the hyperbolic waiting-time PDF Eq. (5.10) and the mean waiting time is given by Eq. (4.10). The mean waiting time diverges at the border $\mu = 2$, between $2 < \mu$, which is compatible with the existence of an infinitely aged condition [18] and the region $2 > \mu$, which is characterized by perennial aging and ergodicity breakdown [78, 79, 80, 81]. It is straightforward to show that Eq. (5.21) yields Eq. (5.19) with $\beta = \mu - 2$. In the case $2 < \mu < 3$, adopting the Wiener–Khintchine theorem [554] suggests that this renewal time series is a form of $1/f$-noise.

Thus, even though we often find references to CWR theories to explain the origin of $1/f$-noise, we cannot rule out RAC as a plausible alternative approach to describing the effects of nonsimplicity. In the next section we point out that with an approach to the hyperbolic PDF Eq. (5.10) based on the slow modulation of a Poisson process [24], that is, on a form of *superstatistics* approach to nonsimplicity, the aging effect is strongly reduced thereby allowing us to establish if a given physical process corresponds to the CWR, or to the RAC, perspective. Furthermore, we have to point out that the practical realization of the CWR prescription generates unexpected renewal events. The slower the modulation, the rarer are the emerging renewal events, thereby making them difficult to detect. The renewal events play the crucial role of determining the scaling of the time asymptotic limit but this anomalous scaling may show up at times so large as to make it almost impossible to reveal the existence of these renewal events. In this case we have to consider the process as belonging to the CWR category.

In the field of single molecule spectroscopy and BQDs, investigators [93, 303, 304, 398, 491] established that the BQD data undergo renewal aging thereby ruling out the CWR, in the description of intermittent switching in BQD. In the field of econophysics, the work of Scalas [471] does not take a position on the two categories being considered, thereby leaving open the possibility that econophysics may be properly described by either SOC or superstatistics. Bianco and Grigolini [94], on the other hand, using the idea of aging, establish that the econophysics processes they study obey renewal theory, even if the critical events underlying trade action are not ostensible and become observable only as a result of additional analysis.

It is important to notice that the BQD phenomenon is related to the origin of $1/f$-noise, with α slightly larger than one, rather than slightly smaller than one. In fact, the BQD phenomenon is known [303, 304, 398, 491] to correspond to Eq. (5.10) with $\mu < 2$. This condition on the IPL index, in turn, is known [357] to violate the ergodic condition, thereby violating the condition for the existence of stationarity. Some analysis [21] affords a way to evaluate the non-stationary correlation function, and, thus using dimensional arguments [554] we reach the conclusion:

$$\alpha = 3 - \mu, \tag{5.22}$$

in accordance with experimental observation [419].

Thus, not only can we not rule out RAC as a possible theory for the origin of $1/f$-noise, but we are tempted to advocate RAC as an approach to $1/f$-noise, as being even more attractive than earlier theories. Notice that Eq. (5.22) locates the ideal condition of $1/f$-noise at $\mu = 2$, which is the border between the ergodic ($\mu > 2$) and non-ergodic ($\mu < 2$) conditions. This interesting aspect seems to have been overlooked by early investigations into the origins for $1/f$-noise and this is probably a consequence of the fact that these earlier approaches were based essentially on the CWR perspective.

5.3 Renewal Aging and Crucial Events

Let us introduce a processing method to generate, as well as to determine if a time series is renewal through an application of an aging technique. We have dealt with a number of similarities between two approaches to hyperbolic relaxation: a non-homogeneous Poisson process and a homogeneous renewal process. The most familiar manifestation of aging is given by the two-time autocorrelation function that does not depend on the time interval between the two times t_1 and $t_2 > t_1$, but depends on each time separately. The relaxation of the correlation, as a function of $t_2 - t_1$, becomes slower for larger t_1. Aging is thought to be determined by the fact that the network under study is out of equilibrium and the regression to equilibrium involves times longer than the observation time [128].

In this section, we have in mind *renewal aging* [18, 66, 217]. The renewal model studied herein yields renewal aging, whose immediate manifestation is revealed by the dependence of $\psi(t)$ on the observation time. The quantity $\psi(t)dt$ is the probability that a laminar region begins at $t = 0$ and ends in the small time interval $[t, t + dt]$. If we have at our disposal a Gibbs ensemble of time sequences, all of which are initiated in the first laminar region, located at $t = 0$, to derive experimentally a $\psi(t)$, we have to observe the time at which the first laminar region of each sequence of the sample ends. This is equivalent to making the *preparation* and *observation* times coincide.

Let us imagine that after preparing the network at $t = 0$, with an event occurring at that time in any of the Gibbs networks, we postpone the beginning of the waiting process for the occurrence of a new event at time $t' > 0$. In this case, if the process is renewal, the probability of an event occurrence at time t is given by the convolution equation:

$$\psi(t, t') = \psi(t) + \sum_{n=1}^{\infty} \int_0^{t'} dt'' \psi_n(t'') \psi(t - t''). \tag{5.23}$$

The physical meaning of this prescription is as follows. At time $t = 0$ an event occurs. However, since the observation process begins later at time $t' > 0$, the probability of observing a new event after t' depends on the last event occurring prior to t', at time t'', $\psi(t - t'')dt''$. The event occurring at $t'' < t'$ is in general the last of a sequence of n events, occurring exactly at t'', while the earlier events can occur at any earlier time. The probability for this last event is $\psi_n(t'')$, with $n \geq 1$; the first term in the integrand of Eq. (5.23). Note that Eq. (5.23) is an exact relation and is the foundation of the new FDT of the first kind discussed in Chapter 6.

The corresponding non-stationary survival probability reads:

$$\Psi(t, t') = \int_t^\infty dt'' \psi(t'', t').\tag{5.24}$$

Note that by taking the t'-derivative of Eq. (5.24) and inserting Eq. (5.23) into Eq. (5.24) we obtain:

$$\frac{d}{dt'}\Psi(t, t') = R(t')\Psi(t - t'),\tag{5.25}$$

where:

$$R(t) = \sum_{n=0}^\infty \psi_n(t).\tag{5.26}$$

The quantity $R(t)$ is the rate of events occurring at time t, under the condition that an event occurs at $t = 0$.

In the case $\mu > 2$, when the stationary condition is adopted, this event generation rate Eq. (5.26) is constant. On the other hand, using Eq. (5.26) and taking advantage of the time convoluted nature of the waiting-time PDF, yields for the rate of generation of events:

$$\widehat{R}(u) = \sum_{n=0}^\infty [\widehat{\psi}(u)]^n = \frac{1}{1 - \widehat{\psi}(u)}.\tag{5.27}$$

Of special interest herein is the case $\mu < 2$, where the Laplace transform of the hyperbolic PDF given by Eq. (5.10), for $u \to 0$ is:

$$\widehat{\psi}(u) = 1 - \Gamma(2 - \mu)(Tu)^{\mu-1}\tag{5.28}$$

so that inserting Eq. (5.28) into this expression for the Laplace transform of the rate of generation of events, we obtain as $u \to 0$:

$$\widehat{R}(u) = \frac{1}{1 - \widehat{\psi}(u)} \approx \frac{1}{\Gamma(2 - \mu)(Tu)^{\mu-1}},\tag{5.29}$$

which, applying a Tauberian theorem, yields for the rate of events being generated asymptotically in time:

$$R(t) \propto \frac{1}{t^{2-\mu}}. \tag{5.30}$$

In other words, the renewal aging, in the case $\mu < 2$ is characterized by the property that the rate of event generation tends to decrease as an IPL in time with index $2 - \mu$.

5.3.1 Modulation and renewal aging

Let us imagine that preparation of a nonsimple network is made at time $t = -t_a < 0$. The measured waiting time to the first event is denoted by τ_1. The first waiting time, distinct from the observation of the successive waiting times, does not necessarily correspond to the total time duration of a laminar region. The resulting histogram records time durations that are generally smaller than those corresponding to preparing the network at time $t = 0$. Nevertheless, in the case when the waiting-time PDF is exponential, both long and short time durations are reduced by the same percentage. Thus, the histogram is turned into a normalized waiting-time PDF, which has the effect of recovering the same exponential form. Consequently, renewal exponential processes do not age.

In the non-exponential case, delaying the observation process has the effect of producing a percentage reduction of the short-time laminar regions larger than that of the long-time laminar regions. As a consequence, with the normalization of the PDF, the weight of the short-time laminar regions is reduced and the weight of the long-time laminar regions is enhanced, thereby generating a slower decay of the survival probability $\Psi(t)$.

A numerically generated PDF $\Delta(\lambda)$ was used to generate trajectories, i.e., artificial sequences of random waiting times [24]. These sequences were compared with the results from the aging analysis of trajectories characterized by the same exponent μ of the IPL, but generated from the renewal process and from modulation processes, with different numbers of drawings N_d from the PDF $\lambda \, exp(-\lambda t)$. After N_d drawings, they [24] selected from $\Delta(\lambda)$ a new rate λ. It is evident that if $N_d = 1$, and only one waiting time is drawn from the Poisson distribution, with a given λ and immediately afterward, a different λ is selected from the PDF $\Delta(\lambda)$, the resulting sequence is renewal. Increasing N_d has the effect of realizing the prescriptions of superstatistics [71], which requires a long sojourn-time in a given Poisson condition, for the network to adapt to the local thermodynamic condition.

Allegrini et al. [24] adopted a procedure referred to with a cumbersome name *aging experiment analysis* [93]. Following that analysis, we adopt a mobile window of size t_a, corresponding to the age of the process to be examined. The beginning of the window t_a is located at the time of occurrence of an event and record the time interval between the end of the t_s window and the first event that occurs after emerging from the time window. Note that this procedure selectively drops events from the original sequence dependent on t_a and thereby produces different decision-making PDFs.

These truncated time intervals are used to build up a t_a-aged histogram, which is then used to define the *aged* PDF $\psi_{t_a}(t)$ and the corresponding aged survival probability:

$$\Psi_{t_a}(t) = \int_t^\infty \psi_{t_a}(\tau')d\tau' = 1 - \int_0^t \psi_{t_a}(\tau')d\tau'. \tag{5.31}$$

Note that with this notation the exact prescription from Eq. (5.23) yields:

$$\psi_{t_a}(t) = \psi(t+t_a) + \sum_{n=1}^\infty \int_0^{t_a} dy\psi(y+t)\psi_n(t_a-y). \tag{5.32}$$

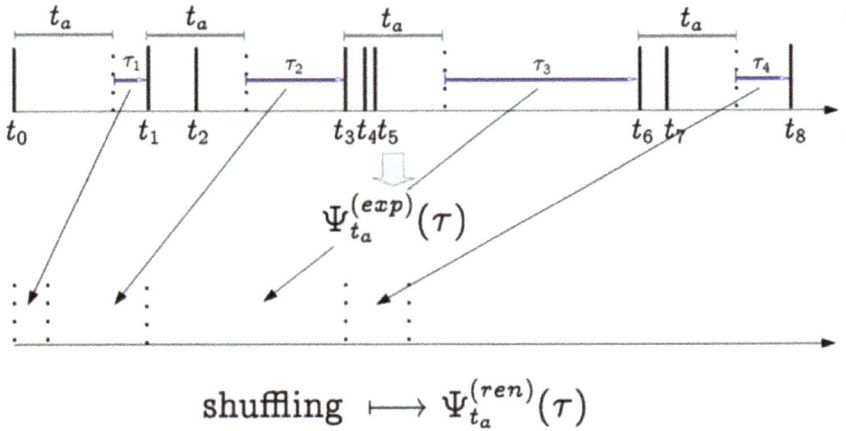

Figure 5.1: An experimental time series $\{t_j\}$ is obtained for analysis. The time intervals in a new time series are selected by choosing an aging time t_a, which is initiated at t_0 and the time $t_1 - t_0 = t_a + \tau_1$ generates the new time interval τ_1. A second time interval initiated at t_1 and the time $t_3 - t_1 = t_a + \tau_2$ generates the new time τ_2, and so on. Note that the time intervals in the experimental time series that are spaced closer than t_a can be lost in the aged time series.

Aging can be established to be renewal by shuffling the experimental sequence $\{\tau_j\}$ obtained in Fig. 5.1 and in this way obtained another time series. Note that the shuffling randomizes the ordering of the time intervals, thereby destroying any correlation contained in the experimental time series. However the change in ordering does not modify the statistics of the time series, since no time intervals have been added to or subtracted from the unshuffled time series. We then apply the aging experiment to the shuffled sequence and if the two survival probabilities coincide, that from the original given by Eq. (5.31) and the shuffled sequences, the process is renewal.

The results of this aging experiment are illustrated in Fig. 5.2. When we draw only one waiting-time τ and then use a different $\psi_\lambda(t) \equiv \lambda \, exp(-\lambda t)$ for the drawing of the next waiting-time, the process is renewal. If we increase the number of waiting times drawn from the same $\psi_\lambda(t)$, the intensity of aging is reduced until the condition of a total lack of aging is reached, at which time the number of waiting times drawn from the same waiting-time PDF becomes very large.

The aging experiment can be used to establish if an experimental data sequence is renewal or not. The aging experiment is applied to an empirical sequence, so as to determine the aged histogram and through that determination, the corresponding survival probability $\Psi_{t_a}^{(exp)}(\tau)$. We also establish a criterion to determine the form of the survival probability produced by the renewal condition. This is not quite straightforward, due to the fact that the exact form of Eq. (5.23) is not a simple functional of $\psi(\tau)$. A simple functional is given by Eq. (5.32) that makes it possible to obtain $\psi_{t_a}(\tau)$ from $\psi(\tau)$ and thus to derive from the form that the renewal theory assigns to the aged survival probability $\Psi_{t_a}(t)$. If these two procedures applied to the same sequence generate the condition:

$$\Psi_{t_a}^{(exp)}(t) = \Psi_{t_a}^{(ren)}(t), \tag{5.33}$$

we consider the process to be renewal. Note that a virtually exact criterion is obtained by shuffling the time intervals between successive events to produce a more reliable $\Psi_{t_a}^{(ren)}(t)$.

5.3.2 Non-stationarity and the GOP

The aging aspect of the PDF is quite important and warrants additional discussion. We note that the above approach [21] can be easily extended to the case where the distribution of first exit times has a finite age. Note that the first exit time differs from a first passage time in that the event is exiting from a region for the first time, which is more general than an

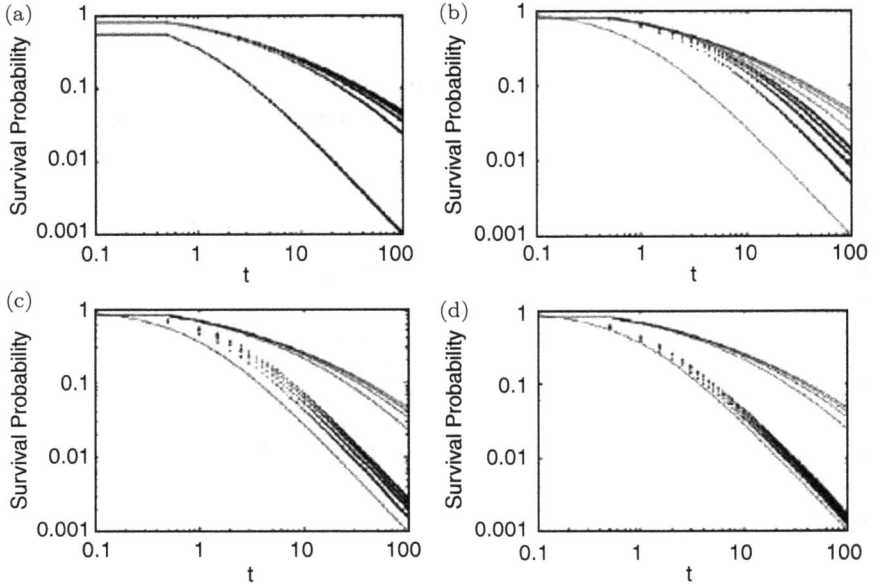

Figure 5.2: Comparison between the function $\Psi_{t_a}(t)$ of a renewal process (continuous lines) and the function $\Psi_{t_a}(t)$ produced by a modulation approach with a changing value of N_d (dotted lines). The curves from bottom to top refer to the ages $t_a = 0, 50, 100, 150, 200$. (a) $N_d = 0$; (b) $N_d = 10$; (c) $N_d = 100$; (d) $N_d = 500$. Taken from [24] with permission.

observable achieving a specific value for the first time. In the case of a two-state system, the conventional master equation is given by:

$$\frac{d}{dt}\mathbf{p}(t) = -\lambda \mathbf{K} \mathbf{p}(t), \tag{5.34}$$

where the probability vector has two components:

$$\mathbf{p}(t) \equiv \begin{pmatrix} p_1(t) \\ p_2(t) \end{pmatrix} \tag{5.35}$$

and the components are coupled by means of the 2×2 matrix:

$$\mathbf{K} \equiv \frac{1}{2} \begin{pmatrix} 1 & -1 \\ -1 & 1 \end{pmatrix}. \tag{5.36}$$

The solution to the deterministic master equation for the probability can be obtained after some algebra to be:

$$p_1(t) = \frac{1}{2}\left(1 + e^{-\lambda t}\Delta p(0)\right),$$

$$p_2(t) = \frac{1}{2}\left(1 - e^{-\lambda t}\Delta p(0)\right),$$

where the initial condition is specified by $\Delta p(0) = p_1(0) - p_2(0)$. Notice that this solution is equivalent to the discrete form given by Eq. (5.133) as $\Delta t \to 0$, such that $\phi(t) = \Delta p(t)$.

We now introduce the concept of subordination by considering the case where the deterministic master equation is subordinated to a stochastic process and subsequently define it in general. Using the CTRW we can write the probability vector as:

$$\mathbf{p}(t) = \sum_{n=0}^{\infty} \int_0^t dt' \psi_n(t')\Psi(t - t')\mathbf{M}^n \cdot \mathbf{p}(0), \qquad (5.37)$$

where the components are coupled by means of the 2×2 matrix:

$$\mathbf{M} \equiv \frac{1}{2}\begin{pmatrix} 1 & 1 \\ 1 & 1 \end{pmatrix}. \qquad (5.38)$$

It is important to understand the meaning of Eq. (5.37). First of all, $\mathbf{p}(0)$ denotes the initial random walker probability vector. The main purpose of this treatment is to determine the probability $\mathbf{p}(t)$ at time t as a function of the initial condition $\mathbf{p}(0)$. Equation (5.37) determines $\mathbf{p}(t)$ through the occurrence of random events. A random event corresponds to the occurrence of jumps from one site to other sites, with a probability described by the matrix $\mathbf{M}$, which in the case indicated is the flipping of a coin. The complexity of this process is indicated by the times of occurrence of these jumps as described by the waiting-time PDF $\psi_n(t)$.

When we observe the network at time t, the probability $\mathbf{p}(t)$ is the result of the occurrence of an arbitrary number n of these random events. The waiting-time PDF $\psi_n(t)$ denotes the probability that at time t, and exactly at time t, the nth member of a sequence of n random events, occurs. Of course, we have to consider the case when no event occurs, as well, this being expressed by $\psi_0(t) = \delta(t)$. It is evident that such a renewal process satisfies the recursion relation Eq. (5.15) and the Laplace transform satisfies Eq. (5.27). Note that the last member of a sequence of n random events in Eq. (5.23) does not necessarily occur at time t, when we observe the timing of events. It might occur at an earlier time $t' < t$.

Thus, for the prescription to become reliable, we set the constraint that no event occurs between time t' and time t. In general, the probability that no event occurs up to a given time t, is given by the survival probability $\Psi(t)$. The Laplace transform of $\Psi(t)$, given by $\widehat{\Psi}(u)$, is related to $\widehat{\psi}(u)$ by:

$$\widehat{\Psi}(u) = \frac{1}{u}\left[1 - \widehat{\psi}(u)\right]. \tag{5.39}$$

Note that between t', the time at which the nth member of a chain of n events occurs, and the observation time t, the distribution keeps the value $\mathbf{M}^n$; the matrix $\mathbf{M}$ being the prescription establishing the jumps that the random walker undergoes when an event occurs. Since n events occurred, this matrix is applied n times to the initial condition $\mathbf{p}(0)$.

Evaluating the sum in Eq. (5.37), using the powers of the Laplace transform of the waiting-time PDF, we write the Laplace transform of $\mathbf{p}(t)$ as $\widehat{\mathbf{p}}(u)$:

$$\widehat{\mathbf{p}}(u) = \frac{1}{u + \widehat{\Phi}(u)(\mathbf{M} - 1)}\mathbf{p}(0), \tag{5.40}$$

where $\widehat{\Phi}(u)$, the Laplace transform of the memory kernel $\Phi(t)$, is related to the Laplace transform of $\psi(t)$, through:

$$\widehat{\Phi}(u) = \frac{u\widehat{\psi}(u)}{1 - \widehat{\psi}(u)}. \tag{5.41}$$

Consider the generalized master equation (GME) given by:

$$\frac{d}{dt}\mathbf{p}(t) = -\int_0^t dt'\Phi(t - t')\mathbf{K} \cdot \mathbf{p}(t'). \tag{5.42}$$

We see that the Laplace transform of Eq. (5.42) yields:

$$\widehat{\mathbf{p}}(u) = \frac{1}{u + \widehat{\Phi}(u)\mathbf{K}}\mathbf{p}(0), \tag{5.43}$$

thereby establishing that the CTRW is equivalent to the GME of Eq. (5.42), with the condition:

$$\mathbf{M} = -\mathbf{K} + 1. \tag{5.44}$$

At this stage, we make the assumption that the random walker starts out in state 1, so that $p_1(0) = 1$ and $p_2(0) = 0$. This is equivalent to establishing an out-of-equilibrium fluctuation at the same moment the network is prepared.

We are interested in determining the regression of the initial fluctuation to equilibrium. Therefore, we define the residual probability:

$$\Delta p(t) \equiv p_1(t) - p_2(t). \tag{5.45}$$

Using the GME of Eq. (5.42) and the explicit expression for $\mathbf{K}$ given by Eq. (5.36), we obtain after some algebra:

$$\frac{d}{dt}\Delta p(t) = -\int_0^t dt'\Phi(t-t')\Delta p(t')dt'. \tag{5.46}$$

It is straightforward to prove that the Laplace transform of $\Delta p(t)$, as determined by Eq. (8.22), is identical to the Laplace transform of $\Psi(t)$ of Eq. (5.42). Consequently their inverse Laplace transforms are also equal:

$$\Delta p(t) = \Psi(t), \tag{5.47}$$

which identifies the regression to equilibrium with the survival probability. The survival probability $\Psi(t)$ is not the equilibrium autocorrelation function, but it can be interpreted as a non-stationary autocorrelation function of age $t_a = 0$.

This approach has been extended [21], to the case when the preparation of the network is done at $t_a < 0$, while the out-of-equilibrium fluctuation is created at $t = 0$. In this case the aged GME reads:

$$\frac{d}{dt}\mathbf{p}(t) = -\int_0^t dt'\Phi_{t_a}(t-t')\mathbf{K}\cdot\mathbf{p}(t'). \tag{5.48}$$

The Laplace transform of the aged memory kernel $\Phi_{t_a}(t)$ is given by:

$$\widehat{\Phi}_{t_a}(u) = \frac{u\widehat{\psi}_{t_a}(u)}{1-\widehat{\psi}_{t_a}(u)}. \tag{5.49}$$

In direct analogy with the $t_a = 0$ case, the time evolution for $\Delta p(t)$ reads:

$$\frac{d}{dt}\Delta p(t) = -\int_0^t dt'\Phi_{t_a}(t-t')\Delta p(t')dt'. \tag{5.50}$$

Thus, we define the <u>generalized OP</u> (GOP):

$$\Delta p(t) = \Phi_\xi^{(t_a)}(t). \tag{5.51}$$

The function $\Phi_\xi^{(t_a)}(t)$ is the aged autocorrelation function of the dichotomous fluctuations $\xi(t)$, whose Laplace transform is given by:

$$\widehat{\Phi}_\xi^{(t_a)}(u) = \frac{1}{u+\widehat{\Phi}_{t_a}(u)}. \tag{5.52}$$

Using Eq. (5.49) we find that Eq. (5.52) can also be written as:

$$\widehat{\Phi}_\xi^{(t_a)}(u) = \frac{1}{u}(1-\widehat{\psi}_{t_a}(u)), \tag{5.53}$$

which establishes the equivalence between the aged autocorrelation function and the aged survival probability:

$$\Phi_\xi^{(t_a)}(t) = \Psi_{t_a}(t). \tag{5.54}$$

It is straightforward to prove that for $t_a = 0$, Eq. (5.49) reduces to Eq. (5.41) so that we recover the relation Eq. (5.51). It is even more important to understand why this procedure is a generalization of the OP discussed in the previous chapter. To realize this important fact, let us rewrite Eq. (5.32) in the equivalent form:

$$\psi_{t_a}(t) = \int_0^{t_a} dy R(y)\psi(y+t), \tag{5.55}$$

where $R(t)$ is the time density (rate) of event production defined in Eq. (5.26). In the case $2 < \mu < 3$, we have in the region $u \to 0$ that Eq. (5.28) becomes:

$$\hat{\psi}(u) = 1 - \langle\tau\rangle u - \Gamma(2-\mu)(uT)^{\mu-1} + \cdots \tag{5.56}$$

In this case Eq. (5.29) yields:

$$R(t) = \frac{1}{\langle\tau\rangle}\left[1 + \left(\frac{T}{t}\right)^{\mu-2}\right]. \tag{5.57}$$

Although $\mu > 2$ ensures the validity of the ergodic condition, as we move closer and closer to $\mu = 2$, the border with the non-ergodic condition, the relaxation towards the constant rate production becomes critically slow. Nevertheless, for $t_a \to \infty$, Eq. (5.55) yields:

$$\psi_\infty(t) \equiv \psi_{t_a=\infty}(t) = \frac{1}{\langle\tau\rangle}\int_t^\infty dt'\psi(t'). \tag{5.58}$$

Then, using Eqs. (5.52) and (5.49), we find that the Laplace transform of the infinitely aged memory kernel:

$$\Phi_\xi^{(\infty)}(t) \equiv \Phi_\xi^{(t_a=\infty)}(t)$$

coincides with the Laplace transform of Eq. (5.21), which is a well-known expression established by renewal theory for the equilibrium autocorrelation function of $\xi(t)$ [211]. This demonstrates, in an explicit way, that Eq. (5.51) is a generalization of the OP. It is, at the same time, a procedure yielding the non-stationary autocorrelation function that coincides [21], with an exact prescription [217].

Unfortunately, the exact expression for the aged autocorrelation function does not have a simple analytical form. However, using the method of conditional probabilities, the following analytical expression was constructed [21]:

$$\Phi_\xi^{(t_a)}(t) = \left(\frac{T}{T+t+t_a}\right)^{\mu-1} + \left[1 - \left(\frac{T}{T+t_a}\right)^{\mu-1}\right] C(t), \qquad (5.59)$$

where:

$$C(t) \equiv \frac{\left(\frac{T}{T+t}\right)^{\mu-2} + \left(\frac{T}{T+t+t_a}\right)^{\mu-2}}{1 - \left(\frac{T}{T+t_a}\right)^{\mu-2}}. \qquad (5.60)$$

In Fig. 5.3 we compare this approximate expression to a numerical treatment yielding the exact autocorrelation function. Moving from the top to

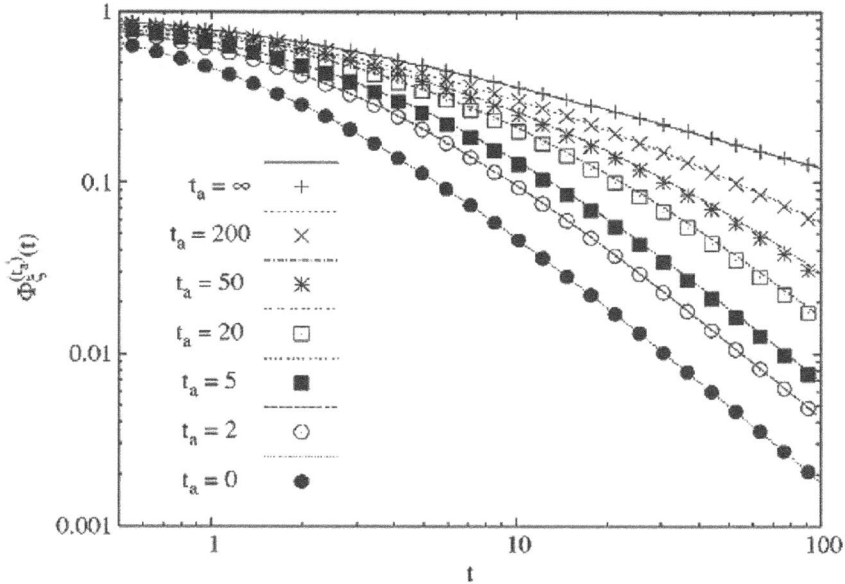

Figure 5.3: The t_a-old correlation function $\Phi_\xi^{(t_a)}(t)$, for different values of the aging time. Curves represent the exact result of Eq. (5.54) for $\mu = 5/2$ and $T = 3/2$, obtained with the numerical procedure described in [21]. The dots correspond to the approximate formula of Eq. (5.59). From [21] with permission.

the bottom curves, the age increases from the brand new kernel ($t_a = 0$) to the infinitely aged, or stationary kernel ($t_a = \infty$). The comparison of the approximate to the exact autocorrelation function is seen to be very good.

5.4   Generalized Langevin Equations (GLE)

There are two prototypical nonsimple physical networks considered in statistical physics, those that are open to the environment and those that are closed to it; the latter being completely described by a Hamiltonian [327]. Of course, simple Hamiltonian procedures have been developed that model both the system of interest and the environment, as a closed universe, as well. The Hamiltonian strategy reduces the infinite set of deterministic dynamic equations to a finite set of stochastic differential equations. The latter set corresponds to the network's physical observables and the eliminated degrees of freedom of the environment. A number of approaches have been developed to show that the influence of the eliminated degrees of freedom (environment) on the observables is to modify the interaction of the network observables among themselves, engender irreversible dissipative forces, and generate random fluctuations. There are mechanical forces produced by energy gradients and information forces produced by nonsimplicity gradients. Here we briefly sketch the direct integration technique [327] to establish the nomenclature for a thermodynamically closed network, that is, networks that eventually achieve thermal equilibrium with their environments.

5.4.1   Explicit integration technique

The method of explicit integration requires obtaining explicit solutions to the equations of motion for the environment, including the back-reaction of the environment to the dynamic effects of the network. Substituting this solution back into the equations of motion for the network variables yields another set of evolution equations, whose only explicit dependence on the environment is through the initial values of the eliminated variables. It is through these initial values that uncertainty is introduced into the network dynamics by means of the initial values.

Again consider the Hamiltonian $H_N(Q, P)$ for a system described by the displacement Q and its conjugate momentum P. We couple this system to the environment described by a Hamiltonian $H_B(\mathbf{q}, \mathbf{p})$ for a heat bath of displacements q_n and conjugate momenta p_n, $n = 1, 2, \ldots, N$, where the bath coordinate vectors are defined by $\mathbf{q} = (q_1, q_2, \ldots, q_N)$ and $\mathbf{p} = (p_1, p_2, \ldots, p_N)$. Finally, an interactive Hamiltonian $H_{NB}(Q, \mathbf{q})$ is considered for the coupling between the system of interest and heat bath.

The total Hamiltonian for the closed composite system is then given by [151, 193]:

$$H = H_N(Q, P) + H_B(\mathbf{q}, \mathbf{p}) + H_{NB}(Q, \mathbf{q}). \tag{5.61}$$

Hamilton's equations of motion for the network, when the dynamics of the bath variables are eliminated, are shown by Lindenberg and West to reduce to [327]:

$$\frac{dP}{dt} = -V'_m(Q) - \int_0^t K(t - t')P(t')dt' + \xi(t), \tag{5.62}$$

which with the proper interpretation of the driving force $\xi(t)$ and memory kernel $K(t)$, is the GLE. If the initial conditions of the bath variables are given by a canonical distribution:

$$W(\mathbf{p}, \mathbf{q}) = Z^{-1}e^{-\beta H_B^{(m)}} \tag{5.63}$$

where Z is the partition function, the inverse temperature is given by $\beta = 1/k_B\Theta$ and the bath comes to equilibrium at a temperature Θ, in the presence of the network, giving rise to a modified bath Hamiltonian [327]:

$$H_B(\mathbf{p}, \mathbf{q}) + H_{NB}(Q, \mathbf{q}) \rightarrow H_B^{(m)}.$$

In this situation the function $\xi(t)$ obtained from solving the dynamic equations for the heat bath (environment) can be interpreted as a random force with Gaussian statistics.

Moreover, using the explicit form of the stochastic force and the memory kernel obtained by solving the Hamiltonian equations yields:

$$\langle [\xi(t) - \langle \xi(t) \rangle_B] [\xi(t') - \langle \xi(t') \rangle_B] \rangle_B = k_B\Theta K(t - t') \tag{5.64}$$

where the subscripted brackets $\langle \cdot \rangle_B$ denote an average over the bath canonical distribution, or said differently, an average over an ensemble of realizations of the thermal fluctuations of the bath. Equation (5.64) is a fluctuation–dissipation relation (FDR) of the second kind [301]. Thus, the memory kernel is dissipative and the mean value of the stochastic force decays on a time scale much shorter than the characteristic time scale of the network dynamics of the bath. Of course, this is not the complete story. Additional conditions on the coupling parameters and the frequency spectrum of the bath are required to complete the identification of Eq. (5.62) as the GLE [193, 327].

In the limit where the number of variables necessary to describe the bath becomes infinite and the bath correlation time approaches zero, the memory kernel approaches a delta function in time:

$$K(t) \rightarrow 2\lambda\delta(t), \tag{5.65}$$

and the GLE reduces to the ordinary Langevin equation:

$$\frac{dP}{dt} = -V'_m(Q) - \lambda P(t) + \xi(t). \tag{5.66}$$

The zero-centered Gaussian random force is subsequently shown to yield the FDR of the first kind:

$$\langle \xi(t)\xi(t') \rangle = 2D\delta(t - t'). \tag{5.67}$$

These equations were introduced into physics by Paul Langevin [312] to describe the erratic motion of a heavy particle immersed in a fluid of lighter particles, such as the pollen mote in water that the botanist Robert Brown [119] observed through his microscope, or the powdered charcoal suspended in alcohol that the physician Jan Ingen-Housz observed a half century earlier. Langevin reasoned that Newton's Laws could describe the motion of the Brownian particle through its being pushed around by the ambient fluid, giving rise to dissipative and random forces. The dissipative force arises from the effects of Stokes' friction for a particle moving through a viscous fluid, whereas the fluctuating (random) force arises from an instantaneous imbalance in the number of fluid particles impacting the surface of the much larger Brownian particle.

Consequently, the arguments leading to Eqs. (5.66) and (5.67) provide a dynamic model for the Einstein [180] form of the FDT in the relation among temperature Θ, dissipation λ and diffusion D parameters:

$$\frac{D}{\lambda} = k_B\Theta, \tag{5.68}$$

thereby making the ratio of the strength of the fluctuations to that of the dissipation proportional to the temperature of the fluid, the Einstein relation. The Langevin equation was thought to be exact for bilinear coupling, independently of the magnitude of the coupling coefficients. This turns out not to be the case when the back-reaction of the ambient fluid to the motion of the heavy particle is properly taken into account using hydrodynamics and gives rise to the aptly named Basset force [68]. Let us be clear, the equation of motion for a Brownian particle with local linear dissipation found in all the textbooks has been shown to be wrong! [600] The Bassett force must be included in Eq. (5.66) to account for the actual back-reaction of the ambient fluid to the particle's motion. This new force introduces a fractional derivative into the dynamics of the Brownian particle [602].

5.4.2 Fractional Langevin Equation (FLE)

Note that a strictly exponential relaxation of a perturbation is inconsistent with a Hamiltonian (reversible) description of a dynamic network, since this would imply irreversible dynamics. Consequently, it is not possible to establish a satisfactory connection between microdynamics and macrody-namics in order to provide a mechanical foundation for stochastic physics [319]. However, there is an alternate derivation of the FLE based on the Smoluchowsky approximation to the GLE, where the inertial force is ne-glected so that Eq. (5.62) reduces to:

$$\int_0^t K(t-t')P(t')dt' = -V_m'(Q) + \xi(t).$$ (5.69)

Integrating the LHS of Eq. (5.69) by parts and using the identity:

$$\frac{\partial}{\partial t}\int_0^t K(t-t')Q(t')dt' = K(0)Q(t) + \int_0^t \frac{\partial K(t-t')}{\partial t}Q(t')dt'$$
$$= K(t)Q(0) + \int_0^t K(t-t')P(t')dt'$$ (5.70)

and the second line on the right in Eq. (5.70) is obtained by integrating the first line by parts. Introducing the IPL memory kernel:

$$K(t) = \frac{t^{-\alpha}}{\Gamma(1-\alpha)},$$ (5.71)

into Eq. (5.70) and rearranging terms, we rewrite (5.69) as:

$$D_t^\alpha[Q(t)] - \frac{t^{-\alpha}}{\Gamma(1-\alpha)}Q(0) = -V_m'(Q) + \xi(t).$$ (5.72)

Here we have introduced the operator:

$$D_t^\alpha[f(t)] \equiv \frac{1}{\Gamma(1-\alpha)}\frac{d}{dt}\int_0^t \frac{f(t')dt'}{(t-t')^\alpha},$$ (5.73)

where $D_t^\alpha[\cdot]$ is a Riemann–Liouville (RL) fractional derivative and $0 < \alpha \leq 1$. In the same way, we write the RL fractional integral as [581]:

$$D_t^{-\alpha}[f(t)] = \frac{1}{\Gamma(\alpha)}\int_0^t \frac{f(t')dt'}{(t-t')^{1-\alpha}}.$$ (5.74)

The FLE given by Eq. (5.72) is a generalization of Newton's force law that incorporates long-term memory into the dynamics of a network coupled to the environment. In addition to the environment inducing memory, it modifies the potential as well. The network would, in isolation, evolve according to the dynamics prescribed by the potential $V(Q)$. Consider the case where the gradient of the modified potential $V'_m(Q) = Q/\tau_c^\alpha$ reduces Eq. (5.72) to the linear FLE (LFLE):

$$D_t^\alpha[Q(t)] - \frac{t^{-\alpha}}{\Gamma(1-\alpha)}Q(0) = -\frac{Q}{\tau_c^\alpha} + \xi(t). \tag{5.75}$$

The solution for the average value of the initial value problem defined by the LFLE, when the fluctuations are zero-centered, is:

$$\langle Q(t) \rangle = Q(0)E_\alpha\left(-\left[\frac{t}{\tau_c}\right]^\alpha\right), \tag{5.76}$$

in terms of the Mittag–Leffler function (MLF), whose fractional Taylor series expansion is:

$$E_\alpha\left(-\left[\frac{t}{\tau_c}\right]^\alpha\right) = \sum_{n=0}^{\infty}\frac{(-1)^n}{\Gamma(n\alpha+1)}\left(\frac{t}{\tau_c}\right)^{n\alpha}. \tag{5.77}$$

Note that the MLF series for $\alpha = 1$ yields the exponential:

$$E_1\left(-\left[\frac{t}{\tau_c}\right]^\alpha\right) = \exp\left[-\frac{t}{\tau_c}\right], \tag{5.78}$$

and consequently the MLF is often referred to as a generalized exponential function.

It is important to stress that the analytical MLF given by Eq. (5.77) has been used to fit the relaxation curves of stress experiments on glassy materials with remarkable accuracy [214, 215] over the entire dynamic range. This success suggests that dynamic randomness, without time scale separation, is manifest through fractional time derivatives and becomes experimentally detectable at the macrolevel. Consequently, the GLE can be further extended to the FLE in which the fluctuations of the heat bath contain IPL memory, leading to a fractional derivative in time for the network dynamics.

GLE and MLF: The introduction of fractional derivatives is an elegant procedure to describe the memory effect of the GLE. We meet here the important MLF, which has been derived using the Mori GLE [377]. An illuminating derivation of the GLE from the interaction of an oscillator with an infinite bath of oscillators can be found in the excellent book by Weiss [568], which in turn inspired another attractive derivation of the MLF relaxation using the GLE [436]. Fractional relaxation in terms of the MLF is attracting an increasing number of investigators and, if for no other reason, deserves further discussion.

We have seen that the MLF can be introduced as a generalization of the ordinary exponential relaxation in time through Eq. (5.77). In the Laplace transform representation the MLE is:

$$\widehat{E}_\alpha(u) \equiv \mathcal{LT}\{E_\alpha(-(\lambda t)^\alpha); u\} = \int_0^\infty e^{-ut} E_\alpha(-(\lambda t)^\alpha)dt, \qquad (5.79)$$

which inserting the series expansion of the MLF into Eq. (5.79) takes the form:

$$\widehat{E}_\alpha(u) = \frac{1}{u + \lambda^\alpha u^{1-\alpha}}, \qquad (5.80)$$

where the microrelaxation rate is given by $\lambda = \frac{1}{\tau_c}$. In the early-time domain $u \to \infty$, the Laplace transform of the MLF is:

$$\lim_{u \to \infty} \widehat{E}_\alpha(u) = \frac{1}{u}\left[1 - \left(\frac{\lambda}{u}\right)^\alpha + \cdots\right]$$

or equivalently in the early-time regime $t \ll 1/\lambda$, the lowest-order terms in the MLF series become:

$$\lim_{t \to 0} E_\alpha(-(\lambda t)^\alpha) \approx 1 - \frac{(\lambda t)^\alpha}{\Gamma(\alpha+1)} \approx \exp\left[-\frac{(\lambda t)^\alpha}{\Gamma(\alpha+1)}\right], \qquad (5.81)$$

where the MLF has the stretched exponential form at early times. In the asymptotic time domain $u \to 0$ the expansion of Eq. (5.80) yields to lowest-order:

$$\lim_{u \to 0} \widehat{E}_\alpha(u) \approx \frac{u^{\alpha-1}}{\lambda^\alpha}$$

or equivalently in long-time regime $t \gg 1/\lambda$, this transforms to:

$$\lim_{t \to \infty} E_\alpha(-(\lambda t)^\alpha) \approx \frac{1}{\Gamma(\alpha)(\lambda t)^\alpha}, \qquad (5.82)$$

so that the MLF has the IPL form asymptotically in time.

Therefore, the condition $\lambda \to \infty$ corresponds to making the stretched exponential regime vanish so as to fill the entire macroscopic time scale with an IPL regime. This is in the spirit of the fractional calculus treatment of the dynamics and is equivalent to ignoring the regime of transition from microdynamics to macrodynamics. In Chapter 7 we shall see, however, that in some cases of neurophysiological and physiological interest, this regime of transition is very extended and offers the only possibility to infer the cooperative nature of the process under study from experimental observation.

5.4.3 MLF entails intermittency

In Sec. 5.4.2, we determine that the MLF was the solution to a simple fractional rate equation (FRE). Assuming that the waiting-time PDF is given by the MLF we can write:

$$E_\mu(-(\gamma t)^\mu) = \int_0^\infty d\lambda e^{-\lambda t}\Pi(\lambda), \qquad (5.83)$$

where we can also interpret the MLF as the ML survival probability. From one perspective the MLF is here given by the average over an ensemble of rates for the exponential distribution, whereas from another perspective the PDF of rates is the inverse Laplace transform of the MLF:

$$\Pi(\lambda) = \mathcal{LT}^{-1}\{E_\mu(-(\gamma t)^\mu); \lambda\}. \qquad (5.84)$$

After some simple algebra, based on the definition of the MLF, it is determined that the PDF of rates is given by the Lamperti distribution [308]:

$$\Pi(\lambda) = \frac{\lambda^\mu \sin\mu\pi}{\pi}\frac{\gamma^{\mu-1}}{\gamma^{2\mu} + 2(\gamma\lambda)^\mu \cos\mu\pi + \lambda^{2\mu}}. \qquad (5.85)$$

It is worth emphasizing that Eq. (5.85) defines a normalized PDF, even though it diverges as $\gamma \to 0$ and goes to zero slowly as $\gamma \to \infty$. The quantity $\Pi(\lambda)$ is integrable in both asymptotic regions and consequently satisfies all the conditions necessary to be interpreted as a PDF.

Lambert *et al.* [307] provided an interesting interpretation of Eq. (5.83) as a ML survival probability function. They observed that Bianco *et al.* [95] had interpreted the variable as the rate of generation of events, for example, the rate of photon emission by chromophore [192]. They then assumed a Gibbs ensemble corresponding to $\Pi(\lambda)$, which produces the desired crucial events and go on to solve a nonlinear Langevin equation whose solution is the equilibrium Gibbs ensemble distribution function. We leave it as an exercise for the student to determine the analytic form of the potential for which the equilibrium solution to the corresponding Fokker–Planck equation yields:

$$p_{eq}(\gamma) = Z^{-1}\exp\left[-\frac{V(\gamma)}{D}\right] \propto \Pi(\gamma). \qquad (5.86)$$

The asymptotic form of the MLF is an IPL, as is that of Eq. (5.85), which is a characteristic of intermittence. In the beginning of this section, we determined that PM map, so-called weak chaos, can be used to provide a theoretical foundation for this type of bursty stochastic process. Noise induced intermittence is a distinct source of intermittency and is given by the Langevin argument presented by Lambert *et al.* [307].

5.5 Phase-space Modeling

Two distinct methods have been developed in the physical sciences to describe the phase space evolution of nonsimple phenomena: the master equation introduced by Pauli [416] to describe the time rate of change of the probability of a system occupying a set of discrete states and the CTRW approach of Montroll and Weiss [382]. The CTRW formalism describes a random walk in which a walker pauses after each jump, selected from a PDF of step lengths, for a sojourn specified by a waiting-time PDF. The Markov master equation is equivalent to CTRW, if the waiting-time process is Poisson [73]. However, when the waiting-time PDF is not exponential, the case we consider herein, the equivalence between the two approaches is only maintained by generalizing the technique to the non-Markov master equation, the so-called GME [286].

It has been argued [372, 483] that the GME unifies the fractional calculus and CTRW. Herein we question the possibility of going beyond a formal equivalence between the two approaches. The reason is, as we shall see, that the CTRW rests on discrete events, while the GME, as derived from a Liouville or Liouville-like approach, relies on continuous trajectories and does not contain events.

Allegrini *et al.* [18] have shown that to create a stationary master equation and therefore one that is compatible with the OP, discussed in the previous chapter, requires that the network be entangled with the heat bath in such a way that the bath does not regress to equilibrium infinitely fast. The network-bath entanglement is the result of rearrangement processes that may take infinitely long to complete, and as a result of this slowing down leads to the memory kernel in the GME [286].

5.5.1 Fokker–Planck Equation (FPE)

The phase-space equation of motion for the phase-space distribution function and subsequently that of the PDF can be determined using a variety of familiar techniques that are readily available in the literature [327] and will not be reproduced here. One strategy for constructing the phase-space equations for the system variables $\mathbf{X}(t)$ in which the explicit dependence on the bath variables $\mathbf{Y}(t)$ has been eliminated, consists of three steps: (1) Formally solve the equations of motion for $\mathbf{Y}(t)$, including the dynamic effects of the system on the bath. (2) Substitute the formal solutions for the bath variables into the equations of motion for the system variables and obtain evolution equations whose only dependence on the bath variables is through the initial state $\mathbf{Y}(0)$. (3) Introduce a PDF for the initial conditions for the bath variables, thereby making the Liouville equations, for the phase-space distribution, stochastic.

Evaluating the average of the product of the stochastic Liouville operator and the phase-space density is not a simple task and is not reproduced here. We merely note that averaging over the $\boldsymbol{\xi}$-fluctuations leads to the introduction of a new operator that allows the phase space equation to be written in the simplified form [327]:

$$\frac{\partial P(\mathbf{x},t \mid \mathbf{x}_0)}{\partial t} = \mathcal{L}_N P(\mathbf{x},t \mid \mathbf{x}_0) + \Delta\mathcal{L} P(\mathbf{x},t \mid \mathbf{x}_0), \qquad (5.87)$$

where $\Delta\mathcal{L}$ is an operator, usually in the form of an infinite series. If the $\boldsymbol{\xi}$-fluctuations have Gaussian statistics, with mean-square cross-correlation strength given by the elements of the matrix $\mathbf{D}$, the operator series truncates at second-order and if the fluctuations are delta correlated in time:

$$\langle \xi_i(t)\xi_j(t') \rangle = 2D_{ij}\delta(t - t'), \qquad (5.88)$$

and the operator reduces to:

$$\Delta\mathcal{L} = \sum_{i,j} D_{ij} \frac{\partial^2}{\partial p_i \partial p_j}, \qquad (5.89)$$

such that Eq. (5.87) becomes the familiar multi-dimensional FPE.

In a homogeneous one-dimensional network, we can write Eq. (5.87) as:

$$\frac{\partial}{\partial t} P(q,t \mid p_0) = \frac{\partial}{\partial q}\left[G(q) + Dg(q)\frac{\partial}{\partial q}g(q) \right] P(q,t \mid q_0). \qquad (5.90)$$

The FPE describes the phase-space evolution of a nonsimple network having two properties; nonlinear dynamics and a multiplicative random force. The steady-state solution to the FPE is given by:

$$P_{ss}(q) = \frac{1}{g(q)}\exp\left[-\int \frac{G(q')dq'}{Dg(q')^2} \right]. \qquad (5.91)$$

This reduces to the canonical distribution when $g(q) = 1$, $\beta = 1/D$, $G(q) = V'(q)$ and $V(q)$ is the one-dimensional potential. The multiparticle–particle generalization of this analysis is given by Lindenberg and West [327].

It is probably useful to interrupt this torrent of formalism in order to provide a sense of how it can be applied to nonsimple phenomena. One recent application for which ample observational data was available, as well alternative theoretical analyses, was to the distribution of animal size in terms of their mass in an ecosystem.

Stochastic ontogenetic growth model: West and West [595] generalized an ontogenetic growth model (OGM) for a single organism that had been constructed by West *et al.* [609] using a conservation of energy argument. The equations were essentially of the same form as those introduced by von Bertalanffy [91] to model metabolism and growth a half century earlier. We [595] generalized the OGM to incorporate the disordering influence of entropy on the growth process, by including statistical fluctuations into the dynamic equations, resulting in the stochastic ontogenetic growth model (SOGM).

In the SOGM we interpret the dynamic variable $Q(t)$ to be the total body mass (TBM) of an individual animal from a particular species and the OGM to be given by [595]:

$$E_m \frac{dQ}{dt} = aQ^b - B_m Q. \tag{5.92}$$

The first term on the RHS of the equation is the basal metabolic rate, E_m is the metabolic energy required to create a unit biomass, and B_m is the metabolic rate required to maintain an existing unit of biomass. Dividing this equation by E_m and introducing the parameters $\bar{a} = a/E_m$, the allometry coefficient normalized to the metabolic rate required to create a unit biomass, and $\kappa = B_m/E_m$, the rate of creation of a unit biomass, we obtain an equation in one independent variable in which the last rate coefficient is considered to be random.

The stochastic component of the SOGM is introduced into Eq. (5.92) by assuming the phenomenological rate of generation of a unit biomass that has a random component $\kappa \to \kappa_0 + \xi(t)$, where κ_0 is a constant and $\xi(t)$ is a zero-centered stochastic process. With this replacement SOGM can be identified to have the deterministic function:

$$G(q) = \bar{a}q^b - \kappa_0 q, \tag{5.93}$$

and the multiplicative function:

$$g(q) = q, \tag{5.94}$$

and the dynamics of the PDF are described by the FPE, whose general solution remains a mystery. However, the asymptotic solution given by Eq. (5.91) can be integrated to obtain [595]:

$$W_{ss}(q) = \frac{\vartheta \gamma^{\frac{\mu-1}{\vartheta}}}{\Gamma\left(\frac{\mu-1}{\vartheta}\right)} \frac{\exp\left[-\gamma q^{-\vartheta}\right]}{q^\mu}, \tag{5.95}$$

normalized on the positive mass interval $[0, \infty]$, with the parameter values:

$$\gamma = \frac{\bar{a}}{\vartheta D}, \quad \mu = 1 + \frac{\kappa_0}{D} \quad \text{and} \quad \vartheta = 1 - b. \tag{5.96}$$

West and West [595] do a least-squares fit of the parameters in the SOGM steady-state PDF to a dataset of mammalian species tabulated by Heusner [251], as depicted in Fig. 5.4. These data were processed by constructing a histogram of interspecies TBM for the 391 mammalian species tabulated by Heusner, by partitioning the mass axis into intervals of 20 grams and counting the number of species within each interval. The vertical axis is the relative number of species and the horizontal axis is in increments of TBM. The allometry exponent is fixed at $b = 3/4$, so that $\vartheta = 1/4$ and the remaining parameters are determined to have the mean-square, best values of $\gamma = 13.4$ and $\mu = 2.04$, with an overall quality of fit measured by $R^2 = 0.96$.

Figure 5.4 shows the fit to the low-mass species out to 500 grams. The remaining fit out to three million grams is included in the parameter determination, but is not shown in this figure, due to constraints of scale. The asymptotic mass region is, in fact, the more interesting part of the distribution. To capture this information on a single graph we [595] constructed a second histogram, this one for the asymptotic region, from approximately one thousand to three million grams. Figure 5.5 depicts the fit to the logarithmic histogram data points, indicated by the dots, starting at a TBM of 1.1 kilogram. An IPL PDF would be a straight line with a negative

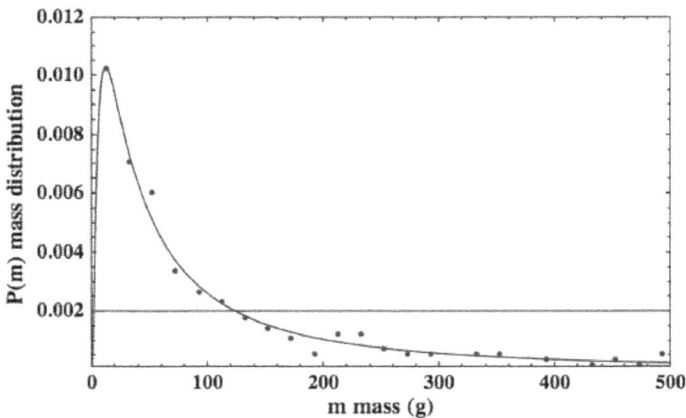

Figure 5.4: The histogram constructed from the average TBM data for 391 mammalian species [251] is given by the dots. The TBM data has been divided into intervals of 20 g each and the number of species within each such interval is counted. The vertical axis is the relative frequency in each interval and the solid line segment is the least squares fit of Eq. (5.95) to the data points. The quality of fit is $R^2 = 0.96$. From [595] with permission.

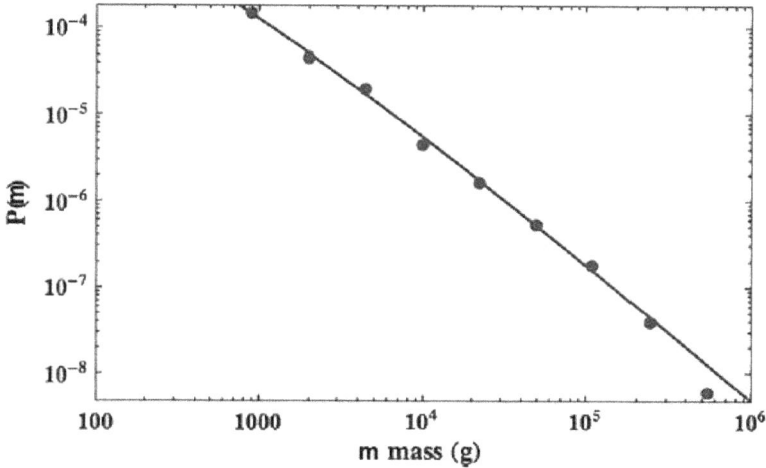

Figure 5.5: The average TBM data for 391 mammalian species [251] are used to construct a histogram. The mass interval is divided into twenty equally spaced intervals on a logarithm scale and the number of species within each interval is counted. The least-square fit to the nine data points is then made using logarithmically transformed distribution, see [595] for details. The quality of the fit is $R^2 = 0.998$. From [595] with permission.

slope on this log–log graph paper. Inserting the fitted parameter values $\mu = 1.67$ and $\gamma = 8.96$ into the steady-state TBM PDF, given by Eq. (5.95), yields the solid curve in Fig. 5.5, which fits the data extremely well. The straight line curve is quite clearly an IPL in the interspecies TBM. This coarse-grained description of the interspecies TBM statistics is indicative of extreme variability, particularly since $\mu < 2$, indicating that the variance of the interspecies TBM over an unbounded domain would diverge.

A diverging second moment of the TBM indicates extreme variability in the data, which in turn provides a measure of nonsimplicity of the under-lying ecosystem. This PDF also provides a way to quantify the influence of one species on another, by means of an information gradient. Assuming that there is an optimal range of TBM for which a given species can survive, this analysis indicates that the variability of the average TBM quantifies the notion of biological fitness. This notion of fitness means the potential to survive to the age of reproduction, find a mate, and have an offspring; the greater the number of offspring, the greater the fitness.

5.5.2 Lévy processes

The non-Markov nature of the GME is consistent with the notion of Lévy RWs, a nomenclature first used in [494]. These walks were introduced into the physical sciences in the context of turbulent fluid flow, where it was recognized that the relative velocity, between parcels of fluid, separated by a given distance, depends on that distance. Therefore, a random walk model of such a process requires the physical assumption that steps to nearby points in the fluid require different times than for steps to more distant points. In simple systems, nearby points require less time to traverse, but in the case of turbulent cascades involving eddies, the relative velocity of fluid particles increases with spatial separation, so that more distant points, counterintuitively, may require less time to traverse a given separation distance, since they are driven by the larger eddies.

A Lévy PDF is the exact solution to the integration chain condition for the PDF, which is essential for a Markov process [383]. We show that it is possible, using certain assumptions, to construct a Lévy process from the GME, even though the latter is explicitly non-Markovian. We note that it is possible to make these two seemingly contradictory processes compatible if we take the non-Markovian properties, stemming from the IPL behavior of the velocity autocorrelation function, to become transition probabilities with an IPL dependence on the length of the jump in space [494]. Consequently, we change the time non-locality into a space non-locality, and show that the latter is the space non-locality of the Lévy process expressed by a fractional diffusion equation [572]. This space non-locality cannot be eliminated by observing suitably large distances because of the IPL structure of the corresponding space transition.

The connection between non-locality in space and non-locality in time is established by observing that, throughout the entire period of time spent by the particle within one of its two velocity states, the particle makes a jump of length $|q| = Wt$. This allows us to interpret the probability of making a jump from one site to another, a distance $|q|$ from the initial site, as being proportional to the waiting-time PDF, obtained by integrating the transition probability overall time:

$$\int_0^\infty \kappa(q,t)dt \propto \frac{1}{(T + |q|)^\mu}. \tag{5.97}$$

Consequently, a transition of length $|q|$ takes place in time $t = |q|/W$, where the walker steps either to the right or left at the constant velocity W and the memory kernel has the factored form:

$$\kappa(q,t) = \psi(t)\delta(|q| - Wt). \tag{5.98}$$

Trefán *et al.* [530] studied the asymptotic regime of Eq. (5.97) incorporating the factored memory kernel Eq. (5.98) and yielding the following explicit form for the doubly transformed kernel:

$$K^*(k, u) = \psi^*(k, u) - \widehat{\psi}(u). \tag{5.99}$$

To derive the Lévy distribution only one basic step remains and that is to properly incorporate the Markov condition into Eq. (5.99). In the asymptotic space-time limit $u \to 0$ and $k \to 0$, we obtain:

$$P^*(k, u) = \frac{1}{u + b\,|k|^\beta}; \tag{5.100}$$

with the constant:

$$b = \frac{(WT)^\beta}{T}\beta\Gamma(1 - \beta)\cos\left[\beta\frac{\pi}{2}\right]; \quad \beta = \mu - 1. \tag{5.101}$$

The inverse Laplace transform of Eq. (5.100) yields the partial differential equation of motion for the characteristic function $\widetilde{P}(k, t)$:

$$\frac{\partial \widetilde{P}(k, t)}{\partial t} = -b\,|k|^\beta\,\widetilde{P}(k, t). \tag{5.102}$$

The properly normalized solution to Eq. (5.102) is simply the exponential:

$$\widetilde{P}(k, t) = e^{-b|k|^\beta t}, \tag{5.103}$$

which is the characteristic function for the symmetric Lévy PDF. Note that the characteristic function is the Fourier transform of the PDF:

$$\widetilde{P}(k, t) = \mathcal{FT}\{P(q, t); k\}. \tag{5.104}$$

The theory developed in this section is tailored to the parameter region $2 < \mu < 3$. The domain $\mu < 2(\beta < 1)$ is excluded by the fact that for $\mu < 2$ the first moment of the waiting-time PDF diverges, thereby preventing us from adopting a characteristic time scale for the process. Furthermore, even if the average sojourn time $\langle t \rangle$ were arbitrarily given a finite value, and β were still identified with $\mu - 1$, in the asymptotic limit of vanishing k and u, the resulting value of b would vanish. Furthermore, the region $\mu > 3$ is excluded by the fact that the Markov approximation leads to the structure Eq. (5.100) with $\beta > 2$, in which case the inverse transform is not positive definite and therefore cannot be interpreted as a PDF.

There is a difficulty associated with the above solution given by the inverse Fourier transform of the characteristic function and that is an apparent inconsistency in scaling. The second moment is known to scale as [530]:

$$\langle Q(t)^2 \rangle \propto t^{4-\mu}, \tag{5.105}$$

implying that the Hurst exponent is:

$$2H = 4 - \mu. \tag{5.106}$$

On the other hand, the Lévy PDF has a scaling behavior, suggesting that the Hurst exponent ought to be given by:

$$H = \frac{1}{\beta} = \frac{1}{\mu - 1}, \tag{5.107}$$

which is different from Eq. (5.106). This difference in scaling is significant and warrants additional explanation. The diffusion process described by Eq. (5.97) consists of a central part and a propagation front, signalled by two sharp peaks [12, 631]. At time t a particle leaving the origin $q = 0$ at time $t = 0$ cannot be found at a distance from the origin larger than Wt. The restriction on the random walker's maximum velocity has the effect of producing an accumulation of particles at the propagation front of the diffusion process, namely at $q = \pm Wt$. This is the origin of ballistic peaks at the respective propagation fronts. At earlier times the initial distribution, concentrated at $q = 0$, splits into these two ballistic peaks and the region between the two peaks is empty.

Due to the effect of sporadic randomness, some trajectories leave the propagation front and the population of the central spatial region steadily increases in time, while the peak intensity, proportional to the autocorrelation function Φ_ξ, slowly decreases. Note that this implies that the diffusion process cannot be described by a single rescaling, as depicted in Fig. 5.6. The peaks of the propagation front rescale with $H = 1$, a fact implying a diffusion faster than that predicted by the rescaling of Eq. (5.106). The rescaling of the central part is properly expressed by Eq. (5.107). The calculations leading to Eq. (5.102) refer to a physical condition where the intensity of the ballistic peaks is negligible, so that the rescaling given by Eq. (5.107) only reflects the diffusion properties of the central part of the distribution.

On the contrary, the rescaling Eq. (5.106) is a balance between the fast rescaling of the propagation front and the rescaling of the central part of the distribution, which is in fact slower than the rescaling in Eq. (5.107). These two rescalings reflect a conflict between the dynamic properties still present within the Lévy walk perspective and a merely probabilistic treatment.

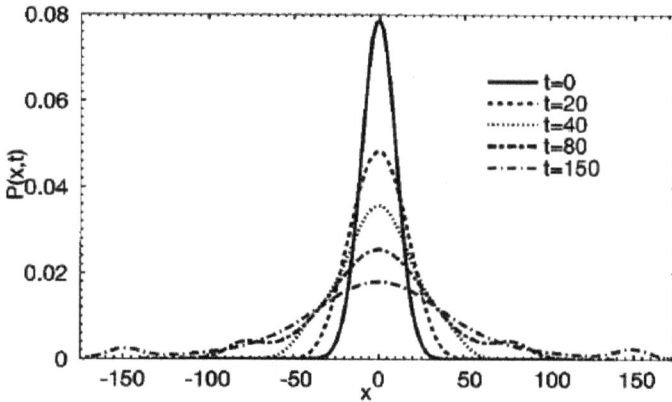

Figure 5.6: The initial Gaussian represents the result of an earlier diffusion process generated by adichotomous, but totally uncorrelated, fluctuations. The subsequent time evolution, responsible for the birth of ballistic peaks, is pursued by a Lévy stochastic generator as explained in [13].

Bologna *et al.* [104] made the observation, concerning the derivation of a Lévy process from the GME, that the memory in the GME must be erased during the network's evolution, so that asymptotically the Markov Lévy PDF is realized. Thus, in spite of the non-Markov structure of the GME, we show in the next section that an asymptotic regime appears whose statistics are determined by this process of memory erasure.

Fractional Lévy motion (FLM): The properties of scaling media are often described by fractal functions and their space-time evolution. However, such functions contain hierarchies of singularities and are typically non-differentiable [354, 366]. Thus, the understanding of such phenomena as fractional wave propagation and fractional diffusion comes about through the development and implementation of alternate modeling strategies, which do not explicitly include the usual differential equations of motion. In recent years, there have been a number of attempts to model the phase space evolution of the PDF for anomalous transport processes, one class of which has led to evolution equations with fractional derivatives [253, 433, 581, 600].

The first successful generalization to an anomalous diffusive process started from a Langevin equation describing a dissipative dynamical process

$Q(t)$, driven by non-Gaussian fluctuations:

$$\frac{dQ(t)}{dt} + \lambda_0 Q(t) = \xi_\beta(t), \tag{5.108}$$

where the stochastic driver $\xi_\beta(t)$ is a delta-correlated stable Lévy process. The formal solution to Eq. (5.108) is given by the quadrature expression:

$$Q(t) = e^{-\lambda_0 t} Q_0 + \int_0^t e^{-\lambda_0(t-t')} \xi_\beta(t') dt', \tag{5.109}$$

such that $Q(0) = Q_0$ and for notational ease, we introduce:

$$f_\beta(t) \equiv \int_0^t e^{-\lambda_0(t-t')} \xi_\beta(t') dt', \tag{5.110}$$

to replace the integral term.

The PDF conditional on the initial value of the dynamic variable is expressed using the notation such that $P(q, t|q_0)dq$ is the probability that the dynamic variable $Q(t)$ lies in the interval $(q, q + dq)$ at time t given the initial value $q_0 = Q(0)$ at $t = 0$. The conditional PDF is defined by the average of the phase space density function $\rho(q, t|q_0)$ over an ensemble of realizations of $\xi_\beta(t)$ as indicated by the subscripted bracket:

$$P(q, t|q_0) \equiv \langle \rho(q, t|q_0) \rangle_{\xi_\beta}. \tag{5.111}$$

The phase space density function is defined as:

$$\rho(q, t|q_0) = \delta(q - Q(t)), \tag{5.112}$$

where $Q(t)$ is the solution to the Langevin equation for a particular realization of $\xi_\beta(t)$, in this case given by Eq. (5.109). The formal expression for the PDF is then given by:

$$P(q, t|q_0) \equiv \langle \delta(q - Q(t)) \rangle_{\xi_\beta}. \tag{5.113}$$

The Fourier transform of the PDF defines the characteristic function, which by inserting Eq. (5.113) into Eq. (5.104) yields:

$$\widetilde{P}(k, t|q_0) = \langle e^{ikQ(t)} \rangle_{\xi_\beta} = \exp[ikq_0 e^{-\lambda_0 t}] \langle e^{ikf_\beta(t)} \rangle_{\xi_\beta}. \tag{5.114}$$

For an arbitrary analytic function $g(\tau)$, Doob [171] showed that for a differential Lévy process $dB(\tau)$ of strength b, with an index β and which is delta-correlated in time:

$$\left\langle \exp\left(\int_0^t ig(\tau)dB(\tau) \right) \right\rangle_B = \exp\left(-b \int_0^t |g(\tau)|^\beta d\tau \right). \qquad (5.115)$$

Consequently, with the assumption that the stochastic driver $\xi_\beta(t)$ is such a Lévy process, after further analysis [572], the characteristic function is determined to be:

$$\widetilde{P}(k,t) = \exp[ikq_0 e^{-\lambda_0 t}] \exp\left[-\sigma_\beta^2(t)|k|^\beta\right], \qquad (5.116)$$

and the time-dependent 'width' of the distribution is:

$$\sigma_\beta^2(t) = \frac{b}{\lambda_0\beta}(1 - e^{-\beta\lambda_0 t}), \qquad (5.117)$$

which agrees with the result obtained by Doob [171] and also by West and Seshadri [572]. The conditional PDF generated by the linear Langevin equation is therefore:

$$P(q,t|q_0) = \frac{1}{\pi} \int_0^\infty dk \cos(k[q - q_0 e^{-\lambda_0 t}]) \exp\left[-\sigma_\beta^2(t)k^\beta\right], \qquad (5.118)$$

where the Fourier transform is taken with respect to the displaced variable $q - q_0 e^{-\lambda_0 t}$. Hence, the solution to the linear Langevin equation driven by a random force with Lévy statistics with scaling index β and constant parameter b, itself has Lévy statistics in the variable $q - q_0 e^{-\lambda_0 t}$ with scaling index β and time-dependent parameter $\sigma_\beta^2(t)$.

The PDF given by Eq. (5.118) can also be determined as the solution to the FFPE corresponding to the Langevin equation Eq. (5.108):

$$\frac{\partial P(q,t|q_0)}{\partial t} = \frac{\partial}{\partial q}[\lambda_0 q P(q,t|q_0)] + K_\beta \partial^\beta_{|q|}[P(q,t|q_0)], \qquad (5.119)$$

whose phase-space derivative is fractional:

$$\partial^\beta_{|q|}[P(q,t;\beta)] = \int_{-\infty}^\infty \frac{P(q',t;\beta)}{|q - q'|^{\beta+1}}dq', \qquad (5.120)$$

where $\partial^\beta_{|q|}[\cdot]$ is a Reisz–Feller fractional derivative with $0 < \beta \leq 2$ [469] resulting in:

$$K_\beta = \frac{b}{\pi}\Gamma(\beta + 1)\sin[\pi\beta/2]. \qquad (5.121)$$

Thus, the phase space equation of evolution of the Lévy PDF, rather than being a partial differential equation of the Fokker–Planck type, is an integro-differential equation from the fractional calculus. The discontinuous form of the Lévy statistics makes the description of the evolution of the underlying process by a fractional diffusion equation not only desirable, but necessary. Solutions to such FFPEs are discussed in [581].

A generalization of this early work to disordered systems with external force fields has been given by Vlad et al. [552]. It was shown [231] that such a fractional transport equation could also be obtained using the IPL transition probabilities in the GME. A similar approach was considered by Metzler et al. [369] at essentially the same time.

A second process of interest is one in which the inhomogeneities are in time, in which case the physical observable is described by an equation of evolution that is second-order in the phase-space variable, but whose time derivative is fractional:

$$D_t^\alpha[F(t)] = \frac{1}{\Gamma(n-\alpha)} \frac{d^n}{dt^n} \int_0^t \frac{F(t')}{|t-t'|^{1+\alpha-n}} dt' \qquad (5.122)$$

where $D_t^\alpha[\cdot]$ is the Riemann–Liouville fractional derivative [433, 469]; where n is the smallest integer such that $\alpha < n < \alpha + 1$. The evolution equation with this fractional time derivative, for $0 < \alpha \leq 1$, was obtained using a CTRW formalism [147] and independently using a stochastic two-state process with memory [12, 578]. In this parameter domain, the equation describes fractional transport. In the parameter domain $1 < \alpha \leq 2$ the fractional equation can describe the propagation of waves in fractal media.

We include both the long-time memory effects that manifest in the fractional time derivative Eq. (5.122) and the long-range spatial effects that manifest in the fractional phase-space derivative Eq. (5.120) in the same fractional evolution equation:

$$D_t^\alpha[P(q,t|q_0)] - P_0(q)\frac{t^{-\alpha}}{\Gamma(1-\alpha)} = K_\beta \partial_{|q|}^\beta [P(q,t|q_0)]. \qquad (5.123)$$

The initial PDF is given by $P_0(q) = P(q, t = 0|q_0)$ and for a point source is taken as a delta function in space $P_0(q) = \delta(q - q_0)$ with the dependence on the initial condition suppressed. The formulation of fractional equations, with fractional time derivatives, correspond to initial value problems, where the initial values are given by certain fractional derivatives. Furthermore, Eq. (5.123) not only describes a more general kind of fractional diffusion, when $0 < \alpha \leq 1$; $0 < \beta \leq 2$, but also describes wave propagation in a non-dissipative fractal medium when $1 < \alpha \leq 2$; $0 < \beta \leq 2$. In addition, when $\beta = 0$ there are no spatial effects in the dynamics, and the resulting equation has been used to describe shear relaxation in viscoelastic materials [216].

Of course, Eq. (5.123) would be of little value if it could not be solved to reveal the physical nature of the complex phenomena being modeled. The eigenfunction expansion technique was used to solve the fractional transport equation in the case $\beta = 2$, for a variety of boundary values [369, 370, 371]. Another method of solution was introduced [216], which involves taking the Laplace–Mellin transform of Eq. (5.123) and in the inversion process writing $P(q, t|q_0)$ in terms of Fox's H-functions. In this way the normalized solution to the fractional evolution equation can be expressed in terms of the inverse Fourier transform of the characteristic function, with the initial condition suppressed:

$$\widetilde{P}(k, t) = \sum_{n=0}^{\infty} \frac{(-1)^n}{\Gamma(1 + \alpha n)} (bt|k|^{\beta/\alpha})^{n\alpha} = E_\alpha(-(bt)^\alpha|k|^\beta), \tag{5.124}$$

where $E_\alpha(z)$ is the MLF [581]. Note that Eq. (5.124) is the characteristic function for this transport–propagation process. The inverse Fourier transform of Eq. (5.124) is the solution to the fractional evolution equation:

$$P(q, t; \alpha, \beta) = \int_{-\infty}^{\infty} E_\alpha(-(bt)^\alpha|k|^\beta)e^{ikq}\frac{dk}{2\pi}, \tag{5.125}$$

where to facilitate discussion we have adopted a new nomenclature for the characteristic function. Thus, we are able to encapsulate in a single dynamical equation all the transport–propagation properties contained within a given complex phenomenon and describe its behavior with a single function.

For example, the Lévy PDF observed in heartbeat dynamics [420], was reviewed and extended to a broad range of other physiological phenomena in [576], and can be obtained from Eq. (5.124) using $\alpha = 1$. Notice that in this limit the MLF sums up to an exponential, so that Eq. (5.124) becomes:

$$\widetilde{P}(k, t; 1, \beta) = \exp[-bt|k|^\beta], \tag{5.126}$$

which is the characteristic function for the centrosymmetric Lévy stable PDF. Thus, using Eq. (5.124) we see that the statistics of such inhomogeneous processes with $\alpha = 1$ are Lévy stable.

Another application of Eq. (5.124) is given for small values of the argument of the MLF. At early times the MLF gives rise to the stretched exponential:

$$\widetilde{P}(k, t; \alpha, \beta) \approx \exp\left[-\frac{(bt)^\alpha|k|^\beta}{\Gamma(\alpha + 1)}\right]. \tag{5.127}$$

If $\beta = 2$, the inverse Fourier transform of Eq. (5.127) is a Gaussian distribution, with a variance that increases as a power-law in time, that is t^α, with $\alpha = 2\mu$, corresponding to FBM for $\alpha \neq 1$. Finally, the solution Eq. (5.127), with $\beta \neq 2$, corresponds to a new kind of statistical process, one having Lévy statistics, but with a power-law time dependence. West and Nonnenmacher [606] called this processes *fractional Lévy motion* (FLM), as a generalization of FBM. Laskin *et al.* [314], a year later independently, used a more formal argument to arrive at an exact expression of the form (5.127) and gave it the same name.

The motivation for combining transport and propagation into a single equation is related to the fact that the diffusion equation is unphysical, since in the classical diffusion equation perturbations propagate with infinite speed to all regions of space. We know however that heat consists of the propagation of phonons, and heat transport results from incoherent scatterings of phonons within a material. As the temperature of the material is lowered there are fewer and fewer scatterings so the transport of heat becomes less diffusive and more wave-like. Maxwell noted the unphysical nature of the heat equation and included a ballistic term in the kinetic theory of gases to incorporate into the description the finite propagation speed of heat. This term resulted in the heat transport equation being replaced with the telegrapher'sequation [383]. The solution to the telegrapher's equation is wave-like at early times and diffusion-like at late times, including as it does in the finite propagation speed of heat, which is the speed of sound in the material.

Thus, the transport equation is only valid asymptotically in time. An alternative argument for the telegrapher's equation can be made starting from the wave equation and considering the influence of absorption. This phenomenon is a manifestation of the resistive loses in the medium supporting the wave motion due to the inelastic scattering of waves. Like the telegrapher's equation, the fractional evolution equation, is physically realizable whereas the diffusion equation is not.

Because of the self-similar character of spatial inhomogeneities in fractal media, the wavelength of waves propagating in such media always take on some of the hierarchical structure of the medium. Berry [90] dubbed such waves *diffractals*, a new wave regime characterized by a short-wave limit in which ever finer levels of structure are explored by the propagating waves and geometrical optics is never applicable. The physical arguments of Berry regarding the influence of fractal media on scalar waves were replaced with more formal arguments [479]. The latter investigators interpreted Eq. (5.123), in the limit $\beta \to 2$, as a fractional diffusion equation when $0 < \alpha \leq 1$ and as a fractional wave equation when $1 < \alpha \leq 2$. In the limit $\beta \to 2$ the general solution to the fractional evolution equation, the Fourier transform of Eq. (5.124), agrees with the Schneider–Wyss solution [479], their Eq. (2.33) with the initial value function indexed with $k = 0$. Theirs is an exact solution to the propagation of a scalar wave through a fractal medium and is therefore an exact diffractal in Berry's sense.

Note that the multiple scattering of regular waves, traditionally used in describing the propagation of waves in heterogeneous media, is here replaced with a fractional time derivative. This fractional time derivative, in the wave context, models the time delay associated with multiple scattering events. In addition, the value of α is a direct measure of the degree of coherence in such scattering events. Scalar waves in homogeneous media have a value $\alpha = 2$ and are completely coherent. Acoustic and light waves in turbulent fluid flow, on the other hand, have a smaller value of α, becoming less and less coherent as the level of turbulence increases and the value of α becomes smaller. This is evidenced by the fact that the physical observable, the intensity of sound or light, is described by a transport equation rather than a wave equation [580].

In summary, it should be clear that complex physical processes, Newtonian mechanics notwithstanding, are non-Markovian in that they contain memory, and the above fractional time derivative incorporates that memory into the description of the evolution of the process. In a similar way, the fractional spatial derivative incorporates the long-range spatial interactions into the evolution of the process. Thus, the fractional evolution equation is non-local in both space and time. This structure has dramatic effects on the transport of information is such media, as we shall see in our discussion of nonsimple networks.

Network traffic models: Information traffic in communication networks, whether telephone networks, the body's central nervous system, or the Internet, requires an understanding of the statistical behavior of information in order to maximize the utility of the metrics of information transfer. The traffic theory devised in the last century to predict delay and blocking of telephone messages, relied on the Markov nature of Poisson statistics and played little role in the design of the Internet. Laskin *et al.* [314] argue that this situation regarding traffic theory will change in the future, as we extend traffic models to properly take into account the observed fractal nature of data traffic [317, 417]. The latter authors recognize that the broadband signals carried by complex networks are significantly different from their more traditional narrowband cousins, in that the design, control and analysis of such networks are strongly dependent on their non-Markov fractal nature.

For example, two models that have been proposed to approximate broadband network traffic are FBM and stable Lévy motion. Mikosch *et al.* [376]

have shown that if the rate at which transmissions are initiated (connection rate) are low, relative to heavy-tailed connection length distribution tails, then the cumulative traffic is reasonably well approximated by stable Lévy processes. If, on the other hand, the connection rate is large, relative to heavy-tailed connection length distribution tails, then FBM is the better approximation. The key to distinguishing between the two models is whether or not the varying connection rates inhibit, as in the Lévy case, or induce, as the FBM case, long-range dependence (memory).

As another example in the queuing analysis of communication networks input processes with infinite variance, such as may arise in self-similar time series, lead to infinite moments in the queuing processes, which translate into extremely long waiting times. Consequently the FLM, discussed previously, can be used to model traffic intensities for rates that diverge [317]. But this is not the only model to describe the self-similarity of broadband traffic [452].

We have discussed Brownian motion, FBM, Poisson and IPL PDFs. Each type of statistic was introduced to describe a well-defined physical process. In Brownian motion the fluctuating velocity variable determines the position of an erratically moving particle in time. On the other hand, in a Poisson process the focus changes from that of a dynamic variable to the occurrence of events. Thus, we have stochastic variables and stochastic point processes and in the case of network traffic these can be the intensity of information, or the occurrence, or non-occurrence, of messages [195]. Some results [185] reveal that message traffic on complex networks do not have the regular statistics of Poisson, but instead have a kind of self-similar bursting, which occurs in fractal time series and fractal modeling has been applied to broadband network traffic; more specifically to Internet traffic [616].

Laskin *et al.* [314] use the FLM to construct a teletraffic model, which takes into account both the Hurst exponent H and the Lévy parameter α by means of the Riemann–Liouville integral Eq. (5.122) with:

$$\alpha \equiv H + 1 - 1/\beta$$

and H is confined to the interval $1/\beta < H \leq 1$. It is straightforward to show that the dynamic variable defined by the integral Eq. (5.122), when inserted into the definition of the characteristic function Eq. (5.124), yields the exact expression:

$$\widetilde{P}(k,t;H,\beta) = e^{-\sigma|k|^{\beta}t^{\beta H}}, \qquad (5.128)$$

which has the same functional form as the approximate expression

Eq. (5.127) with the parameter:

$$\sigma = \frac{b}{\Gamma^\beta(H+1-1/\beta)\beta H}.$$ (5.129)

This teletraffic model is an extension of the application of FBM to the same problem and, as was emphasized [314], it is suitable for traffic modeling in modern broadband networks. It was pointed out that empirical data, collected for a variety of communication networks, exhibit self-similarity and heavy-tailed distributions, so it is reasonable to apply the FLM traffic model to capture these characteristics. Moreover, preliminary results support these arguments indicating that FLM can describe the behavior of today's teletraffic, if not tomorrow's.

It is also important to note that the discussion of the self-similarity in the statistics literature on network traffic takes a different form from that adopted herein. Resnick and Rootzén [452] point out that the focus in the literature has been on heavy-tailed transmission times of sources sending data to one or more servers, assuming a cumulative distribution of transmission times $F(\tau)$ with a Pareto tail:

$$1 - F(\tau) \sim \frac{L(\tau)}{\tau^\alpha} \text{ , as } \tau \to \infty$$

where $L(\tau)$ is a slowly varying function of time. Unlike previous investigators, these authors were concerned with the case $0 < \alpha < 1$ in which the models are statistically unstable and the amount of ongoing traffic increases without limit. Although such models are unrealistic over asymptotically long times they do provide useful descriptions of the uncontrolled network behavior associated with high traffic flows that might be expected over restricted time intervals.

5.6 Models of Nonsimple Networks

The macrovariables observed in complex networks, such as the mean field in physical systems, events in biological systems, and the population density in social or ecological systems, display emergent properties of spatial and/or temporal scale-invariance, manifest in IPLs of connectivity of individuals and waiting-time PDFs between events. Typically these emergent property IPLs cannot be directly inferred from the equations describing the nonlinear dynamics of the individual elements of the network. Despite the advances made by the renormalization group approach and self-organized criticality theories that have shown how scale-free phenomena emerge at critical points, the issue of determining how the emergent properties influence the microdynamics of the single individual at criticality in general remains open.

Herein we follow the arguments contained in [538, 599] and address the problem of quantifying the response of an individual, within a social group, to the dynamics of the collective. This is done by taking advantage of the fractional calculus apparatus, whose utility is demonstrated by capturing the dynamics of the individual elements of a nonsimple network from the information quantifying that network's global behavior. The phase transition of the nonsimple social network, from essentially independent behavior of the individual members, to consensus suggests the wisdom of using a generic model from the Ising universality class to characterize the network dynamics. We do this using the networks of two-state individuals that have an exponential PDF for switching between states, that is in changing their minds from being a Republican to being a Democrat and back again. The influence of the social network on their decision-making habits is determined using a coupled network of two-state master equations from the DMM.

It is then possible to demonstrate that an individual's trajectory responds to the collective (mean field) motion of the network in a way described by a linear fractional differential equation, obtained through a subordination procedure without the necessity of linearizing the underlying dynamics. Following this procedure it is seen that the analytic solution to the linear fractal differential equation retains the influence of the nonlinear network dynamics on the behavior of the individual. Moreover, the solution to the fractional equation of motion suggests a new direction for designing mechanisms with which to control the dynamics of complex networks. But first we examine the statistical properties of the DMM events.

5.6.1   Renewal events

Many nonsimple natural processes are aggregated into events, catastrophic and otherwise. Exemplars from the physical domain are radioactive decay and earthquakes, and from the social domain, elections, peaceful demonstrations and lynchings. No matter how intricate the microdynamics, the detailed interactions between the individual elements, at the macro, or coarse-grained level, the emergent behavior can be viewed as particular events, that can be localized in time, independently of spatial consequences. In this way, the behavior of a macroprocess is characterized by sequence of events described in terms of a waiting-time PDF $\psi(t)$, such as the time interval an individual is a member of one political group before switching to the other.

We are interested in crucial events which are members of the larger class of events called renewal. *Renewal events* instantaneously reset the clock to the initial state of the generating system after their occurrence. Once a renewal event occurs, the system evolves in time independently of whatever occurred earlier. Each event has no memory of previously occurring events. As Turalska and West [538] point out, renewal events found in physical systems include anomalous diffusion of tagged particles inside living cells, blinking quantum dots and defects arising in the weak-turbulence regime of liquid crystals.

The probability of n sequential events being renewal is determined in the following way. First of all, at time $t = 0$ an event occurs, and consequently $\psi_0(t) = \delta(t)$. Secondly, the first renewal event occurs at time later $t > 0$, with a PDF $\psi(t) \equiv \psi_1(t)$. Consequently, the PDF for event n in such a sequence to occur at time t is expressed in terms of PDFs of $n - 1$ earlier events by the convolution chain condition:

$$\psi_n(t) \equiv \psi_{n-1} * \psi(t) = \int_0^t \psi_{n-1}(t')\psi(t - t')dt'. \qquad (5.130)$$

This chain condition is the essence of the renewal process, indicating that for any chain of n successive events the last event is independent of the earlier $n - 1$ events in the chain.

Historically, investigators have often taken waiting-time PDFs to be exponential, with or without supporting experimental data. One reason for this frequently made assumption is that such exponential PDFs define a Poisson process which is renewal and is often the case for simple systems. On the other hand, it is quite often the case that for nonsimple networks the PDFs are IPL. We have found that for the purpose of analysis a waiting-time PDF of the hyperbolic form is convenient in that it has an asymptotic IPL form.

5.6.2 Time subordination

The idea that the time being experienced is not the absolute unchanging measure of duration that Newton hypothesized in the *Principia* is now considered banal in physics and has been since the acceptance of relativity theory in the early 20th century. The latter notion that time depends on the frame of reference of the observer has not been universally embraced outside of physics, however. On the other hand, the recent availability of time-resolved data, in biology, ecology and sociology has also promoted adopting the notion of multiple clocks, each with its own intrinsic mechanism with which to define time. Such clocks distinguish between cell-specific and organ-specific time scales in biology [602] and person-specific and group-specific time scales in sociology [597]. Of course, the notion of subjective and objective time dates back to the middle of the 19th century with the introduction of the empirical Weber–Fechner Law [183] distinguishing between a person's experiential time and their response time to tasks as measured by laboratory clocks.

However, a significant difference between the reference frames of classical physics and natural sciences is that the relations between the clocks in the physical sciences to those in the natural sciences are not linear. While the collective activity of organs, such as the brain or heart could be seen as quite regular, often with periodic fluctuations, the activity of individual neurons are observed to have a bursty kind of noise [61, 587]. In an analogous way, people behave according to their idiosyncratic schedules, not always conforming to the time frame of their social group. Thus, because of the stochastic dynamics of one or both of the clocks being referenced to one another, a transformation between times involving a PDF is necessary and subordination theory has been developed for the construction of such transformations. We provide a semi-quantitative introduction to subordination theory here.

Assume the existence of two clocks, the first of which records a discrete operational time n; measuring the time $\mathcal{T}(n)$ experienced by an individual. The second of which records the ticks of continuous chronological time t, measuring the time $\mathcal{T}(t)$ using a shared device the social group has agreed upon. If each advance of the discrete clock n is experienced as an event, operational time and chronological time can be related by the waiting-time PDF of those events in chronological time $\psi(t)$. If the events are renewal, as given by the chain condition Eq. (5.130), operational and chronological time can be related by:

$$\langle \mathcal{T}(t) \rangle = \sum_{n=1}^{\infty} \int_0^t \Psi(t - t')\psi_n(t')\mathcal{T}(n)dt'. \qquad (5.131)$$

We see that for each tick of the operational clock there is an event that occurs in chronological time. The first tick occurs due to one event occurring in the time interval $t - t'$, the second tick occurs due to two independent events occurring in the same time interval, and so on. The survival probability determines the probability that the last of the n renewal events occurs at the end of the time interval. Every tick of the operational clock is an event, which in the chronological time occurs at time intervals drawn from the renewal waiting-time PDF. Because of this randomness, one needs to sum over all events resulting in an average over many realizations of the transformation.

Consider the ticking of a two-state operational clock, as depicted in Fig. 5.7. In operational or subjective time, the clock switches back and forth between its two states at equal unit time intervals. This is the regularity experienced by a single individual acting alone. In chronological or objective time, however, this regular behavior can be significantly distorted due to the dynamics of the group to which the individual belongs. The time transformation depicted in the figure was taken to be an IPL PDF of waiting times. Thus, a single time step in the operational time corresponds to a time interval being a random number drawn from $\psi(t)$ in chronological time. The long tail of the IPL PDF leads to strong disruptions of the operational time trajectory, since there exist a non-zero likelihood of drawing large time intervals between events. However, since the transformation between the operational and chronological times involves a random process, it is necessary to consider an ensemble of trajectories in chronological time, which leads to the average behavior of the clock in the chronological time denoted in Eq. (5.131) by the bracket.

Turalska and West [538] noted that the time subordination procedure has also been used to model communication delays in the network.

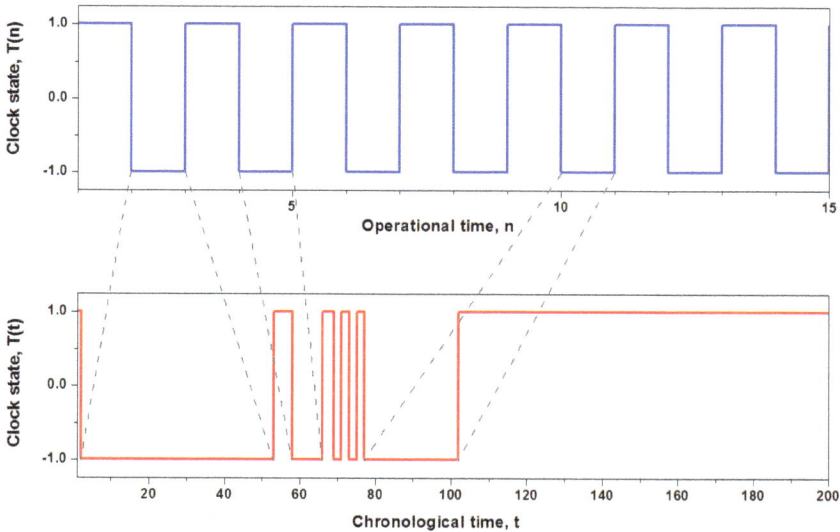

Figure 5.7: The upper curve is the regular transition between the two states of the individual in operational time. The lower curve is the subordination of the transition times to an IPL PDF to obtain the chain of events in chronological time. From [538] with permission.

However, contrary to frequently used approaches, where individuals within the network are subordinated to model an interaction delay, here, we adopt the statistics of the macroscopic variable to derive the behavior of the interacting individuals. The coupling between individuals causes them to deviate from the Poisson behavior of non-interacting individuals. However, using the DMM with ten thousand individuals on a two-dimensional lattice, it was determined that the time scale of interacting individuals is orders of magnitude smaller than the time scale of the macroscopic variable. Thus, we use the statistical properties of the macroscopic variable to provide a first-order estimate of the single individual's dynamics on the network [538, 600].

5.6.3   Nonsimple network subordination

A network's influence on the dynamics of the individual has been determined by adapting the subordination argument of the preceding section, and relating the time scale of the macrovariable to the time scale of the microvariable. The two-state master equation for a single isolated individual in the DMM in discrete time n in steps of $\Delta\tau$ is:

$$\phi(n+1) - \phi(n) = -g_0\Delta\tau\phi(n) \qquad (5.132)$$

in the notation $\phi(n) = \phi(n\Delta\tau)$, g_0 is the rate of switching states for an isolated individual, and $\phi = p_+ - p_-$ is the difference in probabilities for the typical individual to assume one of the two states. The solution to this discrete equation is obtained by iteration:

$$\phi(n) = (1 - g_0\Delta\tau)^n\phi(0), \qquad (5.133)$$

which becomes an exponential in the limit $g_0 \ll 1$. However, the dynamics are not this simple when the individual is part of a network.

A subordination view interprets the discrete index n as an individual's operational time that is stochastically connected to the chronological time t. It is this latter time in which the global behavior is observed. Assuming the chronological time lies in the time interval $(n-1)\Delta\tau \le t \le n\Delta\tau$ and implementing the transformation given by Eq. (5.131), the equation for the average dynamics of the individual probability difference is given by [600]:

$$\langle\phi(t)\rangle = \sum_{n=1}^{\infty} \int_0^t \Psi(t-t')\psi_n(t')\phi(n)dt'. \qquad (5.134)$$

It is evident that the subordinated trajectory involves an ensemble average and Eq. (5.134) replaces the solution to the two-state master equation of the DMM for a single individual [597]. Here the waiting-time PDF $\psi(t)$ is determined from the derivative of the empirical survival probability $\Psi(t)$. The empirically determined analytic expression for the survival probability is the hyperbolic function given by Eq. (5.5) and the waiting-time PDF given by the network as a whole given by Eq. (5.10). The asymptotic behavior ($t \gg T$) of the empirical survival probability is IPL. The extent of the IPL range of the survival probability is determined by the empirical values of T and μ, which are determined by the dynamics of the network. The IPL functional form of the PDF results from the empirical (numerical data generated by the DMM) behavior of the survival probability $\Psi(t)$ with $\mu = 3/2$ [600].

Using a renewal theory argument, Pramukkul *et al.* [438] show that Eq. (5.134) expressed in terms of Laplace transform variables indicated by $\widehat{f}(u)$ for the time-dependent function $f(t)$ has the form:

$$\langle \widehat{\phi}(u) \rangle = \frac{\phi(0)}{u + \lambda_0 \widehat{\Phi}(u)}, \tag{5.135}$$

where $\lambda_0 = g_0 \Delta \tau$. $\widehat{\Phi}(u)$ is the Laplace transform of the Montroll–Weiss memory kernel [382] obtained using the CTRW model:

$$\widehat{\Phi}(u) = \frac{u\widehat{\psi}(u)}{1 - \widehat{\psi}(u)}. \tag{5.136}$$

We give an alternate view of this argument in the next chapter.

The inverse Laplace transform of Eq. (5.135) yields the generalized master equation:

$$\frac{d\langle \phi(t) \rangle}{dt} = -\lambda_0 \int_0^t \Phi(t - t') \langle \phi(t') \rangle dt'. \tag{5.137}$$

The memory kernel $\Phi(t)$ contains the information on how the other members of the network influence the behavior of the individual under study through its dependence on the empirical waiting-time PDF $\psi(t)$.

Past analyses, including DMM calculations, have revealed that the waiting-time PDFs are an IPL. The asymptotic behavior of an individual in time is determined by considering the asymptotic waiting-time PDF as $u \to 0$:

$$\widehat{\psi}(u) \approx 1 - \Gamma(1 - \alpha)T^\alpha u^\alpha, \quad 0 < \alpha = \mu - 1 < 1, \tag{5.138}$$

so that Eq. (5.135) reduces to:

$$\langle \widehat{\phi}(u) \rangle = \frac{\phi(0)}{u + \lambda^\alpha u^{1-\alpha}}. \tag{5.139}$$

The inverse Laplace transform of Eq. (5.139) yields the FRE:

$$\partial_t^\alpha \langle \phi(t) \rangle = -\lambda^\alpha \langle \phi(t) \rangle \qquad (5.140)$$

where the operator $\partial_t^\alpha (\cdot)$ is the Caputo fractional derivative for $0 < \alpha = \mu - 1 < 1$ and

$$\lambda T = \left(\frac{g_0 \Delta \tau}{\Gamma(2 - \mu)} \right)^{\frac{1}{\mu - 1}} . \qquad (5.141)$$

Note that due to the dichotomous nature of the states that $\langle \phi(t) \rangle$ is the average opinion of a typical individual in the network. It is possible to generalize Eq. (5.140) to the FLE [599]:

$$\partial_t^\alpha \phi(t) = -\lambda^\alpha \phi(t) + \eta(t), \qquad (5.142)$$

where $\eta(t)$ is random noise generated by the finite size of the network [586] Equation (5.142) has the form of the FLE discussed earlier in this chapter but with a Caputo rather than a RL fractional derivative and its average reduces to Eq. (5.140) when the noise is zero-centered.

The solution of the asymptotic GME Eq. (5.137), or equivalently the fractional rate equation Eq. (5.140), for a randomly chosen person within the social network is given by the MLF [599]:

$$\langle \phi(t) \rangle = \phi(0) E_\alpha(-(\lambda t)^\alpha). \qquad (5.143)$$

Consequently the full network dynamics discussed in Sec. 3.6 modify the behavior of the individual in an interesting way.

Some numerics: The above theoretical prediction for the average behavior of a typical individual was tested against numerical simulations of a DMM dynamic network on a 100×100 two-dimensional lattice, with nearest-neighbor interactions in the subcritical, critical and supercritical regions of the dynamics. The time-dependent average opinion of a randomly chosen individual is presented in Fig. 5.8, where the average is taken over 10^4 independent realizations of the dynamics in the subcritical, critical and supercritical regimes.

A comparison with the exponential form of $\langle \phi(t) \rangle$ for an isolated individual indicates that the influence of the network on the individual's dynamics clearly persists for increasingly longer times with increasing values of the DMM control parameter within the network. Note that the early time behavior of the MLF is a stretched exponential. In Fig. 5.8(a), the region where the black dashes of the MLF fit diverge from the data is at the onset of the IPL tail of the MLF. An exponential truncation of the MLF would

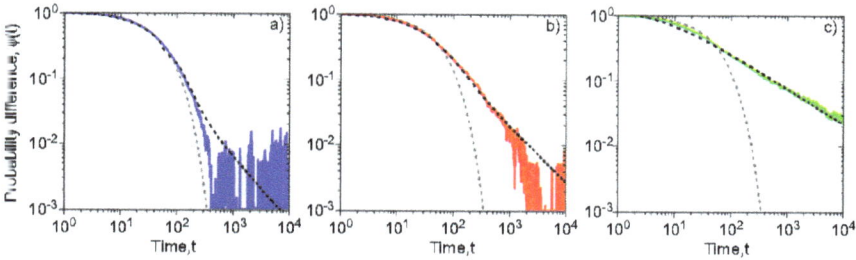

Figure 5.8: The probability difference, estimated as an average over an ensemble of 10^4 independent realizations of single individual trajectory. Each trajectory corresponds to the evolution of a randomly selected individual on an 100×100 lattice with $g_0 = 0.01$ and the same initial condition: (a) subcritical domain, $K = 1.00$; (b) critical domain, $K = 1.70$; and (c) supercritical domain, $K = 2.50$. The gray dashed line denotes the exponential form of the probability difference, that is obtained for a single isolated individual. The black dashed line denotes the optimal fit to the MLF, with α and λ fit to the numerical data. From [599] with permission.

fit the data throughout its domain. The rationale, for an exponentially truncated MLF was taken up elsewhere [538] and will not be repeated here. The fitting of the analytic solution at early times to the DMM's numerically generated curves is certainly very good in the subcritical domain with $R^2 = 0.9968$.

As the control parameter is increased the critical point is approached from the subcritical region, the network undergoes a phase transition and experiences critical slowing down. The plunging stretched exponential that was observed in the subcritical region depicted in Fig. 5.8(a) is replaced with a more gently decreasing function. The time-dependent average solution of a typical individual in the critical regime is depicted in Fig. 5.8(b). It is evident by comparing this fitted curve with the curve in Fig. 5.8(a) that the latter solution is seen to not decrease as quickly in time as the former, and there is less variability asymptotically in time. This behavior is reflected in the value of the IPL index, which is determined to be $\mu = 1.808$ with a quality of fit given by $R^2 = 0.9989$. Note how well separated the solution is from the exponential function, which is given by the gray dashed curve. But here again, a modified MLF, one modulated by an exponential truncation denoting a finite time scale for the dynamics, might provide a better overall fit to the data. Such a modification, following the approach of the tempered fractional derivatives of [367] was made by [538] and supports the conjectures made here.

Increasing the control parameter beyond the critical value we enter the supercritical region of the network dynamics. We can see from the fit of the analytic solution to the data depicted in Fig. 5.8(c) that the solution given by the solid curve extends far beyond that found in either the subcritical or critical domains. By now it is obvious that the long-time behavior of the MLF takes the form of an IPL:

$$\lim_{t \to \infty} \langle \phi(t) \rangle = \frac{1}{\Gamma(\mu)(\lambda t)^{\mu-1}}. \tag{5.144}$$

Here the IPL index is fitted to the value $\mu = 1.535$ with a quality of fit given by $R^2 = 0.989$. The IPL index measured here is very close to that obtained for the global survival probability that is obtained from the numerical calculations on the DMM lattice network.

Chapter 6

Information Exchange Theory

> *This is the essence of intuitive heuristics: when faced with a difficult question, we often answer an easier one instead, usually without noticing the substitution.* — Kahneman [279]

In the middle of the last century, after his successful launching of the new discipline *Cybernetics* [611], as we outline in Chapter 1, the mathematician Norbert Wiener speculated in a popular lecture [612] that a system, high in energy, can be controlled by one that is low in energy. The force necessary to achieve this counter-intuitive result is produced by the low-energy system being high in information content and the high-energy system being low in information content. Consequently, there is an information gradient that creates a force by which the low-energy system controls the high-energy system, through a flow of information against the traditional energy gradient and thereby overcoming the mechanical force. Quantifying the transfer of information from a nonsimple system, high in information, to a second nonsimple system, lower in information, was the first articulation of a universal principle of cybernetics and we referred to this speculation in Chapter 1 as the WH.

In a physical context, the WH can be understood as resulting from an entropic force, used to explain such diverse phenomena as the elasticity of freely jointed polymer molecules [394], oceanic forces [257] and the conscious states in the human brain through neuroimaging [134]. Over the past decade, the nascent field of network science has been applied to determining the conditions under which the WH can be facilitated, or suppressed. After being overlooked for over a half century, the WH was shown to be correct [43, 44], only to be superseded by the more detailed PCM [427], as discussed in the last chapter, which subsequently developed into the POIE described in this chapter.

A number of analyses of information transfer have lead to a result that is continually being rediscovered, that being, nonsimple networks in living systems exist at, or on, the edge of, criticality. In the physical sciences such critical behavior is observed in phase transitions, which optimizes both the transmission of information within a network and the exchange of information between networks [74, 290]. The typical form of an observable's PDF within a large collection of diverse nonsimple phenomena is hyperbolic, or asymptotically IPL, whether modeling the connectivity of the internet, the number of connections within social groups, the frequency or magnitude of earthquakes, the number of solar flares, or the time intervals between solar eruptions, the time intervals in conversational turn taking, and many other nonsimple phenomena [593]. In the absence of a more sophisticated quantitative measure, we adopt the IPL index as the measure of nonsimplicity in each of these phenomena.

As mentioned, resonant or matching dynamics have been among the more useful mechanisms adapted to the extracting of weak signals from noisy backgrounds. A linear dynamic process responds strongly to a harmonic perturbation, with a frequency matching the natural rhythmic behavior of the network. An effective two-state SR phenomena responds strongly when the period of a harmonic perturbation matches the transition rate of the random fluctuations, as well as, in aperiodic SR [146], where the rate of perturbation (Poisson) matches the transition rate of the two-state process. However, as phenomena deviate more strongly from the simple, the matching conditions for resonance and even what is considered to be a resonance becomes more subtle.

Network nonsimplicity increases as interactions become more strongly nonlinear, as the number of degrees-of-freedom increase and as the counting statistics for events deviate more widely from Poisson. These three properties converge in the synchronization of the dynamical elements in a nonsimple network, whether the isolated elements are modeled as linear or nonlinear oscillators, limit cycles, or chaotic attractors [305, 619]. Both the mechanisms of synchronization and SR have been used to provide insight into the behavior of nonsimple networks as, for example, into the dynamics of collections of neurons [389]. In addition, the question of how to transmit information within a nonsimple network in which the simplifying assumptions of traditional statistical physics are no longer valid has been addressed [62, 223, 587]. For example, the assumption that the unperturbed network dynamics can be described by an autocorrelation function, with a finite correlation time [301], has been challenged. A number of investigators have concluded that a nonsimple network, described by intermittent fluctuations with non-Poisson statistics, does not asymptotically respond to external periodic perturbations [62, 504, 505]. These scientists determined that LRT, a cornerstone of non-equilibrium statistical physics, was "dead" when applied to such nonsimple systems. However, it has been established that the determination of the demise of LRT was premature.

5.1   New Ergodic Ensemble-based LRT

Herein we present an extension of the investigations of the perturbation of nonsimple networks [44, 62, 504, 505] to a new resonance-like phenomenon, that is, how nonsimple networks respond to an excitation (a transfer of information) from a second nonsimple network, as a function of the mismatch of the measures of nonsimplicity of the two networks. Consider the nonsimple system to be a non-Poisson renewal (NPR) network and the measure of nonsimplicity to be the IPL index. In assessing the premature death of LRT, our research group [43, 44] considered a NPR network S, with IPL index $\mu_S < 2$, and studied, in analogy with the theory of SR [86], the case where the rate of production of jumps, a kind of renewal event, is modulated by an external excitation. Unlike others, who used a harmonic perturbation, we introduced a stochastic signal as the excitation. The network considered satisfied the NPR condition with $\mu_S < 2$ and the excitation was a second NPR network P, with an IPL index $\mu_P < 2$. We presented heuristic arguments that the transport of information attains maximum efficiency at the matching condition $\mu_S = \mu_P$. This matching condition is called the NME for the obvious reasons that in our theoretical framework, genuinely nonsimple processes satisfy the NPR condition, with $\mu_S < 2$, so that the parameter μ_S becomes a nonsimplicity index and two networks with the same IPL index have the same degree of nonsimplicity.

Note that the parameter range $2 < \mu_S < 3$, although ensuring the existence of a finite mean waiting-time value, generates a divergence of the second moment, as well as, the breakdown of the central limit theorem. It is important to notice further that the response of a nonsimple network of the same nature as the one discussed herein has been studied by other investigators [246, 247, 250, 281, 348, 496, 502, 504, 505, 571]. These scientists did not establish a connection between their results and the LRT of [25, 42, 62, 122] and in some cases, as mentioned, they made the misleading conjecture that their results established the "death of LRT." Actually, these theoretical treatments were asymptotic in time and the only connection made to LRT is through adopting a generalized LRT developed in [43, 44].

The reason why a nonsimple network is the most sensitive to perturbations with comparable nonsimplicity rests on the observation that each non-ergodic, non-Poisson network can be thought of as emerging from the subordination to a network with a fixed time scale. But to see why this is true we must present the story entailing the generalization of LRT in which we establish the conditions for resonance that facilitate information exchange between two networks based on their relative degree of nonsimplicity. We follow the arguments of Aquino *et al.* [44] and present them in due course.

6.1.1   Defining terms

LRT [302] is one of the basic tools used to extract information from fluctuations observed in time series arising in non-equilibrium statistical physical systems [156, 359]. An even more challenging issue is the application of LRT to nonsimple processes outside the physical sciences, such as in social and physiological processes, which we wrap together under the heading of nonsimple systems, or sometimes nonsimple networks. It is now understood [587], if not widely known, that the breakdown of the ergodic condition is caused by the occurrence of *crucial events*. It is interesting that the same perspective applies to nonsimple networks, when the cooperation-induced phase transition produces temporal nonsimplicity and crucial events as first described by Turalska *et al.* [536]. Crucial events are renewal, namely, the time interval between consecutive events does not have any relation whatsoever with the earlier or later events.

Yet, for a Gibbs ensemble, with all the systems prepared such that an event occurs at an initial time, then the rate of event production turns out to be time-dependent, rather than constant, as it would be in the ordinary Poisson case. The time interval between consecutive crucial events is given by a waiting-time PDF $\psi(\tau)$, with the now familiar hyperbolic form and the IPL index μ is assumed to lie in the range $1 < \mu < 3$. It is important to understand that the functional form of the waiting-time PDF given in Eq. (5.10) is not arbitrary, but satisfies a number of important constraints. First of all, it has been widely accepted that only the asymptotic time behavior of the PDF, in which it is an IPL, matters and is given by Eq. (5.12). If we adopt this widely held perspective, it would prevent us from establishing accordance with the experiments on the response of nonsimple systems to external perturbations. Thus, the choice of Eq. (5.10) is dictated by the need to define a border between the asymptotic time regime ($\tau \gg T$) and the microtime regime ($\tau \leq T$). Given the sociophysiology interest of the applications of LRT we refer the interested reader to the work by Grigolini *et al.* [234], where the waiting-time PDF of Eq. (5.10) is obtained by means of a Fechner-transformation from the conventional Poisson PDF.

We showed earlier that the rate of generation of cascading crucial events tends to zero as $1/t^{2-\mu}$, when the IPL index is $\mu < 2$ and to a constant as $1/t^{\mu-2}$, when $\mu > 2$. We note this here because in both cases the time duration of the out-of-equilibrium condition is infinite, thereby raising the challenging task of going beyond conventional LRT to describe the dynamics. In other words, with these IPL PDFs the formal arguments that are the basis of traditional LRT breakdown and the entire discussion must be reformulated from scratch, which we do in this chapter.

Consider a nonsimple network of interest S that is perturbed by a second nonsimple network P. Conventional LRT is given by the following expression:

$$\sigma(t) = \langle \xi_S(t) \rangle \equiv \varepsilon \int_0^t ds \chi(t, s) \xi_P(s) \tag{6.1}$$

where $\xi_S(t)$ is the time series produced by the internal dynamics of the network S. The bracketed symbol $\langle \xi_S(t) \rangle$ denotes the Gibbs average over the fluctuations, which is an average over an ensemble of identically distributed time series and in the absence of perturbations is zero. The variable $\xi_P(t)$ denotes the time-dependent perturbation, ε is its intensity and it is generated by the network P. LRT predicts the response of S on the basis of the unperturbed autocorrelation function of $\xi_S(t)$. In fact, the function $\chi(t, s)$ is a non-stationary kernel in general, traditionally called the linear response function (LRF), and is related to the autocorrelation function $\Psi_S(t, s)$ of the fluctuations $\xi_S(t)$. The quadratic mean value of $\Psi_S(t, s)$ is assumed, for notational ease, to be normalized to unity:

$$\Psi_S(t, s) \equiv \langle \xi_S(t)\xi_S(s) \rangle, \tag{6.2}$$

and is related to the LRF by the following expression:

$$\chi(t, s) = \frac{d\Psi_S(t, s)}{ds}. \tag{6.3}$$

Note that traditional LRT refers to the stationary case, where the autocorrelation function depends only on the time difference and not on the two times separately:

$$\Psi_S(t, s) = \Psi_S(t - s), \tag{6.4}$$

and as a consequence the time derivatives can be interchanged:

$$\frac{d\Psi_S(t, s)}{ds} = -\frac{d\Psi_S(t, s)}{dt}; \tag{6.5}$$

a condition only fulfilled by simple networks. For nonsimple networks the choices of LRFs given by Eq. (6.3) and that obtained using Eq. (6.5):

$$\chi(t, s) = -\frac{d\Psi_S(t, s)}{dt} \tag{6.6}$$

are not equivalent and lead to separate and distinct conclusions when they are independently implemented.

The foundation of both choices, Eqs. (6.3) and (6.6), have been discussed in the literature [25, 42, 62, 122] and established that a new form of LRT is determined by the way perturbations introduce bias. Our research group made the simplifying assumption that $\xi_S(t)$ is a dichotomous signal. Using the language of fluid dynamics they called the time intervals between consecutive crucial events *laminar regions*. At the instant a crucial event occurs, unperturbed dynamics are realized by the random selection of either the positive, $\xi_S = 1$, or the negative value, $\xi_S = -1$. In other words, we assumed that the occurrence of a crucial event entails the tossing of a fair coin, which determines the sign of the next laminar region.

Consequently, the external perturbation can generate a bias in two different ways. The first way rests on affecting the fairness of the coin-tossing process. If $\xi_P(t) > 0$ the choice of the positive sign is more probable than the choice of the negative sign. If $\xi_P(t) < 0$ the opposite is true. This prescription leads to the choice of the LRF given by Eq. (6.3) and we refer to it as the *phenomenological* LRT.

The experiments reported in [27, 497] strongly suggest that nature prefers Eq. (6.6), which we refer to as the *dynamic* LRT. What is the theoretical argument in favor of the dynamic theory? To provide a convincing answer to this important question, let us go back to the special form of Eq. (5.10). We note that we do not know the generator of the dynamics for the nonsimple system being observed, which could be a Hamiltonian in a physical process. However, we do not even know if a satisfactory discussion of the nonsimple dynamics can be made using a Hamiltonian formalism.

Let us assume that the hyperbolic PDF in Eq. (5.10) is a reliable representation for the time durations of the laminar regions. In this case, a reasonable conjecture is that the external perturbation affects either μ or T, or both parameters defining the form of Eq. (5.10). We know that the IPL index μ defines the system's nonsimplicity and emerges from the cooperative interaction among interacting units. A weak external perturbation is not expected to change the system's nonsimplicity and consequently would not affect the IPL index. We shall later consider the effect of a strong perturbation and how it influences the IPL index, but this is beyond the scope of LRT and the present argument.

It is therefore reasonable to assume that the external perturbation only affects T. The time scale is enlarged if ξ_P and ξ_S have the same sign. On the other hand, it is reduced if ξ_P and ξ_S have opposite signs. This assumption leads to the choice of Eq. (6.6), as shown in the literature [25, 42, 62, 122].

We are now in a position to outline the main purpose of this chapter. We focus attention on the results of Aquino *et al.* [43, 44]. These papers address the issue of the response of a nonsimple system to a nonsimple but weak external perturbation, with the surprising result that a nonsimple system does not asymptotically respond to simple stimuli, that is, they do not asymptotically respond to stimuli unless they are nonsimple. It is important to stress that they [43, 44] focused on the cross-correlation between $\xi_S(t)$ and $\xi_P(t)$ in the long-time limit. This is an ideal condition that has the effect of restraining the definition of nonsimplicity to systems with $\mu \leq 2$. In fact, in the long-time limit, a system with $\mu > 2$ reaches the normal condition of a constant rate of event production, thereby recovering the familiar Poisson condition.

The condition $\mu = 2$ is of fundamental importance for brain function. In fact, recent work [27, 29] established that the brain functions in the region of the IPL index $\mu = 2$, which, in turn, is known [337] to correspond to making the brain activity the source of ideal $1/f$-noise. The results of the analysis [43, 44], presented in this chapter, may therefore have important applications to design the most convenient stimuli to drive nonsimple networks, and especially brain dynamics, via what had been defined as "complexity management" [43]. However, an apparent weakness is that these results were derived adopting the phenomenological LRT, thereby raising doubts that those nonsimple systems, which have been proven to obey the dynamical LRT [27, 497], may not obey the PCM, previously established [43] and now called NME. In this chapter we follow the proof presented in [44] that the more realistic dynamical LRT generates NME and subsequently leads to the POIE.

6.2 Non-ergodic Ensemble-based GLRT

Our group [25, 42] proposed a form of LRT that applies to systems (networks), whose dynamics are dominated by crucial events. In the stationary case this LRT becomes indistinguishable from the traditional theoretical prediction [302], as it should, but in the non-stationary, non-ergodic case it is significantly different.

Let us consider the consequences of adopting either of the two choices of the LRF, those being either the *phenomenological* given by Eq. (6.3) or the *dynamical* given by Eq. (6.6), in the special case, where both $\xi_P(t)$ and $\xi_S(t)$ are event-dominated processes. In the case of event-dominated processes we show that the transmission of information from system P to system S is determined by the dialogue between the crucial events of $\xi_S(t)$ and the crucial events of $\xi_P(t)$. Specifically, this discussion focuses on studying the transport of information from P to S, using both forms of generalized LRT (GLRT). Note that there is no limitation on the form of $\xi_P(t)$ provided that the coupling between the two networks is sufficiently weak, so as to be compatible with the emergence of the linear response form of the integral Eq. (6.1).

It is convenient to frame the problem of the transmission of information from P to S, in terms of the ideal case of a Gibbs ensemble of a composite system S + P. Thus, in this chapter we imagine that P generates a fluctuating signal $\xi_P(t)$ and that any such signal evokes a response in $\xi_S(t)$. For each perturbing signal we conduct infinitely many experiments (conceptually) and average over all possible responses. For ease of notation we take both signals $\xi_S(t)$ and $\xi_P(t)$ to be dichotomous, fluctuating between values $+1$ and -1. It is important to remark that Eq. (6.1), for the response of the system S to the perturbation P, is valid when the system is prepared at time $t = 0$ and placed at the beginning of a laminar phase. The interaction of S with the perturbation P is turned on simultaneously with the preparation of S.

In the general case of a dichotomous renewal process $\xi(t)$, generated using a waiting-time PDF $\psi(t)$, the PDF that fixed a time t', the next event observed at time $t > t'$ is given by [21]:

$$\psi(t, t') = \psi(t) + \int_0^{t'} R(t'')\psi(t - t'')dt'', \tag{6.7}$$

with $R(t)$ being the PDF for generating an event to occur exactly at time t:

$$R(t) = \sum_{n=1}^{\infty} \psi_n(t), \tag{6.8}$$

and $\psi_n(t)$ denotes the n-times convolution of $\psi(t)$.

It has been shown [21] that the autocorrelation function of the process is related to two-time waiting-time PDF $\psi(t, t')$ by:

$$\langle \xi(t)\xi(t') \rangle = \int_t^{\infty} \psi(x, t')dx = \Psi(t, t'), \tag{6.9}$$

and therefore coincides with the two-time survival probability $\Psi(t, t')$ for the first event, that is, the probability that, for fixed t', no event is observed until time $t > t'$. We assume that the fluctuating time series $\xi_S(t)$ is the signal generated in S as a dichotomous renewal process defined by the hyperbolic PDF:

$$\psi_S(t) = \frac{(\mu_S - 1)T^{\mu_S-1}}{(T+t)^{\mu_S}}. \tag{6.10}$$

We therefore name, respectively, $\psi_S(t,t')$, $R_S(t)$, and $\Psi_S(t,t')$ the functions obtained by replacing $\psi(t)$ with $\psi_S(t)$ in the earlier equations.

Let us now consider the Gibbs ensemble of composite systems S + P, and evaluate the average $\langle\langle\xi_S(t)\rangle\rangle_{SP}$. Note that the average is over the separate statistics of the two systems S and P:

$$\langle\langle\xi_S(t)\rangle\rangle_{SP} \equiv \langle\langle\xi_S(t)\rangle_S\rangle_P. \tag{6.11}$$

We select all the responses to the same perturbation $\xi_P(t)$; evaluate their average denoted by $\langle\xi_S(t)\rangle_S$; finally we construct the average over all possible perturbations denoted by $\langle...\rangle_P$ so as to obtain the final result denoted by $\langle\langle...\rangle\rangle_{SP}$. This procedure gives us

$$\langle\sigma(t)\rangle = \langle\langle\xi_S(t)\rangle\rangle \equiv \varepsilon \int_0^t ds\chi(t,s)\langle\xi_P(s)\rangle, \tag{6.12}$$

where, for notational convenience, we dropped the subscripts on the brackets, but we understand the averages in the sense described above.

If necessary, the signal $\xi_P(t)$ must share the same properties as $\xi_S(t)$ and they are both assumed to be dichotomous signals, with random renewal fluctuations, between the values $+1$ and -1. Therefore, the signal $\xi_P(t)$ consists of non-Poissonian dichotomous fluctuations with the waiting-time PDF:

$$\psi_P(t) = \frac{(\mu_P - 1)T^{\mu_P-1}}{(T+t)^{\mu_P}}. \tag{6.13}$$

It is therefore convenient to define the additional functions $\psi_P(t,t')$, $R_P(t)$, and $\Psi_P(t,t')$, with the functions obtained by replacing $\psi(t)$ with $\psi_P(t)$ in the earlier equations.

The power spectrum for this type of fluctuating signal, in the absence of perturbations [337, 358], for a time sequence of length L is

$$S_p(\omega) \propto \frac{1}{L^{2-\mu}\omega^{3-\mu}}. \tag{6.14}$$

The dependence of the spectrum on the length of the time series given by Eq. (6.14) occurs because of the breakdown of the Wiener–Khinchine theorem, which relates the spectrum of the process to the Fourier transform of its autocorrelation function. This spectrum is valid for $\mu < 2$, remarkably, even though a stationary autocorrelation function cannot be defined for this IPL index. In the case $\mu > 2$, the Wiener–Khinchine theorem is satisfied and the spectrum becomes

$$S_p(\omega) = \frac{A}{\omega^{3-\mu}}, \tag{6.15}$$

with A independent of L.

At this point, it should be clear that to obtain the important results of this chapter on the transmission of the statistical properties of P to S we must use Eq. (6.12). This suggests employing a prescription to define a value for the average perturbation $\langle \xi_P(t) \rangle$. We assume that the perturbation P is prepared at $t = 0$, to be at the beginning of a laminar phase, with the analogous prescription adopted for the system S. This assumption allows us to replace $\langle \xi_P(t) \rangle$ with $\Psi_P(t)$ in Eq. (6.12).

6.2.1   Dynamic approach

Let us examine the response of a nonsimple network S producing NPR fluctuations to a perturbing signal from a nonsimple network P, with similar properties within the dynamic approach. The phenomenology approach produces qualitatively similar results [43, 44], but in the interest of space, this argument will not be reproduced here. In this section, we analyze both the average response and the input–output correlation, that is, the correlation between the perturbing fluctuating signal produced by the system P (input) and the response signal produced by the system S (output).

Average response to perturbation: We begin the discussion with the linear response expression for the average network response given by Eq. (6.12), obtained by averaging $\xi_S(t)$ over the fluctuations of both the perturbed network S and perturbing network P. The LRF kernel in the integral expression is obtained using the dynamic condition given by Eq. (6.6):

$$\chi(t, t') = -\frac{d\Psi_S(t, t')}{dt} = \psi_S(t, t'). \tag{6.16}$$

Equation (6.12) can then be written as:

$$\langle \sigma(t) \rangle = \varepsilon \int_0^t dt' \psi_S(t, t') \langle \xi_P(t') \rangle. \tag{6.17}$$

We prepare the perturbation $\xi_P(t)$ at $t = 0$, so that in order to observe the influence of P on S we select all the networks of the Gibbs ensemble having $\xi_P(0) = 1$. In this case $\langle \xi_P(t) \rangle$ is given by the survival probability

$\Psi_P(t)$ [21]. Thus, Eq. (6.17) becomes

$$\langle\sigma(t)\rangle = \varepsilon \int_0^t dt' \psi_S(t,t') \Psi_P(t'), \qquad (6.18)$$

where $\Psi_P(t)$ is the survival probability for the process $\xi_P(t)$:

$$\Psi_P(t) = \int_t^\infty dt' \psi_P(t') = \left(\frac{T_P}{T_P + t}\right)^{\mu_P - 1}. \qquad (6.19)$$

This slow decay corresponds to the probability that no perturbation event occurs up to time t. S evolves in time so as to reach a steady-state value that corresponds to a constant perturbation abruptly applied at $t = 0$. However, during this process a perturbation event occurs that has the effect of suddenly changing the external field. Thus, $\langle\sigma(t)\rangle$ does not reach a steady-state value, but after reaching a maxima, it decays. Under the specific conditions discussed in this section, the asymptotic time decay of $\langle\sigma(t)\rangle$ in Eq. (6.18) has the same IPL index as that of survival probability $\Psi_P(t)$.

This process is interpreted as the transmission of the statistics of perturbing network P into the responding network S. In Appendix B we explore the entire range of parameters $1 < \mu_S < 3$ and $1 < \mu_P < 3$, for the responder S and the perturber P, depending on the values of the IPL indexes μ_S, μ_P, bearing in mind that these indices characterize the waiting-time PDFs of their respective networks. The value 2 for the IPL index marks the transition value from a finite to an infinite mean time for these PDFs, that is, the transition from an ergodic, stationary condition to a non-ergodic, non-stationary condition. In fact, for $\mu > 2$ a mean time exists, thereby a finite time scale can be defined for the network and a stationary condition is reached. On the other hand, for $\mu < 2$ such a condition is never achieved, not even in infinite time.

Summary of stationary case, $2 < \mu_S < 3$: Interpreting the equations in Appendix B with the appropriate labels on the quantities in Eq. (B.5), after some algebra, it follows that for $1 < \mu_P < 2$ and $2 < \mu_P < \mu_S$, the time asymptotic behavior of the average response of S is

$$\langle\sigma(t)\rangle \propto t^{1-\mu_P}. \qquad (6.20)$$

This average response is proportional to $\langle\xi_P(t)\rangle$ for large t, meaning that S inherits the relaxation properties of the perturbation P after a sufficiently long time.

On the other hand, for $2 < \mu_S < \mu_P$, the asymptotically dominant term is

$$\langle \sigma(t) \rangle \propto t^{1-\mu_S}. \tag{6.21}$$

In this domain the average response is proportional to the ordinary unperturbed relaxation to equilibrium of $\langle \xi_S(t) \rangle$, when an initial bias for $\xi_S(t)$ is introduced. We see therefore that for $\mu_P < \mu_S$, when $\xi_P(t)$ is slower than $\xi_S(t)$, the perturbation P imposes on the responder S its own relaxation properties, thereby allowing one to *manage* the nonsimplicity of a system or network by using an appropriate stimulus. This is one of the results summarized in the *cube* depicted in Fig. 1.3, as we discussed in Chapter 1.

Summary of non-stationary case, $1 < \mu_S < 2$: The analysis for a non-stationary network S is not as straightforward as the stationary case. For this reason we do not reproduce the details of the calculation in the appendix and refer the reader to the literature. It is worth pointing out, however, that for the calculation leading to the results presented below we made the simplifying assumption that, although at time $t = 0$ half of the S elements of the Gibbs ensemble are in the state $\xi_S = +1$ and half in the state -1, all of them are at the beginning of their sojourn in their corresponding states. This is an out-of-equilibrium condition, corresponding to the initial preparation of the network.

Using the theory of the dynamic LRT we obtain [44]:

$$\frac{\langle \sigma(t) \rangle}{\varepsilon} \approx \frac{k_1(\mu_S, \mu_P)}{t^{\mu_P - 1}} + \frac{k_2(\mu_S, \mu_P)}{t^{\mu_P + 1 - \mu_S}}, \tag{6.22}$$

where the coefficients are known, but inelegant functions of the IPL indices. In the non-stationary range of parameters the dominant term is always the first term in Eq. (6.22) which, if $\mu_P < \mu_S$, is also slower than the unperturbed relaxation to equilibrium of $\xi_S(t)$. In this latter range, therefore, S relaxes to equilibrium again *inheriting* the properties of the perturbation P.

The result in Eq. (6.22) is of special interest, since it discriminates between the dynamic and phenomenologic approaches in the non-stationary regime. It is this difference that enabled the determination that the relaxation of the perturbation of liquid crystals [497] is consistent with the prediction of the dynamic approach. In fact, the phenomenologic approach disregards the influence of the perturbation on the occurrence time of the S events [504, 505], while the dynamic theory does not, thereby affording a criterion for information transport that was judged [44] to be a more appropriate representation of the communication among nonsimple systems with $\mu < 2$. However, the equivalence between the phenomenologic and the dynamic theories in the case when S is infinitely aged, that is, for $\mu_S > 2$, indicates that the generalization of LRT given by Eq. (6.6), namely, the dynamic theory, becomes active only when S is in a far from equilibrium condition and begins drifting toward equilibrium. Although equilibrium is never reached when $\mu_S < 2$, the two-time autocorrelation function $\Psi_S(t,t')$ tends to recover the stationary property [44]:

$$\Psi_S(t,t') = \overline{\Psi}_S(t-t') = \left(\frac{T_S}{T_S + t - t'} \right)^{\mu_S - 2} \qquad (6.23)$$

thereby making the phenomenologic theory formally equivalent to the dynamic theory, when $\mu_S > 2$.

6.2.2   Input–output correlation function

Let us now turn our attention to the asymptotic limit of the input–output correlation (cross-correlation) function between the responder S and stimulator P:

$$\varepsilon \Phi_\infty \equiv \lim_{t \to \infty} \langle \xi_S(t) \xi_P(t) \rangle. \qquad (6.24)$$

It is interesting that the cross-correlation function has also been used as an indicator of aperiodic stochastic resonance [144]. The input–output correlation function, like the other measures used, is defined by the average over the fluctuations in both the responder S and stimulator P.

The asymptotic limit of the cross-correlation function is independent of the way the responder S and the stimulator P are prepared, so we can use the prescription leading to Eq. (6.1), obtained assuming that both S and P are prepared at time $t = 0$, for its construction. Multiplying both sides of Eq. (6.1) by $\xi_P(t)$ and averaging over the fluctuations of the stimulator P, we obtain the input–output correlation function:

$$\Phi(t) = \int_0^t d\tau \psi_S(t,\tau) \Psi_P(t,\tau). \qquad (6.25)$$

Note that we continue to use the dynamic GLRT.

For illustrative purposes, we supplement Fig. 1.4 with Fig. 6.1, showing the functional forms used to construct the three-dimensional plot of the cross-correlation function Φ_∞ in the same parameter range: regions II and III correspond to the condition of minimal and maximal cross-correlation, respectively. Intuitively, these extremes come about because of the difference in time scales between systems S and P in such regions. In region III fluctuating sequences $\xi_S(t)$ and $\xi_P(t)$ have a finite and an infinite time scale, respectively, thereby allowing $\xi_S(t)$ to adapt to the stimulus-induced bias, so as to yield maximal correlation. In region II the roles of the time

$\mu_{S},\mu_{P\rightarrow}$	$1 < \mu_P \leqslant 2$		$2 < \mu_P < 3$	
$1<\mu_S\leqslant2$	$\Phi_\infty = \zeta(\mu_S,\mu_P)^*$	I	$\Phi_\infty = 0$	II
$2<\mu_S<3$	$\Phi_\infty = 1$	III	$\Phi_\infty = \frac{\mu_S-2}{\mu_S+\mu_P-4}$	IV

Figure 6.1: The asymptotic values of the input–output cross-correlation function are recorded as a function of the IPL indices for the S and P systems, based on the dynamic LRT.

scales are interchanged, the bias induced by P on the longer (diverging) time scale of the process $\xi_S(t)$ is asymptotically averaged out, due to the many intervening switching events of $\xi_P(t)$, thereby suppressing any correlation. Figure 6.1 summarizes the asymptotic values of the input–output cross-correlation function as a function of the IPL indices for the S and P systems, based on the dynamic LRT.

The vertex $\mu_S = \mu_P = 2$, representing an exact $1/f$-noise generating system, being stimulated by an exact $1/f$-noise perturbation, marks the abrupt transition from vanishing (region II) to maximal correlation (region III).

6.3   Non-ergodic Time Averages

Let us now turn our attention to an extension of the arguments made concerning the properties of information transfer between networks using ergodic averages to those using non-ergodic averages [427], which reveals some surprising results. For example, if the S network is ergodic and the P network is non-ergodic, the cross-correlation is maximum: this means that there is a flux of information from P to S in accordance with the WH. However, when the P network is ergodic and the S network is non-ergodic, the asymptotic cross-correlation goes to zero. Thus, there is no residual response of the responder S to the stimulator P. Note that this was the domain that earlier investigators prematurely interpreted as the death of LRT. In the case in which both networks are ergodic, there is a partial positive correlation between S and P that changes with μ_S and μ_F; as is the case when both networks are non-ergodic.

These extraordinary results obtained using the asymptotic cross-correlation function have a fundamental limitation, however, because the predictions of this form of NME rely on ensemble averages. Thus, the predictions based on the cross-correlation are not necessarily valid when we have only a single non-ergodic time series for each system, that is, when we cannot apply the equivalence between ensemble averages and time averages. This is a common situation, since many interesting systems cannot be experimentally replicated. Our group [427] has extended the arguments of Sec. 6.2 from ensemble averages to time averages, when the two are not necessarily equal.

The questions now arise: How many of these remarkable properties persist when ensemble averages are replaced with time averages? Does the NME survive the change from ensemble averages to time averages?

5.3.1   What properties do survive?

The network P stimulates the network S as follows: if S has an event at time t and if its next laminar region is assigned a value with the same sign as that of P, then S is perturbed so that its next laminar region tends to be longer, by assigning to its parameter T in Eq. (6.10) the value $T_+ = T(1 + \varepsilon)$. On the other hand, if the next laminar region of S has a value with the opposite sign to that of P, then the value $T_- = T(1 - \varepsilon)$ is used, thus tending to make the next laminar region shorter. In order to assess the influence of P on S for a single time series, using this perturbation procedure, it is natural to consider a time window of size L and analyze the time-averaged cross-correlation function:

$$C(t_0, L) \equiv \frac{1}{L} \int_{t_0}^{t_0+L} \xi_S(\tau)\xi_P(\tau)d\tau. \qquad (6.26)$$

By moving the starting point t_0 of the window and evaluating Eq. (6.26), a density plot for the time-averaged cross-correlation can be created as a function of the IPL indices. A measure of the influence of the P network on the S network is the center of gravity (COG) of this density plot. In the domain $1 < \mu_S, \mu_P < 2$, the COG of the density plot is erratic; in sharp contrast with the smooth behavior found in the calculations of the cross-correlation function in this region obtained using ensemble distribution functions depicted in Fig. 1.3. This variability is clearly shown in the left panel of Fig. 6.2, where the cross-correlation function is plotted as a function of μ_S and μ_P. It is worth noting that different realizations of the figure lead to different landscapes in the non-ergodic quadrant. The reasons behind this behavior will become clear shortly.

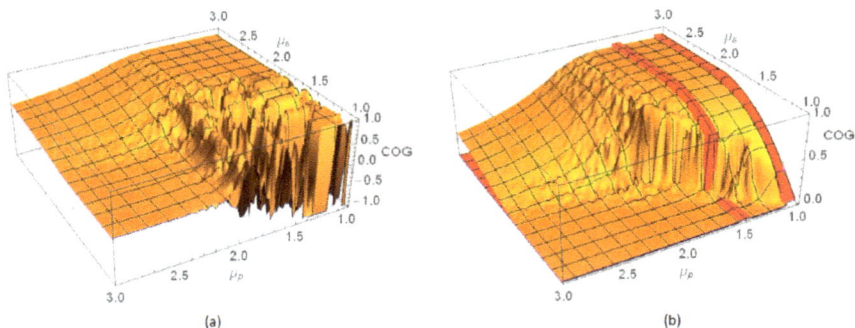

Figure 6.2: The cross-correlation function using the COG as a function of the IPL indexes μ_S and μ_P. (a) The time sequence of length L is divided into N intervals of length T_W. The cross-correlation function of Eq. (6.26) is evaluated for each interval and the landscape is obtained plotting the mean of the resulting distribution of values. (b) The times t_0 of Eq. (6.26) are the times of event occurrence and the landscape is obtained by plotting the mean of the resulting distribution of values. The exact prediction of theory is shown by the two red lines. From [427] with permission.

Piccinini *et al.* [427] resolved this difficulty using an argument consisting of two parts. The first part was a new data processing prescription that enabled the elimination of the erratic behavior observed in the left panel of Fig. 6.2 and produces the smooth behavior of the right panel. The second part provided a theoretical justification for this prescription and was used to calculate the asymptotic cross-correlation function analytically.

The prescription is to locate the beginning t_0 of the window at which each C is evaluated on an event of either the stimulating or responding network. We again note that when $\mu < 2$, the mean length of the laminar regions of a network is infinite. This is the reason for the erratic plot in the left panel of Fig. 6.2; for most of the duration of the time series there are no events, thus the cross-correlation function is either 1 or -1. This fact was exploited to obtain the regular behavior of the right panel of Fig. 6.2; when one of the two networks has an event, it is most probably embedded in a long laminar region of the other network. If the stimulator P has an event, then it is most likely embedded in a long laminar region of the responder S. In this case the resulting value of C follows the statistics of an unperturbed Lamperti distribution given in [427], as the responder has no influence on the stimulator and the latter is non-ergodic. If the S network has an event, then it is most likely embedded in a long laminar region of the P network, which is equivalent to the responder being subject to constant stimulation. In this case the computed value of the cross-correlation function follows the statistics of a perturbed Lamperti PDF also discussed in [427].

The probability $W_S(t)$ of having an event in S at time t is given by:

$$W_S(t) = \frac{R_S(t)}{R_S(t) + R_P(t)}, \tag{6.27}$$

with $R(t)$ given by [15]:

$$R(t) \sim \frac{1}{\langle t \rangle} \left[1 + \frac{1}{3 - \mu} \left(\frac{T}{t} \right)^{\mu - 2} \right], \tag{6.28}$$

when $2 < \mu < 3$, with μ being of the corresponding network. In this case the network is Poisson only in the infinite time limit. The probability $W_P(t)$ can be obtained from Eq. (6.27) by exchanging the roles of S and P. When $t \to \infty$, if $\mu_S > \mu_P$, we have $W_S = 1$ and $W_P = 0$; if $\mu_S < \mu_P$, then $W_S = 0$ and $W_P = 1$. As a side note, we observe that this argument implies that the perturbed system does not respond asymptotically to simple perturbations, which is realized in the phenomenon of habituation.

The red stripes superimposed on the numerical calculations in the right panel of Fig. 6.2 are determined using Eq. (6.27) and the theory developed in [427] shows excellent agreement with numerical simulations. The theory is valid also in the case in which one of the networks is ergodic and the other is non-ergodic; in the long-time limit, only the former has events. This fact and the considerations above entail the CME, the response of an ergodic system to a non-ergodic system is maximal. On the other hand, the response of a non-ergodic system to an ergodic system is minimal, that is, it vanishes asymptotically. In the case in which both systems are ergodic, the above theory is not applicable, but, given the equivalence (by definition) of ensemble averages and time averages, in this case we again recover the results of the CME, as expected.

6.3.2 Some applications

It bears repeating that in the case where both S and P are the generators of non-ergodic fluctuations, the adoption of the prescription provided by the calculation of the cross-correlation function, using time averages, results in the erratic behavior shown in Fig. 6.2(a) for $1 < \mu_S$, $\mu_P < 2$. Our main result is that the beginning of the moving window, t_0, must be chosen to coincide with the occurrence of a renewal event of either S or P. At first sight, this choice is in apparent conflict with the fact that the time average is done only on the crucial events, which are a very small fraction of the total number of events. Naively, one would expect this prescription to make the statistical average less accurate. Instead this restricted outcome, the theoretical approach developed demonstrates that this choice of initial value leads to a smooth function of μ_S and μ_P, in very good agreement with numerical simulation.

To describe the practical applications of these results, following Grigolini *et al.* [235], we establish a connection between two emerging theoretical perspectives that physicists are adopting in their attempts to address fundamental biological issues beyond the limits of reductionism. The former theoretical perspective is illustrated by the debate regarding the effect of finite size on criticality in natural swarms [41]. In fact, the recent experiment [41] is attracting [138] the attention of researchers to the key role of criticality in biology, thereby leading them to look, for instance, at some very interesting proposals [252, 386]. In the conclusion of their article, Mora and Bialek [386] emphasize the phase transition related property of critical slowing down, namely, the infinitely slow regression to equilibrium of processes at criticality. They point out, however, the existence of a possible conflict with the resilience of complex biological systems that are expected to promptly adapt themselves to the changes of their environment. The evasion of predators by flocks of birds [41] is an outstanding example of biological resilience.

An interesting experiment concerning the cognition of living beings (rats) is given by the work done at Duke University by the group of Pais-Vieira [408]. In this experiment, information was transmitted from rat A moving in a box to rat B moving in a different box through a cable connecting the neural network of the brain of rat A to the neural network of the brain of rat B. This experiment is the *in vivo* counterpart of an *in vitro* experiment done in 1999 at the University of North Texas by the group of Gross [237], interpreted by them as a form of chaos synchronization. Thus, we see rapid progress from the 1999 *in vitro* experiment that culminated in the 2014 experiment [447], concerning the same kind of information transfer from the brain of one human subject to that of another.

This form of synchronization seems to be a natural property of the dialogue between two individuals [1], even though the statistical roots of these synchronization processes has remained elusive. The latter remarkable example of biological nonsimplicity is illustrated in a recent review paper [374], reflecting the growing interest regarding anomalous diffusion in biological cells, a paradigm of the special nature of biological processes.

It turns out that the connection between these two forms of biological nonsimplicity requires a deeper understanding of the origin of ergodicity breakdown. We invite the readers to focus their attention on the recent theoretical remarks of [76]. Critical slowing down is a property of critical systems involving interactions among an infinitely large number of units. Beig *et al.* [76] emphasized the importance of "temporal complexity", which must not be confused with critical slowing down, even if in some processes of phase transitions the IPL index of temporal nonsimplicity is the same as that of critical slowing down. At the onset of criticality, a nonsimple system makes a transition from a condition where individuals are essentially statistically independent of one another, to an organized state of highly correlated behavior. This condition, however, is not permanent, and, from time to time, a network, with a finite number of interacting units, undergoes organizational collapse [543]. We have interpreted these intermittent collapses as 'simplicity flicker' states [338]. These simplicity flicker states are crucial events in which the free will of the individuals is reestablished and consequently make the dynamics of criticality non-ergodic, thereby establishing a connection, while retaining the distinction, between temporal nonsimplicity and critical slowing down in nonsimple biological networks.

It is important to stress that, as previously shown [76], ergodicity breaking is confined to the time region $t < T_{eq}$, where $T_{eq} \propto \sqrt{N}$ and N denotes the number of interacting elements within the network. We believe this to be a general property of criticality and that evaluating the transmission of information from one nonsimple network at criticality to another nonsimple network at criticality on the time scale $t > T_{eq}$ [235, 338, 347, 438] gives the misleading impression that the network entrainment found may be a form of chaos synchronization [237]. This is a consequence of the fact that an evaluation of the correlation between the perturbed nonsimple network and the stimulus nonsimple network done on the time scale $t < T_{eq}$ would generate the erratic behavior shown in Fig. 6.2(a) for $1 < \mu_S, \mu_P < 2$, giving the false impression of a lack of correlation.

This leads us to another practical application of the present results, that being brain dynamics. The widely accepted belief that brain activity is not dominated by a characteristic time scale has led investigators in this field of research to make experimental observations that, in turn, have revealed the critical role of crucial events. According to the analysis of EEG of individuals performing a task done by Buiatti *et al.* [125], brain activity reveals the action of crucial events with the waiting-time PDF having an IPL index μ in the region $1 < \mu < 3$. More precisely, moving from one patient to another, the index μ is found to vary from values of $\mu < 2$ to values of $\mu > 2$. As they pointed out the region $\mu < 2$ is not ergodic. As a consequence, there are aging effects that make the statistical evaluation of the effects of perturbation challenging. These authors [125] had to use the theoretical procedure of [280] to evaluate the scaling of the time series generated by a single EEG. The aging procedure [280], as we previously discussed, is equivalent to studying the scaling of the diffusing variable

$Q(t)$ generated by the random steps in a standard random walk obtained using the network time series $\xi_S(t)$ in Eq. (7.25).

Buiatti *et al.* [125] used this procedure to establish for some subjects that the IPL index is $\mu < 2$. It is remarkable that the research work done by the neurophysiologists of the University of Pisa [31] led to the same conclusion, with the additional observation that the anomalous scaling properties revealed by their experimental results are, in fact, a manifestation of temporal nonsimplicity [412]. The discovery that, in the case of the dynamics of the brain, the IPL index of ξ_S influenced by ξ_P may have the non-ergodic value of $\mu_S < 2$ raised the challenging problem of assessing the cross-correlation between S and P. This problem, which is very difficult in the case when both S and P are non-ergodic, is now satisfactorily resolved, both theoretically and numerically, using the methods described in this chapter.

Chapter 7

Detecting Crucial Events

If you can't explain it simply you do not understand it well enough — A. Einstein

In this chapter we retain the commonly used term $1/f$-noise, even though we should more appropriately use the more accurate term $1/f$-signal, since the phenomenon naturally emerges from data produced by nonsimple networks. Leaving such hair-splitting aside in order to avoid ambiguous arguments, we describe the spectrum of variability of the most nonsimple network of all, the human brain, as $1/f$-noise. Following Allegrini *et al.* [27], we note that in the case of physiological processes driven by the brain, some researchers use the form $1/f^{\eta}$, or $1/f$-variability, rather than $1/f$-noise, implying that $1/f$ is a constraint corresponding to $\eta = 1$, while many other authors use the term $1/f$-noise to refer to IPL power spectra whatever be the index. To reduce the confusion produced by using this extended definition of $1/f$-noise, we shall throughout this chapter adopt the latter definition and refer to the case $\eta = 1$ as ideal $1/f$-noise to distinguish it from empirical nonsimple networks, which generate noise in the range $0.50 \leq \eta \leq 1.50$.

Many investigators have studied brain dynamics through the analysis of EEG records by either studying the time decay of the autocorrelation function, or, alternatively, by directly studying its Fourier transform, the power spectral density. Buiatti *et al.* [125] studied the autocorrelation function using a data-generated diffusion technique similar to the one that we develop herein, as proposed by Allegrini *et al.* [27], and evaluated the EEG spectrum under the open-eyes constraint. They found values of η slightly larger than 1, suggesting that the brain lives in a non-stationary non-ergodic condition, the reason for which will subsequently be made clear.

A partial explanation for the differences in the empirical IPL indices obtained for the brain is the fact that no single physiological condition or technique has been used in the data collection. However, from the experimental finding of the condition $\eta > 1$ in [125] one may draw tentative conclusions regarding the origin of $1/f$-noise, which, in turn, clarifies why the determination of the index is so elusive.

In this chapter we consider a sequence of events characterized by the sequence $\{\tau_n\}$ of time intervals between successive events, that is, if an event occurs at the sequence of times $\{t_n\}$ then $\tau_{n+1} = t_{n+1} - t_n$. These events are said to be *renewal* if all of them are mutually statistically independent random variables. Allegrini *et al.* [27] show that this property does not conflict with the existence of a very extended memory [155]. We assume that the interval between consecutive events is described by a waiting-time PDF $\psi(\tau)$, which becomes proportional to $1/\tau^\mu$ for $\tau \to \infty$ and focus our attention on $1 \leq \mu \leq 3$, which ensures that the second moment $\langle \tau^2 \rangle$ diverges. These events are *crucial*, since signals driven by such events show long-time correlations called *intermittent* in the physics literature. Note the condition $\mu < 2$ implies that the first moment diverges as well as the second, which entails an even more striking departure from the conditions of ordinary statistical physics. One property this condition yields is a memory extended to such an extent that the system does not possess an equilibrium, whereas for the condition $2 < \mu < 3$ an equilibrium does exist. However, in the latter case, it takes an infinitely long time to reach equilibrium from an initial out-of-equilibrium condition [586].

Renewal events may be used to generate a random walk (RW) signal $\xi(t)$, with a $1/f^\eta$ power spectral density, in which η depends on the RW rules adopted to generate the signal. The term *laminar region* has been used to refer to the time interval between consecutive events and if $+1$ or -1 is assigned to each laminar region, according to a coin toss, the spectra index is determined to be:

$$\eta = 3 - \mu. \tag{7.1}$$

It is interesting to note that ideal $1/f$-noise is obtained from a RW with $\mu = 2$.

The brain is one of the most nonsimple entities in existence and it is generally agreed that it is a source of $1/f$-noise [587]. This perspective coupled with POIE suggests why $1/f$-spectra are present in the works of the master artists [35] and in the frequency of words in language, since they are both products of the variability and creativity of the brain. In fact, some investigators believe in the "Mozart effect", that in addition to the fact that music generates $1/f$-spectra [272] that certain types of music correlate with higher levels of creativity and intelligence [448]. Allegrini *et al.* [20]

established that Zipf's law of natural human language is connected to scale-free semantic networks [53] entailing that the ideal Zipf's law, like ideal $1/f$-noise, corresponds to the boundary value $\mu = 2$. Allegrini *et al.* [27] observe that the small excursions of the IPL index above the boundary ensure the stability necessary for learnability and those below the boundary make it possible to explore a virtually infinite cognitive space thereby allowing language to evolve. They also comment that there exists a close connection between Zipf's law and phase-transition processes.

7.1 Theory of Crucial Events

In this section we rework the presentation in [415] and study nonsimple processes whose evolution in time rests on the occurrence of a large and random number of events. The mean time interval between two consecutive critical events is infinite, thereby violating the ergodic condition and activating at the same time a stochastic central limit theorem (SCLT) that supports the hypothesis that the MLF is a universal property of nature. The time evolution of these nonsimple systems is properly generated by means of fractional differential equations, thus leading to the interpretation of fractional trajectories as the average over many random trajectories each of which satisfies the SCLT and supports the hypothesis that the Mittag–Leffler PDF is universal.

7.1.1 Invisible crucial events

Crucial events often elude detection for a number of reasons. In the continuous case, detecting a discrete signal embedded in continuous noise may be difficult because the SNR is low and consequently below the sensitivity level of the detector. In the case of discrete events, it is possible that there are two kinds of random events being generated, only one of which is crucial and therefore of interest. If the frequency of occurrence of the unwanted events greatly exceeds that of the crucial events, the more frequent events could mask the sought after rarer events. This was expressed more formally by Pramukkul *et al.* [437], who considered a sensor, with a given sensitivity, intended to monitor events generated from a waiting-time PDF of IPL form. All the various ways the detector may fail to perceive an event are characterized by $(1 - P_S)$, where $P_S < 1$ is the probability of perceiving an event under a given set of conditions.

As a consequence of the various imperfections embodied in P_S, the time interval τ between successively detected (visible) crucial events is the sum of m elementary times. Thus, the probability of m undetected crucial events occurring in the time interval τ between detected events is:

$$P(m) = \mathrm{P}_S(1 - \mathrm{P}_S)^m. \tag{7.2}$$

More precisely, Eq. (7.2) is the probability that the first m events, after the initial preparation event, are not visible, whereas the $(m+1)$st crucial event is visible. When the probability of detection is very small ($\mathrm{P}_S \to 0$) the limit probability distribution is obtained:

$$P(m) = \lim_{\mathrm{P}_S \to 0} \mathrm{P}_S(1 - \mathrm{P}_S)^m = \mathrm{P}_S \exp[-m\mathrm{P}_S]. \tag{7.3}$$

Consequently, the PDF has the explicit form of a simple Poisson process and is renewal. After some straightforward, but tedious algebra, we obtain for the ratio of the standard deviation $\sigma(m) \equiv \sqrt{\langle m^2 \rangle - \langle m \rangle^2}$ to the average $\langle m \rangle$, using the exponential PDF:

$$\frac{\sigma(m)}{\langle m \rangle} \approx 1,$$

yielding the condition

$$\langle m \rangle = \frac{1}{\mathrm{P}_S} \to \infty. \tag{7.4}$$

This result looks similar to the condition $m \to \infty$ imposed in the traditional (Gauss) and generalized (Lévy) CLTs because the random variable m has very large fluctuations around its finite average value $\langle m \rangle$ in those arguments. However, here it is the average that is taken to infinity and not the variate itself. To understand what is entailed by this subtle change in argument, we continue to follow [437] and note that the waiting-time PDF $\psi_m(\tau)$ of the time intervals between visible events can be written:

$$\psi_V(\tau) = \sum_{m=0}^{\infty} \mathrm{P}_S(1 - \mathrm{P}_S)^m \psi_{m+1}(\tau). \tag{7.5}$$

In the limit $\mathrm{P}_S \to 0$, $\psi_V(\tau)$ is the PDF of observing an event at time interval τ after an earlier visible event was generated. Assuming that like the Poisson PDF, the waiting-time PDFs are also renewal we use the relation between the Laplace transforms: $\widehat{\psi}_m(u) = [\widehat{\psi}(u)]^m$, which when inserted into Eq. (7.5) yields, after some algebra:

$$\widehat{\psi}_V(u) = \frac{\widehat{\psi}(u)}{1 - \frac{(1-\mathrm{P}_S)}{\mathrm{P}_S}(\widehat{\psi}(u) - 1)}. \tag{7.6}$$

Note that $\widehat{\psi}(u) \equiv \widehat{\psi}_1(u)$ is the Laplace transform of the waiting-time PDF for a single crucial event.

The asymptotic form of $\widehat{\psi}(u)$ can be formally expressed in the $u \to 0$ limit in terms of the Laplace transform of an auxiliary function $\widehat{\Xi}(u)$:

$$\widehat{\psi}(u) = 1 - \left(\frac{u}{\lambda_0}\right)^{\alpha} + \widehat{\Xi}(u), \tag{7.7}$$

subject to the condition on the auxiliary function:

$$\lim_{u \to 0} \left(\frac{u}{\lambda_0}\right)^{-\alpha} \widehat{\Xi}(u) = 0. \tag{7.8}$$

It needs to be stressed that the auxiliary function must satisfy this constraint because we know that the asymptotic form of the waiting-time PDF is IPL. Therefore the auxiliary function must vanish more rapidly than u^{α} in the limit $u \to 0$ and we can write Eq. (7.6) as:

$$\widehat{\psi}_V(u) = \frac{1 - \left(\frac{u}{\lambda_0}\right)^{\alpha} + \widehat{\Xi}(u)}{1 - \frac{(1-P_S)}{P_S}\widehat{\Xi}(u) + \frac{(1-P_S)}{P_S}\left(\frac{u}{\lambda_0}\right)^{\alpha}}.$$

To fulfill the constraint on the auxiliary function the slowest contribution to $\widehat{\Xi}(u)$ must be:

$$\widehat{\Xi}(u)_{slowest} = k u^{\alpha+\epsilon}, \tag{7.9}$$

with $\epsilon > 0$. Rescaling the Laplace variable u with the detection probability:

$$u = u' P_S^{1/\alpha}, \tag{7.10}$$

transforms the Laplace transform of the waiting-time PDF for the visible crucial events into:

$$\widehat{\psi}_V(u') = \frac{1 - P_S\left(\frac{u'}{\lambda_0}\right)^{\alpha} + \widehat{\Xi}(u'P_S^{1/\alpha})}{1 - \frac{(1-P_S)}{P_S}\widehat{\Xi}(u'P_S^{1/\alpha}) + (1-P_S)\left(\frac{u'}{\lambda_0}\right)^{\alpha}}. \tag{7.11}$$

Now we have the situation that because the rescaled lowest-order contribution to the auxiliary function is proportional to $P_S^{1+\epsilon/\alpha}$ that the contribution from $\widehat{\Xi}(u)$ vanishes in Eq. (7.11) for $P_S \to 0$ and the rescaled expression reduces to:

$$\widehat{\psi}_V(u') = \frac{1}{1 + \left(\frac{u'}{\lambda_0}\right)^{\alpha}}. \tag{7.12}$$

We introduced the concept of survival probability earlier as being connected to the stochastic perspective of a nonsimple system generating events in time. The time interval between consecutive events, the laminar region, is assigned the values $+1$ or -1, according to the outcome of a coin toss. At the initial time $t = 0$ the system is prepared by selecting all the realizations in the ensemble with an event occurring at that time. Consequently, the probability of no crucial event being visible up to time t after observing the original crucial event is called the survival probability $\Psi_V(t)$, and is related to the waiting-time PDF for visible crucial events:

$$\psi_V(t) = -\frac{d\Psi_V(t)}{dt}. \tag{7.13}$$

The Laplace transform of Eq. (7.13) yields:

$$\widehat{\Psi}_V(u) = \frac{1}{u}(1 - \widehat{\psi}_V(u)), \tag{7.14}$$

which after inserting Eq. (7.12) into this expression reduces $\widehat{\Psi}_V(u)$ to the Laplace transform of the Mittag–Leffler function (MLF):

$$\widehat{\Psi}_V(u) = \frac{u^{\alpha-1}}{u^\alpha + \lambda_0^\alpha}. \tag{7.15}$$

Consequently, the inverse Laplace transform of this expression yields the MLF for the survival probability of visible crucial events:

$$\Psi_V(t) = \sum_{n=0}^{\infty} \frac{(-1)^n(\lambda_0 t)^{n\alpha}}{\Gamma(n\alpha + 1)} = E_\alpha(-(\lambda_0 t)^\alpha) \tag{7.16}$$

as obtained in Sec. 5.6.3. Note that this result is a consequence of rescaling $\widehat{\psi}_V(u)$ with P_S and is at the heart of what [438] identified as the SCLT. The SCLT establishes that the limit distribution for visible crucial events in a nonsimple process is a MLF. The important mathematical properties of the MLF in a probability context have been explored and discussed by Mainardi [350]. Now let us make a connection between the SCLT and subordination theory.

7.1.2   Fluctuating trajectories

A typical dynamic description of a nonsimple process is given by a fluctuating trajectory, one often generated by a RW process. Although a RW has discrete steps in space the associated PDF is often continuous in time, which suggests a way to use the SCLT to interpret the fractional trajectories n terms of averages over an ensemble of stochastic realizations. In order to achieve this relation, we use the survival probability of the time interval between successive visible crucial events given by Eq. (7.14). Inserting Eq. (7.6) into this equation, we obtain after some algebra:

$$\widehat{\Psi}_V(u) = \frac{1 - \widehat{\psi}(u)}{u} \frac{1}{1 - (1 - P_S)\widehat{\psi}(u)}, \tag{7.17}$$

then expanding the numerator and taking the inverse Laplace transform of the resulting expression yields:

$$\Psi_V(t) = \sum_{n=1}^{\infty} \int_0^t dt' \Psi(t - t')\psi_n(t')(1 - P_S)^n. \tag{7.18}$$

Equation (7.18) has the well-known structure from the Montroll–Weiss CTRW process [382]. An alternative form to this equation can be obtain by rearranging the terms in Eq. (7.17) a little differently to obtain:

$$\widehat{\Psi}_V(u) = \frac{1}{1 + P_S \widehat{\Phi}(u)} \tag{7.19}$$

where $\widehat{\Phi}(u)$ is the Laplace transform of the Montroll–Weiss memory kernel defined as:

$$\widehat{\Phi}(u) = \frac{u\widehat{\psi}(u)}{1 - \widehat{\psi}(u)}. \tag{7.20}$$

The inverse Laplace transform of the product form of the dynamic equation obtained from Eq. (7.19) yields:

$$\frac{d\Psi_V(t)}{dt} = -P_S \int_0^t dt' \Phi(t - t')\Psi_V(t'), \tag{7.21}$$

which is equivalent to Eq. (7.18) given that they originate with the same equation for the Laplace transform of the survival probability of the visible crucial events. The dynamic form of the GME for visible crucial events given by Eq. (7.21) reinforces the interpretation of $\Phi(t)$ as a memory kernel.

Another identification of the equivalence between two equations, this time between Eqs. (7.5) and (7.17), enables us to employ the scaling argument of the last subsection. Applying the P_S-scaling to Eq. (7.17) and following the argument of the previous subsection leads to the immediate conclusion:

$$\lim_{P_S \to 0} \widehat{\Psi}_V(u) = \frac{u'^{\alpha-1}}{u'^\alpha + \lambda_0^\alpha} \tag{7.22}$$

whose inverse Laplace transform is the MLF. As pointed out by Pramukkul *et al.* [438], this demonstrates why the MLF is ubiquitous in characterizing the survival statistics of the datasets of nonsimple phenomena, because it is the distribution of visible, which is to say measurable, crucial events.

7.2   Empirical Crucial Events

Normal sinus rhythm is one of those phrases known to be an over simplification of the actual rhythmic pattern made by the series of time intervals between successive human heartbeats [420], but is still used in medical schools. This phrase implies that the time interval between beats of the heart is steady and regular, with relatively little variability. However, this is not what is observed. In fact, the frequency spectrum of the time series depicting heart rate variability (HRV) is quite broad. The HRV frequency spectrum is understood to be the result of the firing rate of the heart's natural pacemaker (sinus node) being triggered by a signal from the involuntary (autonomic) portion of the nervous system. The spectrum is the result of a continuous tug-of-war between the parasympathetic and sympathetic components of the autonomic signal to change the firing rate of the sinus node. The first decreases and the second increases the sinus node's pacemaker cells firing rate and it is this balance that produces a healthy HRV spectrum. Ivanov *et al.* [265] established that HRV time series are multifractal and that the frequency spectrum scales.

Tuladhar *et al.* [533] show that HRV time series are closely tied to crucial events through their statistical scaling behavior. As they have shown, crucial events play a fundamental role in the transport of information between nonsimple networks. Relating these events to scaling is accomplished by converting the HRV time series into a diffusion process described by the PDF $P(q,t)$ of the diffusing displacement variable $Q(t)$, as described by Allegrini *et al.* [14]. The PDF generated by the resulting RW process has been proven to satisfy the scaling property:

$$P(q,t) = \frac{1}{t^\delta} F\left(\frac{q}{t^\delta}\right). \tag{7.23}$$

To reiterate, for classical diffusion the scaling index has the value $\delta = 0.5$ and the function $F(\cdot)$ has a Gaussian form. Anamolous diffusion is measured by how far the scaling index deviates from its classical value, which can be measured using the Wiener/Shannon form for information entropy as discussed in Appendix A.0.1.

Note that when the PDF under study departs from the classical Gaussian, the function $F(\cdot)$ can have a long slowly decaying tail of an IPL nature, and the second moment need not be finite. However even a diverging second moment is, for practical purposes finite, due to the unavoidable finite duration of any real world time series. Consequently, a second moment analysis could yield misleading results, determined by statistical inaccuracy, whereas the diffusion entropy analysis (DEA) method, based on Eq. (A.14), would yield correct scaling results.

7.2.1 Distinguishing health from disease

Note that the procedure we introduce here for finding crucial events is not sufficiently accurate to be restricted to detecting only renewal events. It is known that the events revealed by this analysis are a mixture of crucial events and ordinary Poisson events [29]. However, the presence of Poisson events does not prevent us from detecting the anomalous scaling generated by crucial events. The desired scaling is detected in the following way. Grigolini *et al.* [232] use the detected events to generate a diffusion process $Q(t)$ by means of the rule that the random walker jumps ahead when an event, either crucial or Poisson, occurs. The scaling generated by Poisson events has a power-law index $\delta = 0.5$, whereas the scaling IPL index of crucial events is given by the relation between scaling indices:

$$\delta = \frac{1}{\mu - 1}. \tag{7.24}$$

Note that the latter scaling dominates asymptotically in time, due to Eq. (7.24) resulting in $\delta > 0.5$ when the condition $2 < \mu < 3$ applies [232]. When $1 < \mu < 2$ crucial events yield the scaling $\delta = (\mu - 1)$, but the ERV time series studied here uses the subordination theory of Sec. 5.6.2, adopted to explain their nonsimplicity, it shows that we need to focus on $\mu > 2$.

The diffusion process used in the DEA is generated by adopting the procedure proposed by Grigolini *et al.* [232]. We use the empirical data to generate a fluctuation in $\xi(t)$ holding the value 1 when an event occurs, either crucial or Poisson, and a zero value when no event occurs. The diffusion variable $Q(t)$ is obtained from the following equation of motion using the random steps generated by the data:

$$\frac{dQ(t)}{dt} = \xi(t). \tag{7.25}$$

A moving window of size t then generates an ensemble of trajectories from which a histogram produces a PDF $P(q,t)$. This empirical PDF is then used to construct the Wiener/Shannon information entropy, which with empirical diffusion PDF of the form given by Eq. (7.23) yields:

$$S(t) = \delta \log_2 t + \text{constant}. \tag{7.26}$$

Allegrini *et al.* [14] introduced a mathematical model consisting of a superposition of *crucial events*, where the interval statistics are given by the hyperbolic PDF discussed in Sec. 5.2.4, with IPL index $\mu < 3$, and non-crucial or *pseudo-crucial events*, having IPL indices $\mu > 3$. Note that the Poisson PDF has a value $\mu = \infty$ according to Eq. (5.10) for which the waiting-time PDF is constant. Allegrini *et al.* [14] consider the composite time series:

$$\xi(t) = (1 - \varepsilon)\xi_{\mu>3}(t) + \varepsilon\xi_{\mu<3}(t), \tag{7.27}$$

which constitutes a surrogate for HRV time series. The parameter value $\varepsilon = 1$ would indicate that the surrogate time series given by the random walk contains only crucial events. Whereas a value in the interval $0 < \varepsilon < 1$ would imply that the surrogate time series is described entirely by a model containing a mix of crucial and pseudo-crucial events. Thus, in the first case the crucial events are visible, otherwise they are not, implying that ε is the probability that an event is crucial.

The method of empirically determining the parameter ε was given in [14] and was based on evaluating the two-time autocorrelation function:

$$C(i, j) \equiv \frac{\langle \tau_i \tau_j \rangle}{\langle \tau^2 \rangle} = \delta_{i,j}, \tag{7.28}$$

where $\delta_{i,j}$ denotes the Kronecker function; $= 1$ for $i = j$ and $= 0$ otherwise. Whereas, the parameter δ is determined using DEA, which for $\varepsilon = 1$ leads to the detection of the proper scaling given by Eq. (7.24) after an initial transient, during which the nonsimplicity of the process is not yet perceived. We assume the underlying process is stationary, so that when $\tau_i = \tau_j$ we have $C(0) = 1$, when the autocorrelation function is properly normalized, and immediately afterwards there is an abrupt jump down to $C(1) = \varepsilon^2$. This suggested categorizing HRV data from individual patients on the (δ, ε^2)-plane.

The criteria for distinguishing pathological from healthy HRV time series was established by Allegrini *et al.* [14], whose interpretation was extended some time later by Bohara *et al.* [102]. The later investigators noticed that the ideal healthy condition corresponds to $\delta = \varepsilon = 1$, meaning that crucial events should not host any Poisson event and should have $\mu = 2$. This is the border between the domain of perennial aging, $\mu < 2$ and the region where the rate of randomness production becomes constant in the long-time limit, $\mu > 2$ [337].

We use ECG records of the *MIT–BIH Normal Sinus Rhythm Database* and of the *BIDMS Congestive Heart Failure Database,* for healthy and *chf* patients [14, 102]. Bohara *et al.* [102] established that SOTC is the process driving the HRV differences and the detection of crucial events, that is, events with a waiting-time PDF having a diverging second moment. They obtained the results depicted in Fig. 7.1, where the healthy and pathological HRV time series separate into two distinct groups. The healthy patients move toward the diseased as their scaling becomes closer to that of ordinary diffusion $\delta = 0.50$, namely closer to the border separating the region of crucial events $\mu < 3$ and the Gauss basin of attraction $\mu > 3$.

HRV data: Tuladhar *et al.* [533] interpret the mathematical property of the autocorrelation function just given to be that of a crucial event rejuvenating the cardiovascular system. Consequently, the HRV time series evolves towards the occurrence of the next crucial event as if the occurrence of the earlier crucial event made the system brand new. The renewal property facilitates resolving the $1/f$-paradox [563]. This paradox

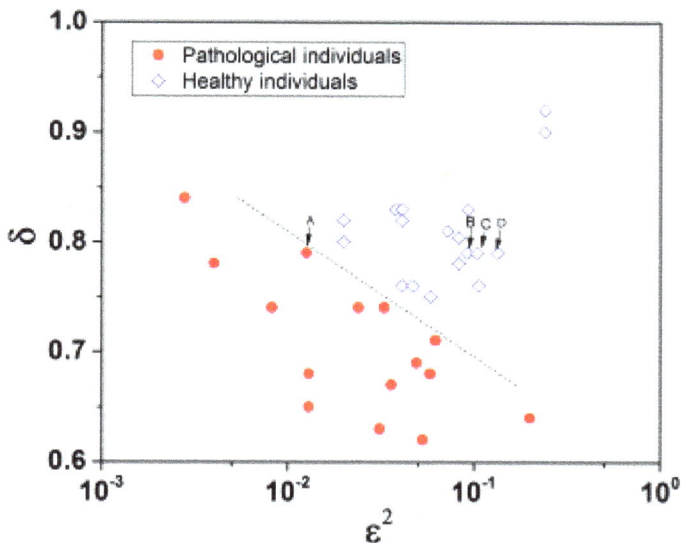

Figure 7.1: Parameters separate healthy from pathological HRV time series: δ is the scaling parameter from the entropy in DEA and ε^2 is the correlation rate. For these data the pathology is congestive heart failure. Adapted from [102] with permission.

has been brilliantly discussed by Niemann *et al.* [397] who noticed that for $\alpha > 1(\mu < 2)$ in Eq. (5.22) the power spectrum $S_p(\omega)$ is not integrable, thereby violating the basis of the Wiener–Khinchine theorem. This theorem is the mathematical foundation necessary for evaluating the spectrum, but has been side-stepped, as physicists often do when the mathematics do not correspond to the observed physical properties, by using a long, but finite, time series of length L to obtain [337]:

$$S_p(\omega) \propto \frac{L^{2-\mu}}{\omega^{3-\mu}}. \tag{7.29}$$

This result is based on the observation [186] that the rate $R(t)$ of production of crucial events, discussed earlier, are all prepared at $t = 0$, and is given by the IPL:

$$R(t) \propto \frac{1}{t^{2-\mu}}. \tag{7.30}$$

Actually, by identifying L in Eq. (7.29) with $1/\omega$, we obtain $S_p(\omega) \propto 1/\omega$, which is interpreted to mean that a time series of length L makes $\omega = 1/L$ the smallest observed frequency. This resolves the paradox with no physical need to impose truncating the frequency distribution, since the empirical truncation is a dynamical effect of the experimental observations.

Electroencephalograms (EEGs):   Research work on brain dynamics has disclosed the existence of crucial events, some of which is revisited here for the purpose of showing that the existence of crucial events is responsible for the $1/f$-variability in brain wave data. Perhaps of equal importance is establishing that although crucial events are generated by criticality they remain compatible with the wave-like nature of the brain processes. We follow Bohara *et al.* [103] in this section and show that although criticality generates large deviations from the regular wave-like behavior, the brain dynamics also host crucial events in regions of nearly coherent oscillations, thereby making many crucial events virtually invisible. Furthermore, the anomalous scaling generated by the crucial events can be established with high accuracy by means of direct analysis of raw data, suggested by a theoretical perspective not requiring the crucial events to yield a visible physical effect. Bohara *et al.* [103] obtain three main results, which we sketch out in this section: (a) confirming the critical role of crucial events in brain activity; (b) demonstrating the theoretical tools necessary to understand the joint action of crucial events and periodicity; (c) shedding light on the nature of the central role of self-organization and thereby contributing to the understanding of cognition.

Periodicity and crucial events: Understanding brain dynamics has forced researchers to go beyond the conventional forms of non-equilibrium statistical physics [409] and is expected to reshape the nascent field of non-simple networks as well [410]. These ideas have been applied to the processing of EEG time series by [103] to better understand the dynamics of the brain by exploiting the notion of homeodynamics [622]. This approach shows how the analysis of biodynamic processes can be done taking into account that although they have a substantial erratic component they are typically driven by rhythms and waves. A separate line of inquiry has recently been developed that addresses the connection between brain dynamics and the phenomenon of criticality [4, 111, 523]. Brain criticality is widely discussed in the neurophysiology literature, for a recent review, see Cocchi *et al.* [142]. We emphasize that criticality is a term invented by physicists and is discussed in Chapter 1 to describe the emergence of collective macrobehavior from nonsimple microdynamics. It is widely thought that an analogous condition is fulfilled by brain dynamics with the consequence of strongly correlating the functionality of different physical regions of the brain. This connection between brain dynamics and phase transition processes at criticality is one of the things that led our research group [103] to introduce the concept of crucial events.

The theoretical connection between the brain waves recorded by EEG time series and crucial events was established in [103] by adopting the SOTC model [342] of individual units having their own periodicity. For example, the SOTC model is applied to a group of individuals spontaneously reaching consensus using the decision-making model (DMM) [342, 344, 345]. However, the DMM is a computationally demanding technique and has not yet been converted into an analytical approach for bridging the gap between waves and crucial events. The research work done in the past on brain dynamics led Allegrini *et al.* [28] to conclude that the crucial events are characterized by values of μ very close to 2, according to the prescription for the EEG power spectrum with an IPL index:

$$\alpha = 3 - \mu. \tag{7.31}$$

The bridge between crucial events and periodicity was theoretically established by Ascolani *et al.* [39] using an extension of the well known CTRW [382]. Their approach was applied to EEG data by Bohara *et al.* [103] using subordination to a coherent process with frequency Ω as a mathematically simple way of simulating a genuine process of self-organization. Assume we have a clock, the hands of which move clockwise with frequency Ω from noon to noon, making T_Ω clicks with the time interval Δt between one click and the next. Thus, the frequency of the timepiece can be expressed as:

$$\Omega = \frac{2\pi}{T_\Omega \Delta t}. \tag{7.32}$$

The crucial events are then imbedded into this regular motion by assuming that the time interval between consecutive clicks is derived from a waiting-time PDF $\psi(t)$ with the temporal nonsimplicity of a hyperbolic PDF.

This procedure of infusing the perfectly coherent original clock with nonsimple randomness was used to establish a bridge between waves and crucial events [103]. This has the effect of turning the frequency Ω into an effective frequency Ω_{eff}, thereby modeling a process of self-organization of interacting oscillators, each of which is characterized by its own frequency, into a collective homeodynamic process. Lambert *et al.* [307] worked out the theory establishing the effective frequency to be:

$$\Omega_{eff} = \frac{\Omega(\mu_s - 2)}{T_S}, \tag{7.33}$$

which is valid for $\mu_s > 2$. This theoretical prediction suggests, in agreement with Fig. 7.2, that the frequency peak is evident for $\mu_s > 2$ and that, in addition, it also depends on the parameter T_S of the hyperbolic waiting-time PDF.

Figure 7.2 illustrates the spectra generated by surrogate sequences obtained using the subordination method with $\Delta t = 1$. We keep the frequency Ω fixed, change the IPL index μ_s in the waiting-time PDF and note that the resulting spectrum consists of three parts: a peak corresponding to the effective frequency Ω_{eff} that shifts to the left with decreasing μ_s and vanishes for $\mu_s < 2$; to the left of the Ω_{eff} peak the slope α of the IPL part of the spectrum is determined to be given by Eq. (7.31) with $\mu = \mu_s$. The spectrum becomes flat when $\mu_s = 3$ and remains flat for higher values of μ_s, as clearly shown in Fig. 7.2. Moreover, to the right of the peak is an IPL spectral slope of $\alpha = 2$, corresponding to that obtained for Brownian motion.

Note that due to the average of many realizations, which is not possible with real EEG time series, the region of low frequency is regular and is not affected by the fluctuations that would appear when evaluating the spectrum with only one time series. For this reason, the adoption of surrogate time series makes it possible for us to prove that, as expected, subordination is compatible with the emergence of $1/f$-variability in the ideal case $\mu_s = 2$.

Empirical spectra: Let us now consider spectra generated by real EEG time series, fluctuations and all, as shown in Fig. 7.3. We see that the region

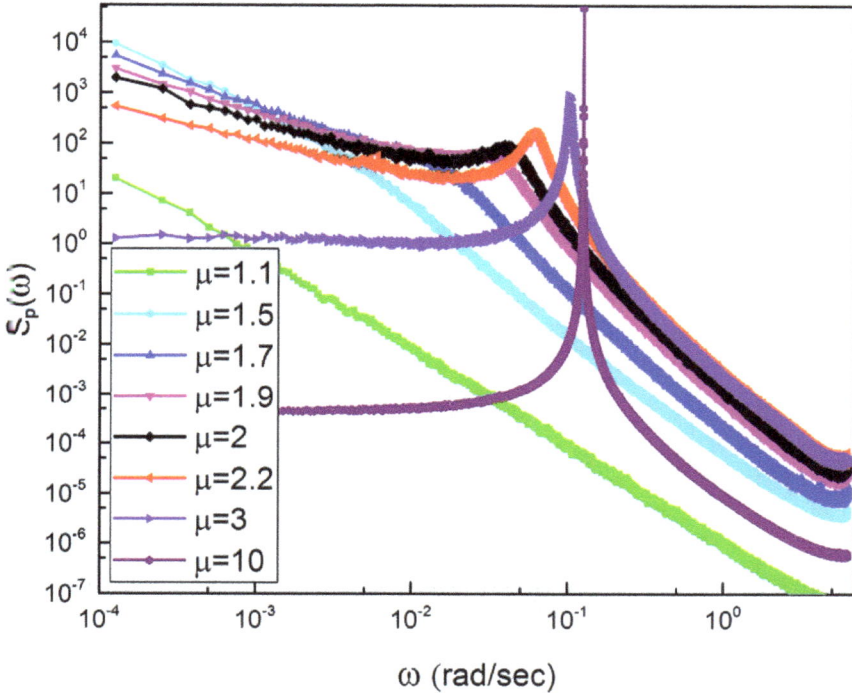

Figure 7.2: Spectra obtained averaging over 300 trajectories with numerical parameters $T_S = 0.5$ and the regular oscillation before subordination has the frequency $\Omega = 0.77$. The inlay denotes the calculations for IPL indices spanning a range $1.1 \leq \mu_s \leq 10$. From [103] with permission.

of low frequencies is very erratic, due to the fact that, as just mentioned, the use of only one time series makes it numerically impossible to generate a smooth curve. In addition there exists an indication of a frequency bump, generated by periodicity, and for frequencies larger than that located by this bump the spectral slope $\alpha = 2$ is rediscovered. This empirical spectrum depends on a wide swath of frequencies, not just those in the vicinity of the bump.

In constructing Fig. 7.3, Bohara *et al.* [103] used the subordination prescription described earlier in this section, with Δt given by an inverse sampling frequency of real data ($\Delta t = \frac{1}{2048}$ sec) to generate the surrogate spectra. They assigned to the monochromatic frequency $\Omega = 62$ hz with six different values of the parameters T_S, mimicking the dominant frequencies

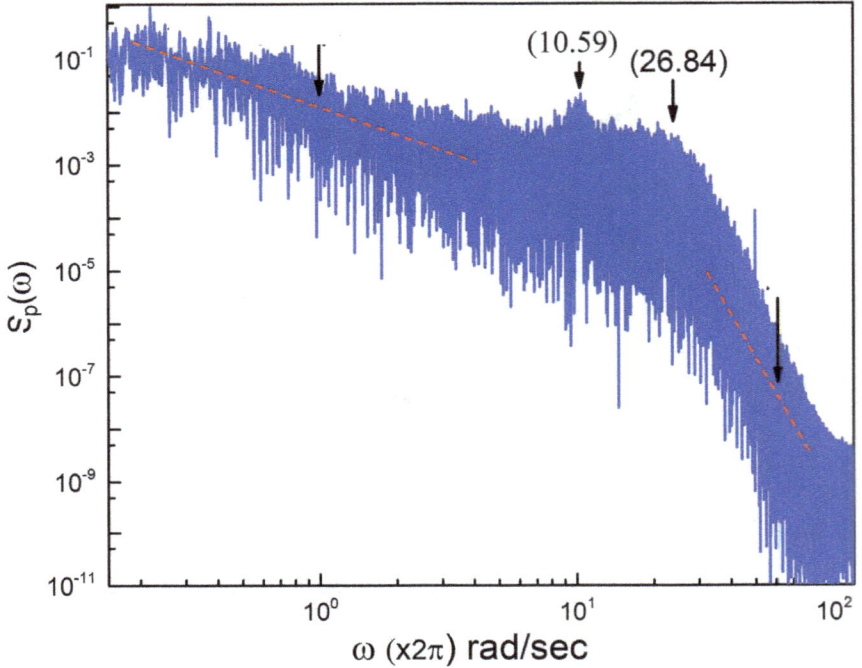

Figure 7.3: Spectrum obtained from raw EEG time series data. From [103] with permission.

of the six spectral components that they superposed to obtain the empirical spectrum of Fig. 7.3. In fact all these unshown spectra share the property $\alpha = 3 - \mu_s$ in the low frequency region, the property $\alpha = 2$ in the high frequency regions, and an intermediate region, where the change of slope occurs, which is significantly broader than in the monochromatic case given in the numerical example depicted in Fig. 7.2.

7.2.2   Method of stripes

The MoS was originally developed to detect crucial events hidden among clusters of Poisson events as previously described. Although the crucial events may be invisible within the time series, the time intervals between the crucial events have an IPL PDF and therefore scale very differently from that of the time intervals between the neighboring Poisson events. The strategy to detect these unseen events was to devise a processing technique which highlighted this difference in scaling and which was adopted to detect the scaling of crucial events hosted by heartbeats [20]. The reason the technique was not applied to EEG time series data earlier remains a mystery, perhaps it was because of the lack of a proper theoretical understanding of the connection between crucial events and periodicity. The same method was more recently applied by Bohara *et al.* [102] to establish a connection between the occurrence of crucial events and multifractality.

In this chapter we have used an intuitive illustration of the process of self-organization, based on subordination, which affords theoretical support for the adoption of the method of stripes. The central idea is that previous methods detect only a small fraction of crucial events, whereas empirical EEG time series and subordination theory with them, host a much larger number of crucial events, even if they do remain invisible.

Figure 7.4 shows how the method of stripes is constructed. As is well known, an EEG measures event related potentials (ERPs) which, in turn,

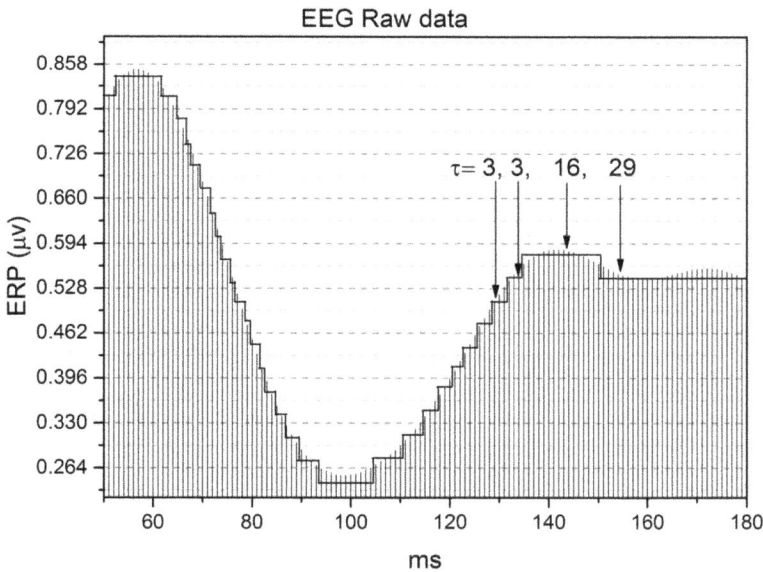

Figure 7.4: The amplitude in μv (microvolts) of the event related potential (ERP) of a EEG time series is plotted versus time in ms (milliseconds). The vertical axis is segmented into a sequence of strides, each of width $\Delta E = 1/30 \ \mu v$ and the method is described in the text. From [102] with permission.

measures the rate of neuron firings. The method divides the vertical axis
into many stripes of size ΔE, here assumed to have the empirical value
$\Delta E = 1/30\ \mu v$, and we record the times at which the unprocessed signal
crosses the line separating adjacent stripes. The level of the stripe is determined by the number of neuron firings at a given time, and we record for
how long that firing rate remains constant.

An event is the change from one firing rate to another and is signified by
crossing from one stripe to another. Of course this event is not necessarily
a crucial one. Consequently, the time interval between consecutive events
cannot be used directly to define the important IPL index μ. As pointed
out earlier, the non-crucial events generates a diffusion process with scaling
$\delta = 0.5$ and the crucial events, on the contrary, generate the scaling index
of Eq. (7.24) that for $\mu > 2$ is larger than $1/2$, thereby making it possible
for DEA to establish the correct scaling of Eq. (7.24) at long times.

The result of applying the method of stripes to an EEG time series is
depicted in Fig. 3.12, which shows that the detected scaling of EEG data
for a healthy individual has $\delta = 0.81$, which using Eq. (7.24) yields the
IPL index $\mu = 2.12$. Bohara *et al.* [103] discuss the details of the data
processing and demonstrate that the complex scaling of Eq. (7.24) appears
in the intermediate time regime. The slope of the straight line fitted to the
processed data covers over three decades of scale, which is a clear indication
that the method of stripes is robust.

Heart-Brain Coupling: Tuladhar *et al.* [533] note that stress, a function of the brain, disrupts normal HRV time interval variability, a function
of the heart, and meditation was historically developed to circumvent its
debilitating influence. The origin of meditation is unclear, but it has been
part of civilized culture for thousands of years. Its earliest records, circa
1500 BCE, can be traced back to Vedantism, a Hindu tradition in India.
From a physiological perspective meditation is a coupling of the brain's
functionality with that of the heart and has been explored and refined for
millennia. It is only recently that we have begun to apply the methods
of science to the study of this brain-heart coupling. The present section
reviews the development of a measure of reduction in the level of stress
provided by meditation. A statistical analysis of HRV time series before,
during and after meditation, is used to quantify precisely how much stress
is alleviated through meditation by means of control of the heart-brain
coupling.

In this section we review a statistical analysis of data based on an approach developed earlier [14, 102] establishing that the dynamics of HRV
time series host both Hamiltonian (deterministic) and crucial (stochastic)
memory. It is also proven that meditation affects both kinds of memory,

transforming the Hamiltonian memory into a coherent process [422]. Surprisingly, meditation also affects crucial memory, as well. In fact, analysis [103] reveals that meditation shifts the index μ of the IPL waiting-time PDF, shifting it from values very close to the perfect $1/f$ condition, $\mu = 2$, to those in the vicinity of the border with the Gaussian basin of attraction, $\mu = 3$. This is a strong perturbation. The meditation-induced coherence establishes that Chi, as well as, Kundalini Yoga meditation reduce stress, although the advanced practice of the latter yields higher levels of reduction.

Two kinds of memory: One form of memory, which includes Laplace determinism, is modeled by the rate equation

$$\frac{d\xi}{dt} = i\Omega\xi, \tag{7.34}$$

and is referred to as Hamiltonian Memory for real Ω. Note the $\xi(t)$ is the notation used for the stochastic driver that is applied to the rate equation earlier in this section. The stochastic nature of this time series is included in the present case by subordinating the harmonic term to a stochastic dynamic process as done in Sec. 7.2.1, resulting in an integrable correlation function. By way of contrast, we refer to the non-integrable correlation function generated by the crucial event fluctuations as crucial memory, which the cross-correlation function of Eq. (7.28) plays an important role in discriminating from the integrable case. In fact, in the case of crucial event infinite memory $C(1) = 0$, whereas in the case of Hamiltonian memory $C(1) \neq 0$.

As pointed out in [533] research has shown that in the case of HRV time series these two forms of memory coexist [14, 102]. It is true that this coexistence can be viewed as being paradoxical, just as that of wave-particle dynamics in quantum mechanics. But its existence is not as surprising in the life sciences as its analog was in the physical sciences when it was discovered over a century ago. In brain dynamics the notion of nested frequencies [244] was adopted to connect the EEG's wave-like nature to the action of crucial events that can be uncovered by the proper method of analysis [32]. We exploit the observation of Peng *et al.* [422] that meditation has the effect of generating coherent behavior in HRV time series. The link between coherence and criticality is a subject of great interest [528] that we address herein using our subordination perspective. This link has been studied by many authors moving from the adoption of positive emotions [115] to different forms of meditation [422, 423].

Harmonic subordination: Recall the MDEA processing technique we introduced to bridge the gap between the harmonic component of EEG signals and crucial events. The computational spectra obtained using harmonic subordination is depicted in Fig. 7.2 and has three major components: (1) a low-frequency IPL structure with an index $\alpha = 3 - \mu$; (2) a high-frequency IPL structure with an index $\alpha = 2$; (3) a mid-frequency broad peak centered on the frequency of the coherent process Ω_{eff} [103]. Aside from some additional noise the EEG time series were found to display the same characteristics. We now apply the same processing technique to individuals engaged in meditation.

We illustrate the average power spectrum $S_p(\omega)$ for an HRV time series $\xi(t)$ of time length L [533]:

$$S_p(\omega) = \frac{1}{L} \left\langle \left| \int_0^L dt e^{i\omega t} \xi(t) \right|^2 \right\rangle. \tag{7.35}$$

To make our procedure clear in Fig. 7.5 we depict samples of the HRV time series analyzed herein, referring to individuals practicing Chi Yoga and others practicing Kundalini Yoga meditation. The data is obtained from Goldberger *et al.* [220], previously analyzed using very different techniques by Peng *et al.* [422]. The Chi meditators are novices and the Kundalini Yoga meditators are advanced practitioners. We see that the main qualitative difference between Kundalini Yoga and Chi meditation is that at the onset of mediation the rate of heartbeats in the former case significantly increases, whereas in the latter it does not. However, during both types of mediation coherent-like oscillations appear.

As discussed using EEG time series, the search for crucial events hidden within real time series is carried out using the method of strips [14, 102]. Here we analyze sequences of heartbeat data, recording the beat numbers in the HRV time series along the horizontal axis and the time interval between successive beats, $T_B(i)$, along the vertical axis. The vertical axis is divided into strips of size $s = \Delta T_B \approx 30$ msec [220] and the crossings of the trajectory from one strip to the nearest neighbor strip are recorded as events, but not necessarily crucial events. It is expected that the time interval between consecutive crucial events is filled with *pseudoevents*, as discussed earlier for surrogate data. Consequently, both pseudoevents or crucial events are assumed to generate a RW in which the walker jumps ahead by a fixed quantity equal to 1, for each event, thus generating a diffusion-like process. In spite of the fact that crucial events are rare, the long-time limit of this diffusion process is dominated by their scaling with index δ [232]. Note that according to the scaling prediction given in Sec. 7.2.1, the IPL index

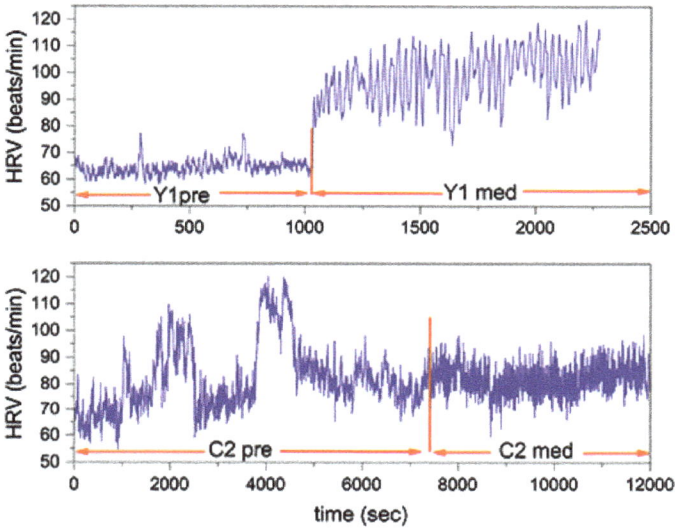

Figure 7.5: HRV time series of Kundalini Yoga meditator, at the top, and the Chi Yoga meditator, at the bottom. The vertical red lines denote the time at which the two meditations start. From [533] with permission.

is given by:

$$\mu = 1 + \frac{1}{\delta}. \tag{7.36}$$

The scaling index δ is evaluated numerically using DEA [232]. The pseudo-events imbedded within the time intervals between crucial events are characterized by the deterministic memory of Laplace and Hamilton, a form of memory different from that produced by the waiting-time of crucial events [14].

The deterministic memory violates the renewal condition, which in terms of the autocorrelation function is expressed as $C(1) = 0$. The earlier analysis of memory [14] assumed that the ideal condition of healthy heartbeats, in terms of crucial events dressed with the additional memory of pseudoevents, yields for the autocorrelation function $C(1) \approx 1$. In addition to the crucial events the HRV time series host totally random Poisson events, with probability $1 - \varepsilon$, as well. Consequently, the determination of the size of ε^2 is accomplished by evaluating $C(1)$ using Eq. (7.28) in the case of real real HRV time series and setting the condition:

$$C(1) = \varepsilon^2. \tag{7.37}$$

Figures 7.6 and 7.7 depict the results obtained through the application of the analysis technique of [102] to the *Physionet* data of [422]. The effects of different forms of meditation are strikingly different. Compare the results shown in Fig. 7.6, that are due to Chi Yoga meditation, to those illustrated in Fig. 7.7, Kundalini Yoga meditation. In both cases there is a clear tendency for meditation to increase the IPL index μ. In the case of Chi meditation only one subject does not change the IPL index μ and that is C7. This is a clear anomaly, since for all the other subjects, δ moves from above to below the dashed line. The change in μ goes from the minimal change, for C1 and C5, having a change of $\Delta\mu = 0.07$ to the maximal change, for C2, of $\Delta\mu = 0.26$.

Yoga meditation, in addition to decreasing δ and consequently increasing μ, has the remarkable effect of increasing ε. Allegrini *et al.* [14] hypothesized that $1 - \varepsilon$ is an indicator of stress that eventually results in heart failure. If this conjecture is valid, we could infer that yoga is demonstrably an efficient method to reduce stress. Furthermore, it would appear the Chi meditation is less efficient at reducing stress than Kundalini Yoga, since for some individuals, C3, C4 and C8, the quantity ε^2 is decreasing rather than increasing. Of course, as of now, this is merely interesting speculation, but these results do suggest a new avenue of research.

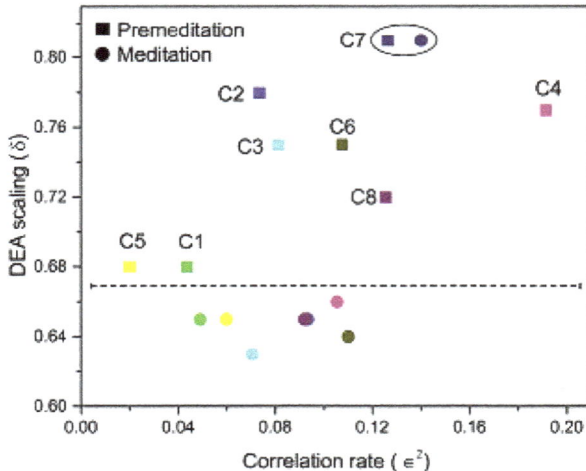

Figure 7.6: DEA scaling δ, IPL index μ and ε^2 of the HRV time series of eight different participants before and during Chi Yoga meditation. From [533] with permission.

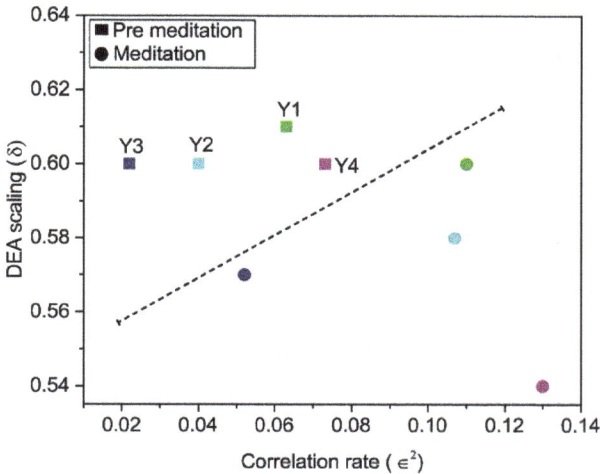

Figure 7.7: DEA scaling δ, IPL index μ and ε^2 of the HRV time series of four different participants before and during Kundalini Yoga meditation. From [533] with permission.

The case of Kundalini Yoga meditation depicted in Fig. 7.7 is quite different, which was to be expected since these were seasoned veteran yoga meditators and those under going Chi Yoga meditation were not. The minimal change is for Y1, having $\Delta\mu = -0.04$. The maximal change is for Y4, with $\Delta\mu = -0.19$.

It is well known that meditation generates "exaggerated heart rate oscillations" [422], and in addition, we also find that meditation hosts crucial events. Bohara *et al.* [102] pointed out these crucial events usually arise surrounded by an entourage of uncorrelated and irrelevant events, which prompted us to introduce the notion of dressing. A dressed crucial event is embedded in a cloud of Poisson events, whereas the time interval between bare crucial events is empty. Meditation triggers a form of dressing manifest by the exaggerated heart rate oscillations. The adoption of the strip technique of analysis [102] made it possible to quantify the level of dressing by means of the measure ε^2. This measure enabled us to interpret $1 - \varepsilon$ as the percent of Poisson events affecting the HRV time series. Allegrini *et al.* [14] conjectured that the Poisson events are generated by stress and that an excessive amount of such non-crucial events as measured by ε^2, that is, excessive amounts of stress, may be the cause of heart failure. The results depicted in Fig. 7.7 show that the Kundalini Yoga meditation yields a significant reduction in the effect of stress, using this interpretation.

Consequently, Tuladhar *et al.* [533] present a technique for analyzing HRV time series that may be used to directly quantify the benefits of meditation, which are currently evaluated indirectly, through the observation of psychological health [208].

Kundalini Yoga is known to be an efficient way to treat certain psychiatric disorders [490], but the therapists using this technique are looking for further improvements [288]. We hope that the adoption of the analysis techniques discussed here may assist progress in this important field of research.

7.2.3   Sun–Earth information transfer

We have to point out that, although the condition $\mu < 2$ is challenging due to the breakdown of the ergodic condition, as we have discussed, the condition $\mu > 2$ can also generate interesting effects. For this we turn to a timely scientific topic, that has unfortunately been drawn into the very visible political arena before the science was unambiguously settled. The topic is, of course, global warming, which has morphed into climate change. The book *Disrupted Networks: From Physics to Climate Change* [592] presented the arguments as to why the nonsimple physics problems associated with determining the behavior of global temperature changes have not been settled, in spite of the hype surrounding the politics of the phenomena. Here we adapt our network information exchange argument to the Sun–Earth system to determine if we obtain anything worth commenting on.

Here we treat the Sun as the perturbing system P, the Earth as the responding system S, and solar radiation as the interaction. The causes of the changes in the average global temperature, the increase in the average global temperature near the Earth's surface of approximately $0.80 \pm 0.10°C$, since 1900, are not as apparent as some recent scientific publications and the popular media would have us believe. We show that the changes in the Earth's average surface temperature are directly linked to two distinctly different aspects of the Sun's dynamics: the short-term statistical fluctuations in the Sun's irradiance and the longer-term solar cycles. This argument for the direct linking of the dynamics of the Sun to the response of the Earth's climate is based on the statistical arguments presented herein and elsewhere [592], augmenting the interpretation of the causes of global warming presented in the Intergovernmental Panel on Climate Change (IPCC) report in 2007 [263].

The 'majority opinion' regarding the causes of global warming is based on the analysis done using large-scale computer codes that incorporate all identified Newtonian-based physical/chemical mechanisms into global circulation models (GCM) in an attempt to recreate and understand the variability in the Earth's average temperature over time. The IPCC report concludes that the contribution of solar variability to global warming is negligible, to a certainty of 95%. It is reported that the 'scientific majority' believes the average warming observed since the beginning of the industrial era is due to the increase in anthropogenic greenhouse gas concentrations in the atmosphere.

The Earth's atmosphere, landmasses and oceans, absorb and redistribute the total solar irradiance (TSI) by means of coupled nonlinear hydrothermal, geochemical and radiative dynamic processes, producing the average global temperature. Versions of these physical mechanisms are included in the GCMs, but what is not addressed in these simulations, over and above the global temperature trend of the last thirty years, are the statistics of the average global temperature anomalies. The statistical variability in the Earth's average temperature is interpreted as noise and as these temperature fluctuations are thought to contain no useful information, they are consequently smoothed to emphasize the presumably more important long-time changes in the average global temperature. Moreover, according to the CLT the statistics of the fluctuations in such large-dimensional networks ought to be Gaussian and the fact that they are not remains unexplained. This prompted the study of temperature fluctuations as a problem in non-equilibrium statistical physics, wherein statistical fluctuations often provide useful information about the transport properties of complex phenomena, for example, through the fluctuation–dissipation theorem, as previously demonstrated.

We use the response of the Earth's air temperature fluctuations to the variability of solar irradiation as an interesting contemporary example. Scafetta *et al.* [473, 475] have found the surprising property that the air temperature anomalies, analyzed with the DEA method [232, 472], reveal the complexity index for the Earth's temperature anomalies $\mu_E > 2$, which, in turn, suggests a possible connection with the statistics of hard-X-ray solar flares studied by Grigolini *et al.* [233]. These investigators found that the statistics of solar flares generates an IPL index $\mu_P = 2.14$. The application of the same DEA techniques to the average global temperature fluctuations, as discussed by Scafetta and West [473], yields $\mu_E = 2.18$ (Northern Hemisphere), $\mu_E = 2.10$ (Southern Hemisphere), $\mu_E = 2.19$ (over land) and $\mu_E = 2.07$ (over ocean). Note that the Earth plays the role of the system S so, we replace the label S with E. Furthermore, the Sun plays the role of the perturbation. Thus, in this case the symbol P denotes the Sun and is denoted by the intensity suffix I.

How can we explain the connection between the activity on the Sun and the temperature anomalies on the Earth? In this case, the system E generates the time sequence of air temperature fluctuations $\{\Delta T\}$ and the system P generates a sequence of flares, described by the time series $\{t_i\}$. We consider the flares as indicators of, or surrogates for, the Sun's dynamic nonsimplicity. Although Scafetta and West [473] did not use the aging technique to ensure the renewal character of the solar flares, they made a careful analysis based on the comparison between the diffusion and second moment scaling that afforded a compelling indication that solar flares are in fact renewal. In other words, they established that the conversion of the time of occurrence of solar flares into a diffusion process is generated by a Lévy walk, see Sec. 5.5.2, and this is only possible if the solar flares are renewal.

Thus, the sequence of the times of occurrence of solar flares is a sequence of renewal events, characterized by the parameter μ_I fulfilling the condition $2 < \mu_I < 3$. The condition $\mu_I > 2$ makes the solar flare process compatible with equilibrium and ergodicity. However, although the mean interval between consecutive flares exists and is finite, the mean-square value is divergent, thereby implying fluctuations from the mean value of large intensity, which is responsible for the scaling:

$$\delta = \frac{1}{\mu_I - 1}, \qquad (7.38)$$

larger than the ordinary diffusive scaling of $\delta = 1/2$. The DEA method is characterized by the property of affordable correct scaling, therefore going beyond the evaluation of the distribution variance that, in principle, should diverge, as a result of Lévy statistics [15].

Let us now discuss the second time series, namely, the sequence of air temperature fluctuations $\{\Delta T\}$. This is a virtually continuous signal, insofar as it is represented by a time scale much larger than the time scale of the former sequence. We [473, 475] derived a surrogate sequence, qualitatively similar to the real sequences of temperature fluctuations, by recording the number of solar flares contained in the unit time scale of temperature fluctuations. As a result of the fractal and self-similar properties of the time series $\{t_i\}$, the resulting fluctuation has the same clustering structure as the original time series, and the interval between consecutive clusters is characterized by the same IPL index μ_I, as the original time series. Due to the coarse graining adopted, the variance of the surrogate fluctuation ΔT is finite, and, as a consequence, the DEA technique is expected to detect the same scaling as that produced by the original time series, namely, the scaling parameter Eq. (7.38).

This argument is attractive, but its weakness is that an energy balance is implied that would suggest an extremely detailed and probably very nonsimple model. Let us see, therefore, how adopting the concept of POIE helps us to bypass these complications. We refer to the air temperature

fluctuations in a hypothetical situation in which there are no interactions
with the solar flares. Also in this hypothetical case, it is plausible to imagine
that a surrogate time series exists corresponding to an unknown nonsimplicity parameter $\mu_E^{(0)}$, see West and Grigolini [586] for a complete discussion.

The problem that we have to address now is that of establishing a
connection between the Sun's irradiance $\Delta I(t)$ and the Earth's temperature
fluctuation $\Delta T(t)$. Let us imagine that both are satisfactorily described
by surrogate time series, such as the one derived in [473] from the solar
flare theory [233]. In the absence of coupling they are characterized by the
nonsimplicity parameters $\mu_E^{(0)}$ (air temperature fluctuations) and $\mu_I = 2.14$.
We do not use the superscript (0) for sunlight, insofar as we assume that
the Sun drives the Earth's temperature with no feedback.

In the absence of linking, the time series $\Delta I(t)$ and $\Delta T(t)$ are uncorrelated. Let us now discuss how the POIE establishes a correlation and
why the DEA approach enables us to detect it, regardless of its weakness.
We imagine an ideal experiment and explain why the application of DEA
is virtually equivalent to realizing this ideal experiment in practice. Imagine that it is possible, in principle, to record infinitely many time series of
the former kind, corresponding to the same time series of the latter kind.
According to the LRT of Eq. (6.1) we have

$$\langle \Delta T(t) \rangle = \varepsilon \int_0^t dt' \chi(t, t') \Delta I(t'). \tag{7.39}$$

Moreover, imagine that this excitation experiment can be done infinitely
many times so as to turn Eq. (7.39) into

$$\langle \Delta T(t) \rangle = \varepsilon \int_0^t dt' \chi(t, t') \langle \Delta I(t') \rangle. \tag{7.40}$$

This experiment is impossible in practice, but the application of the
DEA technique corresponds to its realization. In fact, this technique rests
on the use of a mobile window of length l that explores all the possible
positions of the time series. The left end of the window corresponds to the
beginning of the observation process, the time origin $t = 0$ of Eq. (7.40).
The right end of the window corresponds to the evaluation of the response
of E after a time l. Among all possible conditions concerning $\Delta I(t)$ we
have to select only those fulfilling the condition $\Delta I(t) > 0$. The negative
contributions would cancel with the positive, thereby preventing us from
estimating the transfer of complexity from the Sun to the Earth's atmosphere. Under these conditions, we obtain the response function

$$\chi(t, t') \propto \frac{1}{(t - t')^{\mu_E^{(0)} - 1}}.$$

In fact, the linear response function is proportional to the time derivative of the equilibrium autocorrelation function, which is proportional to IPL with an index $\mu_E^{(0)} - 2$, as required by the condition of infinite age. We assume that $\langle \Delta I(t') \rangle$ is prepared at $t = 0$ to be $\langle \Delta I(t') \rangle \propto 1/t^{\mu_I}$, and thereby obtain the condition

$$\mu_E^{(0)} > \mu_I.$$

Note that the solar IPL index is empirically determined to be $\mu_I \approx 2.10$ and the unknown average global temperature IPL index is supposed to fit the ergodic condition $\mu_E^{(0)} > 2$. Thus, the Earth assumes the anomalous solar flare statistics, independently of whatever ergodic value $\mu_E^{(0)} > 2$ it had previously. This explains why the Earth's air temperature fluctuations inherit the solar flare statistics [473] as discussed in the DEA context [586].

To recapitulate, we argued that the Earth's air temperature anomalies inherit the Sun's complexity by means of the CME, which we now understand using POIE, as well. The POIE regulates the transfer of information between nonsimple systems and was realized here through a linking of extremely small intensity, which is, in fact, the condition on which the key prescription of Eq. (7.40) rests. As mentioned, this phenomenon is a general property of information transmission from P to S, thanks to the role of crucial events: the influence of music on the human brain [94] is explained in a similar manner.

7.3   Crucial Events Simplified Review

As quoted at the beginning of this chapter, Einstein once said something to the effect that if you cannot explain a thing simply then you do not fully understand it. So in this section is where we determine whether we understand crucial events, which is to say whether we can make the idea comprehensible to the mildly interested investigator without resorting to equations, jargon or citations. However, in various drafts of this effort, which are not reproduced here, we determined that most equations could be dispensed with along with citations, but a minimum of jargon was necessary in order to avoid writing a companion volume containing complete discussions of the jargon being replaced. Consequently, we will simplify the review of critical events keeping jargon to a minimum.

Information transport within a nonsimple dynamic system is of singular importance and has frequently been taken to be a measure of a system's nonsimplicity. In fact, nonsimplicity as we have used it herein emerges in systems containing many elements and having nonlinear interactions that generate collective modes that dominate the system's dynamics. These

collective observable modes are known as phase transitions and are the basis for understanding thermodynamic processes in physical systems. The
nonsimple dynamics of observables of this kind become independent of the
underlying microscopic equations. Consequently, depending on some macroscopic parameter, such as temperature, a critical point exists marking the
border between macroscopic determinism and macroscopic indeterminacy.

Said differently, while the macroscopic control parameter is below the
critical point that a single solution exists, whereas above the critical point
multiple solutions are available. Consequently, as the system-dynamics explores one large-scale solution, the spectrum of possible microscopic states
entailed by the alternative large-scale solutions (not activated) remain un-
explored. This happens because as the critical point is approached from
above the macroscopic fluctuations are amplified, resulting in increasingly
extended correlations in space and in time. The amplification continues
until at the critical point the correlation scales diverge to infinity. Below
the critical point the large-scale solutions separate and the system becomes
non-ergodic.

Here the mathematical divergence means that, for example, the correlation of the fluctuation in time no longer decays exponentially with a
finite time scale. Instead, at the critical point, the decay of the correlation
function becomes scale invariant and is represented by an IPL in the time
between critical events $\tau^{-\beta}$. Of course, the same is true for the system
relaxation after a large fluctuation. This critical phenomenon is normally
denoted as $1/f$-noise, or more specifically $1/f^{\eta}$-noise, with the spectral index in the interval $0 < \eta < 2$ and is related to the correlation IPL index
$\beta = 3 - \eta$.

It is important to emphasize that not all IPL indexes are produced by
systems having the same properties. They may all satisfy the condition
for the emergence of collective states, with constrained dynamics, IPL correlation functions that may, or may not, be integrable, depending on the
correlation index β, but it is important to establish whether the system fluctuations are intermittent. If the system's statistics are intermittent then all
properties, including the correlation functions, only depend on the distribution of time intervals τ between events (waiting times). If intermittent, the
waiting-time PDF is $\tau^{-\mu}$ and the intermittency IPL index μ can be used
as a measure of the system's nonsimplicity. As discussed in Chapter 1,
for non-ergodic systems, where $1 < \mu < 3$ a novel form of LRT has been
discovered, which suggests a quenching of information transfer for systems
with $\mu < 2$ to regular stimuli, non-quenched information transfer for systems sharing the same or similar μ, with maximal information transfer for
$\mu \approx 2$.

A great many phenomena are characterized by the frequency content of their spectra, so that simple diffusion, also referred to as Brownian motion, has an asymptotic inverse square frequency spectrum. This we all understand and agree upon. It is the next step in which simplicity is lost and the behavior deviates from ordinary to anomalous diffusion that the phenomena are called $1/f$-noise again referring to the spectrum. Such $1/f$-noise was first observed in physical systems at the beginning of the last century and as the disciplines in which such $1/f$-phenomena were observed increased over the intervening years, so too did the number of theories necessary for its understanding. The phenomenological theory that has gained the most followers is that proposed in the last century by Mandelbrot and which he named fractional Brownian motion (FBM) and we call Type I $1/f$-noise.

The historic difficulty with categorizing phenomena by their frequency content is that this large-scale characterization does not have a unique reductionist origin. There exists at least one other kind of $1/f$-noise that is separate and distinct from Type I $1/f$-noise and it is generated by crucial events, which we refer to as Type II $1/f$-noise. The confusion over these types of $1/f$-noise arises because the mechanisms generating these two types of noise can appear separately, or they can appear together, depending on the nonsimplicity of the phenomenon being considered. Moreover, whenever the two appear together the critical events can be masked as we have seen.

The Type I process has also been labeled stationary-FBM (SFBM) and is characterized by stationary autocorrelation functions. The Type II process is generated by crucial events having both non-stationary as well as aging effects and is labeled aging-FBM (AFBM). The crucial events are a consequence of criticality, which is to say, as the control parameter in a nonsimple dynamical network approaches its critical value the correlation function changes from exponential to IPL. The phase transition is accompanied by a transition of a Gaussian process with short-term exponential memory to a hyperbolic statistical process with an IPL long-term memory. The critical behavior of the microdynamics entails emergent macrodynamics whose statistics are crucial events, that is, the statistics of the macroprocess is non-stationary, renewal, and IPL.

Recall that Type II $1/f$-noise is characterized by non-stationary correlation functions. This form of $1/f$-noise is due to the action of crucial events, and the distance between consecutive crucial events is described by a waiting-time IPL PDF, with an IPL index μ in the range $[1, 3]$. These crucial events generate perfect $1/f$-noise at $\mu = 2$. For $\mu > 2$ the non-stationary correlation function becomes stationary in the long-time limit and for this reason is often confused with Type I $1/f$-noise.

The dynamic variable $Z(t)$ scales if the time is scaled by a parameter λ the variable becomes a scaled version of itself with a multiplicative factor λ^η. This scaling is familiar from the displacement in simple diffusion where the statistics are Gaussian and the scaling index is $\eta = 1/2$. In the case of infinite memory SFBM the scaling is given by $\eta = H$ where H is the Hurst exponent which is confined to the interval $0 < H \leq 1$. However, even though the AFBM has the same scaling behavior as SFBM, its non-stationary nature makes its other properties quite different. For example, if the time interval between successive events is generated by a hyperbolic PDF with an IPL index in the interval $1 < \mu < 3$, it is possible to show that the effect of aging is to change the IPL index for the survival probability from μ to $\mu - 1$.

We emphasize here that $1/f$-noise is not in fact noise, but rather is the temporal signature of the self-similar properties of a critical dynamic state and is more properly denoted as a $1/f$-signal. However, it is necessary to recognize that there are many ways that critical states can be achieved. We have already discussed the most familiar way, that physical phase transitions exist in which an outside control parameter determines the onset of criticality at a specific value of that parameter. Historically, the first theory of critical behavior without dependence on an external parameter was self-induced criticality (SOC) in which the internal dynamics of the system attracts the system into a critical state. The SOC $1/f$-signal occurs in such catastrophic phenomena as earthquakes, forest fires, brainquakes, punctuated equilibrium in biological evolution, to name a few. SOC is the defining property of a class of dynamic systems that has a critical point as an attractor and generates spatial self-similarity producing the fractal statistics of self-organized critical events, whose self-similar and scale-invariant features lack natural length and time scales. Note that this is accomplished without the tuning of an external parameter to a critical value.

The existence of a SOC phenomenon is usually signaled by the onset of anomalous avalanches, whether the avalanches are in terms of the number of tremors in an earthquake or neurons in a brainquake. SOC is based on the dynamic search for the critical value of the control parameter K, which is selected by the network through a bottom up process, involving the dynamic behavior of the individuals and is not externally imposed. The signature of SOC is the time interval between consecutive critical events, with a IPL waiting-time PDF, a property historically referred to as temporal complexity (nonsimplicity). We therefore refer to the form of SOC as self-organized temporal criticality (SOTC) and identify the critical events as crucial.

SOTC spontaneously generates temporal nonsimplicity. Adopting temporal nonsimplicity as the signal of criticality alerted us to this perhaps being a more convenient indicator than the observation of avalanches with an IPL PDF. This assumption was confirmed by finding that two networks in critical states signaled by temporal nonsimplicity exchange information with greater efficiency than do two corresponding networks with critical states signaled by IPL avalanches. The close connection between maximal efficiency of information transport and temporal nonsimplicity is based on the explanation of information exchange using PCM theory discussed in Chapter 1. In extending the PCM to POIE theory, it is noted that criticality generates IPL renewal events characterized by IPL indexes and the efficiency of the exchange of information between two such networks is based on the degree to which the crucial events of the perturbing network influence the crucial events of the perturbed network.

It is evident that SOTC generates crucial events, whereas it is equally clear that the critical events generated by SOC may or may not be crucial, we do not know. More research is needed to determine the necessary and sufficient conditions for critical events to become crucial. On the other hand, there is a great deal on the formalism leading up to crucial events and that is the subject of the next and last chapter.

Chapter 8

Crucial Events in 3 Acts

This chapter is devoted to illustrating three distinct theoretical levels by which to generate within the macroscopic dynamics significant deviations from the prescriptions of ordinary statistical physics derived from the Second Law of Thermodynamics (SLT). As prototypical descriptions of microscopic dynamics compatible with the equilibrium properties established by the SLT, we offer the pioneering works of Einstein [180] and Langevin [312].

As clearly pointed out by the authors of [318], Langevin focused his attentions on trajectories, thereby stressing individual realizations of stochastic processes, while Einstein dealt with PDFs and therefore with ensembles of trajectories. The latter view corresponds to moving from the Langevin equation for stochastic dynamics to the Fokker–Planck equation in phase space describing the evolution of the PDF. Both Einstein and Langevin addressed the problem of characterizing the microscopic dynamics that at the macroscopic level is expected to be consistent with the prescription of thermodynamic equilibrium that according to the work of Boltzmann and Gibbs [449] is compatible with microscopic dynamics. Microscopic dynamics are characterized by fast fluctuations $\xi(t)$ that generate the simple diffusion of the slower fluctuations of the dynamic variable $X(t)$ and in the case of simple diffusion the long-time limit $X(t)$ is known to be proportional to $\sqrt{t}$. We make the assumption that nonsimple processes are characterized by $X(t) \propto t^{\delta}$, where $\delta \neq 1/2$. We shall discuss three distinct sources of such nonsimplicity: (1) the first is due to quantum coherence; (2) the second is generated by *crucial events* that we introduce using DNA sequences and turbulence; (3) the third source is a combined action of the first two sources.

If this were a play, it would have three separate and distinct acts. Each act has its list of characters that behave in identifiable and predictable ways, each contributing to the overall theme of the play in both known and unknown ways. As you are carried along by the plot line you will form attachments for this or that character, those that align with your values and those that do not, but do not let that determine your overall reaction. We here extend Hamlet's meaning of "the play's the thing" to encompass the broader message of the play's theme or message as being of over-riding importance. But enough of metaphor, let us get down to science.

8.1   Act I: Quantum Coherence

The Langevin equation was made popular by supplementing Newton's classical equations with the introduction of linear dissipation and random fluctuations. However it is not nearly as well known that the Langevin equation was also derived using quantum mechanical arguments in 1965 by Hazime Mori [388]. In this section we derive the Langevin equation (LE) following the approach of Grigolini [236] which is equivalent to the Mori treatment, and its quantum mechanical nature is made even more evident by adopting the Dirac notation in its derivation. This nomenclature is reviewed in Appendix C.

8.1.1   Derivation of the Langevin equation

We propose to derive the Langevin equation from the quantum equation of motion for the ket $|f\rangle$:

$$\frac{d}{dt}|f\rangle = \Gamma|f\rangle, \tag{8.1}$$

where the dynamic operator is proportional to the Hamiltonian operator $\widehat{H}$:

$$\Gamma = -\frac{i}{\hbar}\widehat{H},$$

in the case of the Schrödinger equation of Eq. (C.1) and the super-operator of Eq. (C.10) $\Gamma \to \mathcal{L}$ in the case of the time evolution of the observable A:

$$\frac{d}{dt}A(t) = \mathcal{L}A(t).$$

In both cases we need to define the scalar product.

Let us build up a basis set of orthogonal states moving from the initial condition $|f\rangle$ interpreted as the ground state of the basis set that we propose to build up. Let us call the ground state $|f_0\rangle$. To emphasize this choice of the initial state we rewrite Eq. (8.1) as:

$$\frac{d}{dt}|f_0(t)\rangle = \Gamma_0|f_0(t)\rangle, \tag{8.2}$$

stressing that $|f_0(0)\rangle = |f\rangle$ and $\Gamma_0 = \Gamma$. Using this nomenclature, we build up the basis set using the following prescriptions:

$$|f_1\rangle \equiv (1 - P_0)\Gamma_0|f_0\rangle,$$

where we define the projection operator onto the initial state:

$$P_0 \equiv \frac{|f_0\rangle\langle f_0|}{\langle f_0|f_0\rangle}. \qquad (8.3)$$

The meaning of this prescription is that we define the first 'excited' state $|f_1\rangle$ of the basis set by applying the dynamical operator Γ_0 to the initial state $|f_0\rangle$. To make it possible for the first excited state to be orthogonal to the initial state we apply to $\Gamma_0|f_0\rangle$ the projection operator

$$Q_0 = 1 - P_0, \qquad (8.4)$$

which is equivalent to subtracting from $\Gamma_0|f_0\rangle$ the initial state $|f_0\rangle$. Thus, only that part of the state $Q_0\Gamma_0|f_0\rangle$ generated by the dynamics which is orthogonal to the initial state remains as the contribution to $|f_1\rangle$.

This procedure is generalized to the kth-excited state by writing:

$$|f_{k+1}\rangle \equiv (1 - P_k)\cdots(1 - P_0)\Gamma_0|f_k\rangle = \Gamma_k|f_k\rangle,$$

along with the bra conjugate to the ket $|f_{k+1}\rangle$:

$$\langle \widetilde{f}_{k+1}| \equiv \langle \widetilde{f}_k|\Gamma_0^+(1 - P_0)\cdots(1 - P_k) = \langle \widetilde{f}_k|\Gamma_k^+.$$

We emphasize the fact that we are using a bi-orthogonal basis set of vectors, which can be used in general without making any assumption on special symmetries of the operator Γ_{0k}^+, the Hermitian conjugate of the operator Γ_k, yielding the scalar product:

$$\langle \widetilde{f}_k|f_{k'}\rangle = \delta_{kk'}\langle \widetilde{f}_k|f_k\rangle.$$

However, herein we limit our applications to the conventional Hermitian case discussed by Mori [388]. In Appendix C.1 the Mori projection technique is reviewed and the dynamics of coupled correlation functions is derived:

$$\frac{d}{dt}\Phi_0(t) = \lambda_0\Phi_0(t) - \Delta_1^2\int_0^t dt'\Phi_1(t - t')\Phi_0(t'). \qquad (8.5)$$

This interesting result shows that the correlation function of the system of interest depends on the correlation function of its environment. We do not yet see the structure of the Langevin equation, but be patient. As stressed the state $|f_1\rangle$ is the environment of the system of interest $|f_0\rangle$.

More generally, the state $|f_{k+1}\rangle$ is the environment of the state $|f_k\rangle$. Let us show how to create a connection with the LE structure at the level of the state $|f_k\rangle$. Note that using the general properties stressed in Appendix C.1, we prove that Eq. (8.5) is the special case with $k = 0$ of the more general equation:

$$\frac{d}{dt}\Phi_k(t) = \lambda_k \Phi_k(t) - \Delta_{k+1}^2 \int_0^t dt' \Phi_k(t')\Phi_{k+1}(t-t'), \tag{8.6}$$

where λ_k is given by Eqs. (C.22) and the coupling coefficient by:

$$\Delta_k^2 = -\frac{\langle \widetilde{f_k}|f_k\rangle}{\langle f_0|f_0\rangle} > 0. \tag{8.7}$$

We use the Laplace transform of the correlation functions to rewrite Eq. (8.6):

$$z\widehat{\Phi}_k(z) - 1 = \lambda_k \widehat{\Phi}_k(z) - \Delta_{k+1}^2 \widehat{\Phi}_k(z)\widehat{\Phi}_{k+1}(z), \tag{8.8}$$

to express the Laplace transforms of the correlation functions of successive orders in terms of one another:

$$\widehat{\Phi}_k(z) = \frac{1}{z - \lambda_k + \Delta_k^2 \widehat{\Phi}_{k+1}(z)}. \tag{8.9}$$

Note that we use the notation of a carrot over a variable for both the Laplace transform of the variable as in the expression for the relations between correlation functions, as well as for a quantum operator corresponding to a classical variable. The proper interpretation of the carrot should be given by the contexts.

Let us go back to Eq. (C.19) and retracing the steps followed in the Appendix C.1 we prove that Eq. (8.5) is the case $k = 0$ of:

$$Q_k|f_k(t)\rangle = \int_0^t dt'|f_{k+1}(t')\rangle\Phi_k(t-t'). \tag{8.10}$$

Note that

$$|f_k(t)\rangle = (P_k + Q_k)|f_k(t)\rangle = \Phi_k(t)|f_k\rangle + Q_k|f_k(t)\rangle, \tag{8.11}$$

and inserting Eq. (8.10) into Eq. (8.11), we obtain:

$$|f_k(t)\rangle = \Phi_k(t)|f_k\rangle + \int_0^t dt'|f_{k+1}(t')\rangle\Phi_k(t-t'). \tag{8.12}$$

Consequently, taking the Laplace transform of Eq. (8.12) we obtain:

$$|\widehat{f_k}(z)\rangle = \widehat{\Phi}_k(z)(|f_k\rangle + |\widehat{f}_{k+1}(z)\rangle). \tag{8.13}$$

Inserting Eq. (8.9) into this expression and reordering terms yields:

$$|\widehat{f}_k(z)\rangle = \frac{(|\widehat{f}_k\rangle + |\widehat{f}_{k+1}(z)\rangle)}{(z - \lambda_k + \Delta_{k+1}^2 \widehat{\Phi}_{k+1}(z))}. \tag{8.14}$$

The inverse Laplace transform of this expression yields:

$$\frac{d}{dt}|f_k(t)\rangle = \lambda_k|f_k(t)\rangle - \Delta_{k+1}^2 \int_0^t dt' \Phi_{k+1}(t - t')|f_k(t')\rangle + |f_{k+1}(t)\rangle. \tag{8.15}$$

We emphasize here that this equation partitions the states into the one of interest $|f_k(t)\rangle$ and its environment $|f_{k+1}(t)\rangle$. If these were classical quantities and the environment was random this would be the generalized Langevin equation (GLE). It is not yet the GLE but we are nearly there.

Markov approximation

Let us assume that the fluctuating environment $|f_{k+1}\rangle$ is characterized by a correlation function with such rapid decay as to make it virtually equivalent to a Dirac delta function. More precisely, let us make the assumption that the correlation function of the environment has an extremely short correlation time τ_c defined by:

$$\tau_c \equiv \widehat{\Phi}_{k+1}(z = 0).$$

Moreover, let us focus on the result of Eq. (8.9) in the specific case $\lambda_k = 0$:

$$\widehat{\Phi}_k(z) = \frac{1}{z + \Delta_{k+1}^2 \widehat{\Phi}_{k+1}(z)}. \tag{8.16}$$

Note that in this case Eq. (8.15) becomes:

$$\frac{d}{dt}|f_k(t)\rangle = -\Delta_{k+1}^2 \int_0^t dt' \Phi_{k+1}(t - t')|f_k(t')\rangle + |f_{k+1}(t)\rangle. \tag{8.17}$$

Let us study this correlation function for $z \to 0$ being asymptotically in time. Since the correlation function $\widehat{\Phi}_1(z)$ decays much more rapidly than the correlation function $\widehat{\Phi}_0(z)$, we obtain:

$$\widehat{\Phi}_k(z) = \frac{1}{z + \Delta_{k+1}^2 \widehat{\Phi}_{k+1}(z)} \approx \frac{1}{z + \Delta_{k+1}^2 \widehat{\Phi}_{k+1}(0)} = \frac{1}{z + \Delta_{k+1}^2 \tau_c}, \tag{8.18}$$

whose inverse Laplace transform yields an exponential relaxation. It is interesting to note that in this approximation Eq. (8.15) becomes:

$$\frac{d}{dt}|f_k(t)\rangle = -(\Delta_{k+1}^2 \tau_c)|f_k(t)\rangle + |f_{k+1}(t)\rangle. \tag{8.19}$$

To complete the derivation of the Langevin equation from this quantum argument, we interpret $|f_{k+1}\rangle$ as a stochastic force and to make the assumption that its mean value vanishes. Multiplying Eq. (8.17) by $\langle \tilde{f}_k|$ has indeed the effect of killing $|f_{k+1}\rangle$ due to the wise choice of Hilbert basis set adopted by Mori. However, the result of this prescription is:

$$\frac{d}{dt}\langle \tilde{f}_k|f_k(t)\rangle = -(\Delta^2_{k+1}\tau_c)\langle \tilde{f}_k|f_k(t)\rangle,$$

equivalent to the inverse Laplace transform of Eq. (8.18):

$$\frac{d}{dt}\Phi_k(t) = -(\Delta^2_{k+1}\tau_c)\Phi_k(t). \tag{8.20}$$

This quantum-like formalism forces us to interpret $|f_k(t)\rangle$ as a classical variable f_k and f_{k+1} as a totally random noise, under the condition that:

$$\frac{f_k(t)}{f_k(0)} = \Phi_k(t).$$

We refer to this interpretation as the Onsager principle (OP). In conclusion, Eq. (8.17) can be written in the form

$$\frac{dX(t)}{dt} = -\Delta^2 \int_0^t dt'\Phi_\xi(t-t')X(t') + \xi(t), \tag{8.21}$$

where $\Phi_\xi(t)$ is the stationary correlation function of $\xi(t)$ and the dynamic variable $X(t)$, if we assign to it an initial value $X(0)$ significantly larger than the root-mean-square displacement $\sqrt{\langle X^2\rangle_{eq}}$ is expected to regress to zero according to the GOP:

$$\frac{X(t)}{X(0)} = \Phi_\xi(t). \tag{8.22}$$

Equation (8.22) was defined by Allegrini et al. [16] as a generalized form of the OP [405]. In the literature Eq. (8.21) is called the GLE. The regression of Eq. (8.22) applied to the Markov approximation to GLE is the ordinary Onsager regression principle. In this case, as made clear by Eq. (8.20):

$$\frac{X(t)}{X(0)} = \Phi_\xi(t) = e^{-\lambda t}, \tag{8.23}$$

where the relaxation rate is:

$$\lambda \equiv (\Delta^2_1\tau_c).$$

Here we make the simplifying assumption that the Markov approximation holds true at the level of the variable f_0, namely the real variable of interest for the Mori chain.

3.1.2   Generalized form of Onsager principle

The OP plays a fundamental role in defining crucial events. In the original work of Onsager [405] the regression to equilibrium of macroscopic fluctuation of large intensity was defined in a form that is essentially equivalent to the Markov approximation done on the GLE of Mori. Allegrini *et al.* [16] proposed to call Eq. (8.22) the generalized Onsager principle, in a full agreement with the spirit of Mori's GLE, shedding light at the same time on the surprising result of earlier work [11].

Allegrini *et al.* [11] addressed the study of DNA sequences, developing a "dynamical" method based on the assumption that the statistical properties of DNA paths are determined by the joint action of two processes, one deterministic with long-range correlations and the other random with delta-function correlations. It is important to notice that the term "dynamical" implies the assumption that the separation of two molecules of the same DNA sequence can be interpreted as a "time interval". From the point of view of searching for a message carried by a sequence of symbols, the interpretation of distance as a "time interval" is legitimate and it may have a connection with the important issue of the evolution of life [124].

The weakly chaotic deterministic model adopted by these authors will be discussed here in some detail in order to define crucial events. Here we limit ourselves to stressing that biologists found that DNA sequences are characterized by a sequence of molecules of two different classes, purines, denoted by the symbol $+1$ and pyrimidines, denoted by the symbol -1. If we advance along the sequence from the symbol $+1$ there is a high probability that the next symbol is $+1$ as well. As a result of adopting the prescription of weak chaos, we observe an extended sequence of symbols $+1$, that, using the language of turbulence we call *laminar regions*. It will become evident that the survival probability $\Psi(t)$, the probability that at a "time interval" t from the initial value $+1$ the symbol has not changed is:

$$\Psi(t) \propto \frac{1}{t^{\mu-1}}, \tag{8.24}$$

where the IPL index is in the interval $1 < \mu < \infty$.

It is important to notice the significant difference between the survival probability of Eq. (8.24) and the exponential regression of Eq. (8.23). The exponential function is faster that the IPL decay even if in the region $\mu > 3$ the distinction between IPL and exponential regression may require a statistical analysis with increasing precision as μ increases towards ∞ to verify that distinction.

The authors also made the important assumption that the DNA sequence transmitting genetic information is realized as a result of the joint action of two prescriptions, one is the weak chaos prescription generating for instance the IPL index $\mu = 2.5$ and the other is a random prescription corresponding to $\mu = \infty$. This is closely related to an important recent result that will be used to describe the third act mentioned earlier. These authors found that in the case evaluating the equilibrium correlation function is technically difficult and imprecise. The adoption of the GOP, which is equivalent to evaluating the survival probability, leads to satisfactory results. In Act III of this chapter we discuss the case where the sequence under observation is generated adopting at any time step with probability p_C the crucial event generator and with probability p_M a survival probability generated by the memory function of Mori theory.

Quantum coherence and exponential waiting-time PDFs

In this subsection we show that quantum coherence may generate extended laminar regions reminiscent of those created by turbulence. However, their waiting-time PDFs are exponential. Consider the Hamiltonian operator:

$$\widehat{H} = V(|1\rangle\langle 2| + |2\rangle\langle 1|), \tag{8.25}$$

where V is a real number. In this case, the quantum time evolution is given from $|f_0\rangle = |\psi(0)\rangle$ and the unitary operator is given by Eq. (C.5). We also make the assumption that the kets $|1\rangle$ and $|2\rangle$ are an orthonormal and complete basis set, such that:

$$\langle 2|1\rangle = \langle 1|2\rangle = 0,$$

$$\langle 1|1\rangle = \langle 2|2\rangle = 1,$$

and the projection unitary operator on the complete space is:

$$|1\rangle\langle 1| + |2\rangle\langle 2| = \mathbf{I}.$$

Note that $\mathbf{I}$ is the identity operator, a natural consequence of the assumption that this is a complete basis set.

The Hamiltonian operator $\widehat{H}$ is Hermitian, thereby producing real eigenvalues that are obtained by solving the eigenvalue equation:

$$det(\widehat{H} - \lambda\mathbf{I}) = \mathbf{0},$$

the determinant yields the eigenvalue equation:

$$\lambda^2 - V^2 = 0,$$

whose solution yields the eigenvalues:

$$\lambda_\pm = \pm V,$$

along with the eigenstates:

$$|\pm\rangle = \frac{1}{\sqrt{2}}(|1\rangle \pm |2\rangle).$$

The eigenstates are another complete basis set of orthonormal kets and the states $|1\rangle$ and $|2\rangle$ are expressed, after some algebra, in this new basis set as:

$$|1\rangle = \frac{1}{\sqrt{2}}(|+\rangle + |-\rangle),$$

and

$$|2\rangle = \frac{1}{\sqrt{2}}(|+\rangle - |-\rangle).$$

Let us study the time-dependence of the wave function initiated from $|\psi(0)\rangle = |1\rangle$:

$$|\psi(t)\rangle = \frac{1}{\sqrt{2}}e^{-\frac{i\widehat{H}}{\hbar}t}|1\rangle = e^{-\frac{i\widehat{H}}{\hbar}t}\frac{1}{\sqrt{2}}(|+\rangle + |-\rangle)$$
$$= \frac{1}{\sqrt{2}}(e^{-\frac{i}{\hbar}Vt}|+\rangle + e^{\frac{i}{\hbar}Vt}|-\rangle).$$

Let us re-express this result in the original basis set to obtain:

$$|\psi(t)\rangle = \frac{1}{2}(e^{-\frac{i}{\hbar}Vt}(|1\rangle + |2\rangle) + e^{\frac{i}{\hbar}Vt}(|1\rangle - |2\rangle))$$
$$= \frac{1}{2}((e^{-\frac{i}{\hbar}Vt} + e^{\frac{i}{\hbar}Vt})|1\rangle + (e^{-\frac{i}{\hbar}Vt} - e^{\frac{i}{\hbar}Vt})|2\rangle)$$

which simplifies to:

$$|\psi(t)\rangle = \cos\left(\frac{V}{\hbar}t\right)|1\rangle - i\sin\left(\frac{V}{\hbar}t\right)|2\rangle. \tag{8.26}$$

Let us imagine that the states $|1\rangle$ and $|2\rangle$ are two distinct sites of a lattice. The condition described by Eq. (8.26) represents a strange condition if we move from the microscopic to the macroscopic level. An extensive literature is devoted to settle this paradoxical property of quantum mechanics. There is a wide agreement (see [172] for a recent interesting discussion of this fundamental issue) that no quantum system is fully isolated and that its environment yields a measurement process making the system collapse into either the state $|1\rangle$ or the state $|2\rangle$, according to the prescriptions of Copenhagen interpretation [69].

The linear superposition of the site $|1\rangle$ and of site $|2\rangle$ of Eq. (8.26) has the form:

$$|a\rangle = c_1|1\rangle + c_2|2\rangle,$$

which is a normalized state satisfying the condition:

$$\langle a|a \rangle = 1,$$

which due the orthogonalization properties of this basis set yields:

$$c_1^* c_1 + c_2^* c_2 = 1. \tag{8.27}$$

The normalization condition of Eq. (8.27) is closely connected to the important fact that

$$|c_j|^2 = p_j, \quad j = 1, 2,$$

are probabilities. More precisely, p_j is the probability for the system to collapse into the state $|j\rangle$ as a result of a measurement.

Let us make the assumption that the environment of this quantum system has the property of measuring it with a very large frequency $1/\tau$ satisfying the condition $\frac{V}{\hbar}\tau \ll \pi/2$. As a consequence, if the system is prepared in the state $|1\rangle$ it remains in the state $|1\rangle$ for a very long time. In fact, if the first measurement occurs after a time interval τ from preparation the probability of collapsing into the state $|1\rangle$ is

$$p_1 = \left(\cos\left(\frac{V}{\hbar}\tau\right) \right)^2,$$

which is very close $p_1 = 1$, due to the high frequency of the measurement process. When the second measurement occurs, since the first measurement "prepares" the quantum system in the state $|1\rangle$, the probability of collapsing again into the state $|1\rangle$ remains very large. If we assign to the state $|1\rangle$ the symbol $+1$, as we did for purines, we obtain a much extended laminar region consisting of a much sequence of symbols $+1$.

The probability of collapsing into the state $|2\rangle$ is very small, but is not zero. As a consequence, after a large number of measurement events the quantum system will collapse into the state $|2\rangle$. Note that with calculations of the same kind as those that led us to Eq. (8.26) we can establish the time evolution of this quantum system when we prepare it in the state $|2\rangle$. In this case, we obtain:

$$|\psi(t)\rangle = \frac{1}{\sqrt{2}} e^{-\frac{i\hat{H}}{\hbar}t}|2\rangle = \cos\left(\frac{V}{\hbar}t\right)|2\rangle - i\sin\left(\frac{V}{\hbar}t\right)|1\rangle \tag{8.28}$$

and consequently, the measurement process will generate an extended laminar region with the symbol -1.

The definition of a crucial event that we are proposing on the basis of the dynamical properties of the coding DNA sequence, is the occurrence of an abrupt transition from a laminar region filled by the symbols $+1$ to an extended laminar regions filled with the symbols -1. We have already pointed out that the survival probability of a crucial event is an IPL.

Let us evaluate the survival probability in the quantum case that we are illustrating here. The probability of making a transition from the symbol $+1$ to the symbol -1 is:

$$x = \left(\sin \left(\frac{V}{\hbar} \right) \tau \right)^2 \ll 1.$$

Imagine that the measurement frequency is large, namely τ is sufficiently short as to realize the condition:

$$\sqrt{x} = \sin \left(\frac{V\tau}{\hbar} \right) \approx \frac{V\tau}{\hbar} \ll 1.$$

In this case the probability of not making a transition is very close to 1:

$$\left(\cos \left(\frac{V\tau}{\hbar} \right) \right)^2 = 1 - x.$$

At time 2τ the probability of keeping the same symbol will again be $(1-x)$, and so on. In conclusion, the probability of not changing symbols at a generic time t is evaluated by noticing that at time t the number of measurement processes that have been affecting the system is $n = t/\tau$. The probability of keeping the same symbol up to time t is:

$$(1 - x)^n = e^{\ln(1-x)^n} = e^{-nx}.$$

Consequently, the survival probability $\Psi(t)$ is given by the exponential:

$$\Psi(t) = e^{-\frac{t}{\tau}x} = e^{-tr},$$

and the decay rate is:

$$r \equiv \left(\frac{V}{\hbar} \right)^2 \tau.$$

The waiting-time PDF $\psi(\tau)$ of the time duration of the region where the signal is characterized by the same symbol, either $+1$ or -1 is therefore:

$$\psi(\tau) = re^{-r\tau}. \tag{8.29}$$

This is not the IPL necessary to identify this transition as a crucial event.

In conclusion, if the system is frequently observed, it appears to be frozen. This is an interesting property of quantum mechanics established in 1977 by Misra and Sudarshan [378]. The extended laminar regions are characterized by exponential waiting-time PDFs, a condition that prevents us from interpreting the transitions from the symbol $+1$ to the symbol -1 as crucial events. The exponential nature of this measurement effect has been the subject of debates in the last few years. In [105] and [121] an attempt was made to overcome the exponential nature of the measurement induced waiting-time PDFs by using the method of subordination, which is equivalent to using the CTRW theory of Montroll and Weiss [382].

Reduced density matrix and measurement

Here we discuss the entanglement of a quantum system with its environment without making the assumption that the entanglement is classical and we show that the concept of conditional probability is not powerful enough to explain satisfactorily the process of wave function collapse of the Copenhagen interpretation.

We have seen that Mori theory is based on the adoption of projection operators. We have also mentioned the importance of the density matrix formalism. The derivation of GLE is closely related to the adoption of reduced density matrices that are obtained by tracing the total density matrices over the degrees-of-freedom of the corresponding baths.

Let us remind the reader about the trace operation by considering the operator:

$$A = |a\rangle\langle b|.$$

The trace of the operator A is defined:

$$Tr\{A\} \equiv \sum_j \langle v_j|A|v_j\rangle, \tag{8.30}$$

where the kets $|v_i\rangle$ form a complete basis set so that:

$$\sum_j |v_j\rangle\langle v_j| = \mathbf{I}. \tag{8.31}$$

As a consequence of the completeness of the basis set we obtain

$$Tr\{A\} = \sum_j \langle v_j|a\rangle\langle b|v_j\rangle$$
$$= \sum_j \langle b|v_j\rangle\langle v_j|a\rangle = \langle b|a\rangle, \tag{8.32}$$

since in the second term on the RHS of this equation, we can see the identity operator of Eq. (8.31) in action.

Let us consider a quantum system of the same kind as the quantum system with the Hamiltonian operator $\widehat{H}$ given by Eq. (8.25) and the ket:

$$|\psi_A\rangle = \frac{1}{\sqrt{2}}(|a\rangle + |b\rangle), \qquad (8.33)$$

which is a linear combination of two distinct sites, $|a\rangle$ and $|b\rangle$. Using the property given by Eq. (8.32) to evaluate the trace of $|\psi_A\rangle\langle\psi_A|$, we immediately obtain:

$$Tr_A\{|\psi_A\rangle\langle\psi_A|\} = \langle\psi_A|\psi_A\rangle = 1,$$

as a consequence of the fact that the ket $|\psi_A\rangle$ is properly normalized.

Let us now consider a different quantum system of the same kind and a ket $|\psi_B\rangle$ of this quantum system:

$$|\psi_B\rangle = \frac{1}{\sqrt{2}}(|\alpha\rangle + |\beta\rangle). \qquad (8.34)$$

We see that this ket, which will serve as the environment of the ket $|\psi_A\rangle$, interpreted as the system of interest, is the linear superposition of two distinct sites, $|\alpha\rangle$ and $|\beta\rangle$. Note that both systems are characterized by complete basis sets of orthonormal kets:

$$\langle a|a\rangle = 1, \quad \langle a|b\rangle = 0, \quad \langle b|b\rangle = 1, \qquad (8.35)$$

as well as:

$$\langle\alpha|\alpha\rangle = 1, \quad \langle\alpha|\beta\rangle = 0, \quad \langle\beta|\beta\rangle = 1. \qquad (8.36)$$

Let us first study the case when the kets $|\psi_A\rangle$ and $|\psi_B\rangle$ are not entangled. In this case the total ket $|\psi_T\rangle$ has the product form:

$$|\psi_T\rangle = |\psi_A\rangle|\psi_B\rangle. \qquad (8.37)$$

The total density matrix ρ for the system plus environment is:

$$\rho = |\psi_T\rangle\langle\psi_T|.$$

The reduced density matrix σ is obtained by tracing over the information concerning the environment B, and as a consequence is defined by:

$$\sigma = Tr_B\{\rho\}.$$

Thus, in the case of the separable condition given by Eq. (8.37) we obtain:

$$\sigma = Tr_B\{|\psi_T\rangle\langle\psi_T|\} = |\psi_A\rangle\langle\psi_A|Tr_B\{|\psi_B\rangle\langle\psi_B|\}$$
$$= |\psi_A\rangle\langle\psi_A|. \qquad (8.38)$$

Equation (8.38) shows that the trace on the environment in conjunction with the separable condition leaves the density matrix of the system of interest totally unchanged. This is no longer true in the case on entanglement between the system of interest and its environment. More explicitly the density matrix is:

$$\rho = \frac{1}{4}(|a\alpha\rangle + |a\beta\rangle + |b\alpha\rangle + |b\beta\rangle)(\langle a\alpha| + \langle a\beta| + \langle b\alpha| + \langle b\beta|),$$

whose trace over the environment B yields the reduced density matrix:

$$\sigma = \frac{1}{2}(|a\rangle\langle a| + |b\rangle\langle b| + |a\rangle\langle b| + |b\rangle\langle a|), \qquad (8.39)$$

as a consequence of Eq. (8.38). We see that the reduced density matrix is identical to that of the system without the environment and that the density matrix has off-diagonal elements in addition to the diagonal ones that should be the only ones remaining as a result of the measurement process.

Let us now consider the ket for the entangled condition:

$$|\psi_T\rangle = \frac{1}{\sqrt{2}}(|a\beta\rangle + |b\alpha\rangle),$$

along with the corresponding bra:

$$\langle\psi_T| = \frac{1}{\sqrt{2}}(\langle a\beta| + \langle b\alpha|).$$

Thus, we obtain the density matrix:

$$\rho = \frac{1}{4}(|a\beta\rangle\langle a\beta| + |a\beta\rangle\langle b\alpha| + |b\alpha\rangle\langle a\beta| + |b\alpha\rangle\langle b\alpha|).$$

The reduced density matrix is then obtained by taking the trace over the environment:

$$\sigma = \frac{1}{2}(|a\rangle\langle a| + |b\rangle\langle b|). \qquad (8.40)$$

Comparing the density matrix given by Eq. (8.40) to that of Eq. (8.39), we see that as a result of entanglement the off-diagonal elements vanish. We can interpret this result as a realization of the Copenhagen interpretation of quantum mechanics. As a consequence of the measurement done on the state of Eq. (8.33), we find the eigenvalue a with probability $1/2$ and the eigenvalue b with probability $1/2$.

8.1.3   Recovering the tri-diagonal matrix of Mori

Let us assume that the operator A is the dimensionless harmonic displacement operator $\widehat{Q}$ and that the state $|f_0\rangle$ is the ground state $|0\rangle$ in the number representation. In this case:

$$|f_1\rangle = \frac{1}{R_1}(1 - P_0)A|f_0\rangle = \frac{1}{R_1}(1 - P_0)\widehat{Q}|0\rangle. \tag{8.41}$$

Let us focus on $\widehat{Q}|0\rangle$, using Eq. (C.53):

$$\widehat{Q}|0\rangle = \frac{1}{\sqrt{2}}(a + a^+)|0\rangle. \tag{8.42}$$

Using Eqs. (C.66) and (C.67) for the creation and annihilation operators, Eq. (8.42) becomes:

$$A|0\rangle = \widehat{Q}|0\rangle = \frac{1}{\sqrt{2}}(a + a^+)|0\rangle = \frac{1}{\sqrt{2}}|1\rangle. \tag{8.43}$$

As a consequence, from Eq. (8.41) we obtain:

$$|f_1\rangle = \frac{1}{R_1}\frac{1}{\sqrt{2}}|1\rangle. \tag{8.44}$$

The normalization factor R_1 must therefore be given by $R_1 = 1/\sqrt{2}$ so that:

$$|f_1\rangle = |1\rangle. \tag{8.45}$$

Let us now move to evaluate $|f_2\rangle$ which according to the earlier discussion can be written as:

$$|f_2\rangle = \frac{1}{R_2}(1 - P_0)(1 - P_1)A|f_1\rangle. \tag{8.46}$$

We know that $|f_1\rangle = |1\rangle$ and $A = \widehat{Q}$, so that:

$$A|f_1\rangle = \widehat{Q}|1\rangle = \frac{1}{\sqrt{2}}(a + a^+)|1\rangle$$

$$= \frac{1}{\sqrt{2}}(|0\rangle + \sqrt{2}|2\rangle).$$

The projection operator $1 - P_0$ annihilates $|0\rangle$ thereby yielding:

$$|f_2\rangle = \frac{1}{R_2}|2\rangle.$$

Since both $|2\rangle$ and $|f_2\rangle$ are normalized, we obtain $R_2 = 1$ so that:

$$|f_2\rangle = |2\rangle.$$

It is now evident that by induction one can show that, in general, we can relate the nth Mori state to the oscillator state n:

$$|f_n\rangle = |n\rangle.$$

Let us examine the general prescription:

$$|f_{n+1}\rangle \equiv \frac{1}{R_{n+1}}(1 - P_0)(1 - P_1)\cdots(1 - P_n)A|f_n\rangle, \qquad (8.47)$$

where using the formal definition of the normalization factor we have:

$$R_{n+1} \equiv \sqrt{\langle f_n|A(1 - \mathbf{I_n})A|f_n\rangle}.$$

In this case:

$$A|f_n\rangle = \widehat{Q}|n\rangle = \frac{1}{\sqrt{2}}(a + a^+)|n\rangle$$

$$= \frac{1}{\sqrt{2}}a|n\rangle + \frac{1}{\sqrt{2}}a^+|n\rangle.$$

Using Eqs. (C.66) and (C.67) we obtain:

$$A|f_n\rangle = \frac{1}{\sqrt{2}}(n)^{\frac{1}{2}}|n - 1\rangle + \frac{1}{\sqrt{2}}(n + 1)^{\frac{1}{2}}|n + 1\rangle. \qquad (8.48)$$

The projection operator $1 - P_{n-1}$ annihilates the first term on the right-hand side of Eq. (8.48). For this reason Eq. (8.47) reduces to:

$$|f_{n+1}\rangle \equiv \frac{1}{R_{n+1}}\frac{1}{\sqrt{2}}(n + 1)^{\frac{1}{2}}|n + 1\rangle. \qquad (8.49)$$

Since both $|f_{n+1}\rangle$ and $|n + 1\rangle$ are normalized (are the same ket) we obtain:

$$R_{n+1} = \sqrt{\frac{(n + 1)}{2}}, \qquad (8.50)$$

and from which we can conclude:

$$\langle f_{n+1}|\widehat{Q}|f_n\rangle = R_{n+1} = \sqrt{\frac{(n + 1)}{2}}.$$

Note that for the oscillator states:

$$\langle f_n|\widehat{Q}|f_n\rangle = 0,$$

because $A|n\rangle$ generates either $|n + 1\rangle$ or $|n - 1\rangle$.

The arguments given generate the following tri-diagonal matrix expression:

$$\widehat{Q} = \begin{pmatrix} 0 & \frac{1}{\sqrt{2}} & 0 & 0 & 0 & 0 & \dots \\ \frac{1}{\sqrt{2}} & 0 & 1 & 0 & 0 & 0 & \dots \\ 0 & 1 & 0 & \sqrt{\frac{3}{2}} & 0 & 0 & \dots \\ 0 & 0 & \sqrt{\frac{3}{2}} & 0 & 2 & 0 & \dots \\ 0 & 0 & 0 & 2 & 0 & \sqrt{\frac{5}{2}} & \dots \\ 0 & 0 & 0 & 0 & \sqrt{\frac{5}{2}} & 0 & \dots \\ \dots & \dots & \dots & \dots & \dots & \dots & \dots \end{pmatrix},$$

which is the matrix obtained from the sum of matrixes (C.76) and (C.77) multiplied by $1/\sqrt{2}$.

8.2 Non-integrable Correlation Functions and Localization

This section is devoted to discussing the link between the Mori GLE and the theory of quantum dissipative systems illustrated in the important 1999 book by Weiss [568]. Weiss considered a bath of non-interacting quantum oscillators, in which case we must replace the time evolution generated by the Schrödinger equation Eq. (C.1) with the quantum Liouville equation Eq. (C.7). More precisely, we adopt the corresponding Heisenberg equation, where an observable A, corresponding for instance to the displacement $\widehat{Q}$ of the Weiss oscillator changes in time and the averages are done using the initial condition of the density matrix ρ given by Eq. (C.6).

8.2.1 Heisenberg time evolution

According to Heisenberg, an observable A driven by the equation of motion Eq. (C.10) changes in time according to:

$$A(t) = e^{\frac{i\widehat{H}}{\hbar}t} A e^{-\frac{i\widehat{H}}{\hbar}t}, \tag{8.51}$$

which is Eq. (C.9), written here again to facilitate reading. The mean value of the observable is:

$$\langle A(t) \rangle = Tr\{e^{\frac{i\widehat{H}}{\hbar}t} A e^{-\frac{i\widehat{H}}{\hbar}t} \rho\}.$$

Let us make the assumption that $A = \widehat{q}$. Using the notation of Appendix C.3 we express $\widehat{q}$ in terms of the creation and destruction operators:

$$\widehat{q} = \left(\frac{\hbar}{2m\omega}\right)^{\frac{1}{2}}(a + a^+).$$

Returning to the generalized expression for the density matrix ρ given by Eq. (C.6) for $t = 0$ we have in the Heisenberg picture:

$$\rho(0) = \sum_j p_j |\psi_j\rangle\langle\psi_j|.$$

Let us make the assumption that the wave function $|\psi_j\rangle$, denoting the ket of index j, is proportional to $|n\rangle$. We establish a connection with the temperature of an oscillator heat bath T by replacing the PDF with the Boltzmann distribution:

$$\rho(0) = \sum_{n=0}^{\infty} |n\rangle\langle n| \frac{e^{-\beta E_n}}{Z},$$

in which case $\beta = \frac{1}{k_B T}$, where k_B is the Boltzmann constant and Z is the canonical normalization factor.

Note that here with a slight notation change compared to Appendix C.3 we define the Hamiltonian operator as:

$$\widehat{H} = \hbar\omega\left(\widehat{N} + \frac{1}{2}\right),$$

where $\widehat{N}$, as in Appendix C.3, is the number operator:

$$\widehat{N} = a^+ a,$$

and the energy eigenvalues are:

$$E_n = \hbar\omega\left(n + \frac{1}{2}\right).$$

For $T = 0$, we have the projection onto the ground state:

$$\rho(0) = |0\rangle\langle 0|,$$

and for the other asymptote $T = \infty$:

$$\rho(0) = \sum_{n=0}^{\infty} |n\rangle\langle n|.$$

Using the property established in Appendix C.3:

$$\langle n|(a + a^+)|n\rangle = 0,$$

we conclude that the mean value of the displacement operator $\widehat{q}$ vanishes. This is the reason why we have to rest on the OP. A reasonable assumption to make is that the fluctuations of $\widehat{q}$ from time to time may generate values larger than the size of the root-mean-square value $\sqrt{\langle \widehat{q}^2\rangle - \langle \widehat{q}\rangle^2}$, thereby leading us to evaluate the time interval between consecutive large fluctuations. According to the OP this is equivalent to evaluating the correlation function:

$$\langle \widehat{q}\,\widehat{q}(t)\rangle = Tr\{\widehat{q}e^{\frac{i\widehat{H}}{\hbar}t}\widehat{q}e^{-\frac{i\widehat{H}}{\hbar}t}\rho(0)\}. \tag{8.52}$$

To establish a connection with the Weiss quantum mechanical approach we set the condition $T = 0$ and we make the assumption that the subscript j refers to oscillators with the frequency ω_j. The evaluation of the Trace of Eq. (8.52) yields:

$$\langle \widehat{q}\,\widehat{q}(t)\rangle = \sum_{n=0}\langle n|\widehat{q}e^{\frac{i\widehat{H}}{\hbar}t}\widehat{q}e^{-\frac{i\widehat{H}}{\hbar}t}\rho(0)|n\rangle. \tag{8.53}$$

Taking into account the sum over oscillators with different frequencies, we obtain:

$$\langle \widehat{q}\,\widehat{q}(t)\rangle = \sum_{j}p_j\sum_{n=0}\langle n|0_j\rangle\langle 0_j|n\rangle, \tag{8.54}$$

where $|0_j\rangle$ denotes the oscillator with index j in the ground state. We rewrite the last equation in the form:

$$\langle \widehat{q}\,\widehat{q}(t)\rangle = \sum_{j}p_j\sum_{n=0}\langle 0_j|n\rangle\langle n|\widehat{q}e^{\frac{i\widehat{H}}{\hbar}t}\widehat{q}e^{-\frac{i\widehat{H}}{\hbar}t}|0_j\rangle, \tag{8.55}$$

which leads us to the more convenient expression:

$$\langle \widehat{q}\,\widehat{q}(t)\rangle = \sum_{j}p_j\langle 0|\widehat{q}e^{\frac{i\widehat{H}_j}{\hbar}t}\widehat{q}e^{-\frac{i\widehat{H}_j}{\hbar}t}|0\rangle, \tag{8.56}$$

with $\widehat{H}_j$ denoting the Hamiltonian operator with frequency ω_j.

Note that expressing the two $\widehat{q}$'s in terms of the creation and destruction operators, we obtain:

$$\langle 0|\widehat{q}e^{\frac{i\widehat{H}_j}{\hbar}t}\widehat{q}e^{-\frac{i\widehat{H}_j}{\hbar}t}|0\rangle = k_j\langle 0|(a + a^+)e^{\frac{i\widehat{H}_j}{\hbar}t}(a + a^+)e^{-\frac{i\widehat{H}_j}{\hbar}t}|0\rangle, \tag{8.57}$$

where we have introduced:

$$k_j \equiv \frac{\hbar}{2m\omega_j}.$$

We see that:

$$\langle 0|(a + a^+)e^{\frac{i\widehat{H}_j}{\hbar}t}(a + a^+)e^{-\frac{i\widehat{H}_j}{\hbar}t}|0\rangle$$

$$= \langle 0|(a + a^+)e^{\frac{i\widehat{H}_j}{\hbar}t}(a + a^+)|0\rangle e^{-\frac{i\omega_j}{2}t}$$

$$= \langle 0|(a + a^+)e^{\frac{i\widehat{H}_j}{\hbar}t}|1\rangle e^{-\frac{i\omega_j}{2}t},$$

and the RHS of this equality simplifies to:

$$\langle 0|(a + a^+)e^{\frac{i\widehat{H}_j}{\hbar}t}|1\rangle e^{-\frac{i\omega_j}{2}t}$$

$$= \langle 0|(a + a^+)|1\rangle e^{-\frac{i\omega_j t}{2}} e^{i\omega_j t(1+\frac{1}{2})}$$

$$= \langle 0|0\rangle e^{i\omega_j t} = e^{i\omega_j t}.$$

Consequently, the normalized correlation function reads:

$$\frac{\langle \widehat{q}\,\widehat{q}(t)\rangle}{\langle \widehat{q}^2\rangle} = \sum_j p_j e^{i\omega_j t},$$

because the pre-factor k_j appears in both numerator and denominator.

To establish a connection between the formalisms of Weiss and Mori, we notice that the adoption of the bi-orthogonal basis set leads to a continued fraction structure equivalent to a Hermitian matrix, implying that the coefficient Δ_j^2 is a real number. As a consequence, we set:

$$\Phi_0(t) = \frac{\langle f_0|f_0(t)\rangle}{\langle f_0|f_0\rangle} = Re\left(\frac{\langle \widehat{q}\,\widehat{q}(t)\rangle}{\langle \widehat{q}^2\rangle}\right).$$

These arguments lead us to the important result:

$$\Phi_0(t) = \sum_j p_j \cos(\omega_j t),$$

so that turning the discrete sum into an integral and replacing the factor p_j with the frequency density $g(\omega)$ we obtain:

$$\Phi_0(t) = \int d\omega g(\omega) \cos(\omega t). \tag{8.58}$$

A closer connection with Weiss theory is obtained by setting:

$$g(s) \propto \omega^{s-1}, \tag{8.59}$$

with $s = 1$ corresponding to the ohmic condition, $s > 1$ to the sub-ohmic and $s < 1$ to the super-ohmic condition.[1] Weiss focuses on the condition

---

[1] We note that our s is related to s_W of Weiss through relation $s = 2 - s_W$.

$0 < s < 2$. Here we show that this region corresponds to the scaling index H used by Mandelbrot to define the form of anomalous diffusion known as fractional Brownian motion (FBM) [353]. It shall become evident that:

$$s = 2 - 2H, \qquad (8.60)$$

or rearranging terms:

$$H = 1 - \frac{s}{2}. \qquad (8.61)$$

The next subsection is devoted to establishing a connection between the Weiss theory and FBM. We shall see that the super-ohmic regime generates super-diffusion ($H > 1/2$) and the sub-ohmic regime generates sub-diffusion ($H < 1/2$). But before we address that connection, it is important to stress another important aspect of the present results. In principle, quantum mechanical observables at different times do not commute:

$$\langle \widehat{q}(t_1)\widehat{q}(t_2) \rangle \neq \langle \widehat{q}(t_2)\widehat{q}(t_1) \rangle,$$

if $t_1 \neq t_2$. However, the connection between the Mori and Weiss theories that we have been discussing leads to:

$$\langle \widehat{q}(t_1)\widehat{q}(t_2) \rangle = \sum_j p_j \, \cos(\omega_j[t_1 - t_2]),$$

which is equivalent to making the ordinary classical assumption that $q(t_1)$ commutes with $q(t_2)$. The assumption of Gaussian statistics, compatible with FBM, yields, for instance:

$$\langle \widehat{q}(t_1)\widehat{q}(t_2)\widehat{q}(t_3)\widehat{q}(t_4) \rangle = \langle \widehat{q}(t_1)\widehat{q}(t_2) \rangle \langle \widehat{q}(t_3)\widehat{q}(t_4) \rangle$$
$$+ \langle \widehat{q}(t_1)\widehat{q}(t_3) \rangle \widehat{q}(t_2)\widehat{q}(t_4) > + \langle \widehat{q}(t_1)\widehat{q}(t_4) \rangle \langle \widehat{q}(t_2)\widehat{q}(t_3) \rangle,$$

showing that all 24 permutations of the four labeled times yield the same result. This is in line with the Onsager theory requiring this condition to be ensured by the condition of equilibrium realized as a result of infinite aging.

Long and short-time properties of $\Phi_0(t)$

Let us use Eqs. (8.58) and (8.59) to study the long-time limit of $\Phi_0(t)$:

$$\Phi_0(t) = \frac{\int_0^\infty \omega^{s-1} \cos(\omega t) d\omega}{\int_0^\infty \omega^{s-1} d\omega}.$$

The change of integration variable $y \equiv \omega t$ makes it possible to extract the scaling behavior and write:

$$\Phi_0(t) = \frac{constant}{t^s} \int_0^\infty y^{s-1} \cos(y) dy \tag{8.62}$$

and that [224]:

$$\int_0^\infty y^{s-1} \cos y dy = \Gamma(s) \cos\left(\frac{s\pi}{2}\right).$$

Note additionally that due to the Tauberian theorem:

$$\mathcal{L}\left(\frac{1}{t^s}\right) = z^{s-1}\Gamma(1-s),$$

the Laplace transform of the correlation function Eq. (8.62) becomes negative for $s > 1$, in which case we would be tempted to conclude that:

$$\widehat{\Phi}_0(z) \propto \mathrm{sign}(1-s)z^{s-1}. \tag{8.63}$$

This result is correct, but cannot be derived from the integral of Eq. (8.62), because this integral is correct only for $s < 1$.

To prove that Eq. (8.63) is correct, we notice that the normalized correlation function $\Phi_0(t)$ must satisfy the condition:

$$\Phi_0(0) = 1.$$

For this reason, we have to establish a bridge that spans the long-time and short-time values of the correlation function. Following Tuladhar *et al.* [534] we assign to $\Phi_0(t)$ the following analytical form:

$$\Phi_0(t) = \frac{1}{2-s}e^{-\gamma t} - \frac{s-1}{2-s}\frac{1}{(1+\gamma t)^s}. \tag{8.64}$$

Again using the Tauberian theorem for the IPL on the RHS of this equation, we obtain for the Laplace transform of $\Phi_0(t)$:

$$\widehat{\Phi}_0(z) = \frac{1}{2-s}\frac{1}{\gamma+z} - \frac{s-1}{2-s}\frac{1}{\gamma^s}\Gamma(1-s)z^{s-1}. \tag{8.65}$$

This is an exact result, even if it is based on an arbitrary choice we made to establish the bridge between the short and the long-time regions of this process. We note that the first term on the RHS of Eq. (8.65) is neglected because it generates a singularity to the left of the integration path used when inverting the Laplace transform through integration on the Bromwich contour. This bright observation was made by Nöelle Pottier [436].

This leads us to the important conclusion that Eq. (8.63) is correct. Of course, in the case $s < 1$, we adopt the less controversial form:

$$\Phi_0(t) = \frac{1}{(1+\gamma t)^s}. \tag{8.66}$$

Preparing the theory for FBM

To make contact with the results of the previous subsection let us apply the result of the Mori analysis giving rise to Eq. (8.9) to $k = 0$:

$$\widehat{\Phi}_0(z) = \frac{1}{z - \lambda_0 + \Delta_1^2 \widehat{\Phi}_1(z)}. \tag{8.67}$$

Using Eq. (8.9) for $k = 1$ and applying this prescription to Eq. (8.67), we obtain:

$$\widehat{\Phi}_0(z) = \frac{1}{z - \lambda_0 + \Delta_1^2 \frac{1}{z - \lambda_1 + \Delta_2^2 \widehat{\Phi}_2(z)}}. \tag{8.68}$$

We can apply this procedure again and obtain:

$$\widehat{\Phi}_0(z) = \frac{1}{z - \lambda_0 + \Delta_1^2 \frac{1}{z - \lambda_1 + \Delta_2^2 \frac{1}{z - \lambda_3 + \Delta_3^2 \widehat{\Phi}_3(z)}}}. \tag{8.69}$$

This nesting procedure can be applied with no limit so as to obtain an infinitely extended continued fraction.

Let us discuss the application of this procedure to the case of Weiss processes and assume that $s < 1$, the super-ohmic condition. In this case, we know that:

$$\widehat{\Phi}(z = 0) = \infty, \tag{8.70}$$

as a consequence of the fact that the correlation function in this case is not integrable. This is the prediction of Eq. (8.63) when $s < 1$. Note that in the case of Weiss processes $\lambda_k = 0$. Thus Eq. (8.67) yields:

$$\widehat{\Phi}_0(z = 0) = \frac{1}{\Delta_1^2 \widehat{\Phi}_1(z = 0)}. \tag{8.71}$$

To obtain a result compatible with Eq. (8.70) it is necessary that $\widehat{\Phi}_1(z = 0) = 0$. As a consequence the function $f_1(t)$ must belong to the sub-ohmic condition. Of course the variable $f_2(t)$ will be super-ohmic again, and so on.

8.3 Fractional Brownian Motion

The dynamical approach to the Mandelbrot FBM was established in 1995 [11] to study DNA sequences. It is important to stress that 25 years ago there was consensus among a group of outstanding researchers that a DNA sequence can be interpreted/modeled as a diffusional process with infinite memory, a conviction reflecting the belief that DNA sequences share the same kind of infinite memory. Herein, after using turbulence to introduce crucial events, we show that DNA sequences are renewal processes. This perspective seems to conflict with the concept of anomalous diffusion generated by stationary but not integrable correlation FBM, which is not as popular as the Mandelbrot FBM theory.

Let us study the diffusion process:

$$\frac{dX}{dt} = \xi. \tag{8.72}$$

We can interpret $\xi(t)$ as the fluctuation $f_0(t)$ discussed in the earlier sections. In this way $\xi(t)$ can be either the displacement $q(t)$ or the momentum $p(t)$, interpreted in the statistical sense discussed in Sec. 8.2.1. As a consequence, they may refer to either $s < 1$, the Weiss super-ohmic condition or to $s > 1$, the Weiss sub-ohmic condition. Let us integrate Eq. (8.72):

$$X(t) = \int_0^t dt' \xi(t'), \tag{8.73}$$

and assuming that all our random walkers in the classical picture of diffusion are located at the origin at time $t = 0$. We also make the assumption that the fluctuations are zero centered $\langle \xi(t) \rangle = 0$, compatible with our arguments that we need to focus on correlation functions rather than on single intense fluctuations. For this reason, we evaluate the second moment

$$\langle X^2(t) \rangle = \int_0^t dt_1 \int_0^t dt_2 \langle \xi(t_1) \xi(t_2) \rangle. \tag{8.74}$$

The assumption that the fluctuations $\xi(t_1)$ and $\xi(t_2)$ commute allows us to write Eq. (8.74) as:

$$\langle X^2(t) \rangle = \langle \xi^2 \rangle \int_0^t dt_1 \int_0^{t_1} dt_2 \Phi_\xi(t_2), \tag{8.75}$$

where

$$\Phi_\xi(t) \equiv \frac{\langle \xi \xi(t) \rangle}{\langle \xi^2 \rangle} \tag{8.76}$$

and coincides with the autocorrelation function $\Phi_0(t)$ discussed earlier.

Making the assumption that this dynamical derivation of anomalous diffusion is equivalent to the FBM of Mandelbrot, we may set;

$$\langle X^2(t) \rangle \propto t^{2H}. \tag{8.77}$$

Differentiating Eq. (8.75) twice with respect to time, yields:

$$\frac{d^2}{dt^2} \langle X^2(t) \rangle = \langle \xi^2 \rangle \Phi_\xi(t) \tag{8.78}$$

and using the FBM assumption of Eq. (8.77) we find the following asymptotic in time property:

$$2H(2H-1)t^{2H-2} \propto \langle \xi^2 \rangle \Phi_\xi(t) \propto \frac{1}{t^s}, \tag{8.79}$$

yielding the important relations given by Eqs. (8.60) and (8.61).

It is also remarkable that Eq. (8.79) shows that for H < 1/2 the long-time tail of $\Phi_\xi(t)$ must be negative in full agreement with the Mori theory yielding Eq. (8.64).

Infinite memory of FBM

Cakir et al. [127] stress that the dynamical approach to FBM that we illustrate here to be compatible with the well-known FBM property [353]:

$$\frac{\langle X(-t)X(t)\rangle}{2\langle X^2(t)\rangle} = 1 - 2^{2H-1}, \tag{8.80}$$

which is an impressive way of stressing the infinite memory of FBM. The dynamical approach to FBM affords the same result, which is proven using Eq. (8.73) and the backward expression:

$$X(-t) = - \int_0^{-t} dt' \xi(t') dt'. \tag{8.81}$$

Thus, we can write:

$$\langle X(-t)X(t)\rangle = - \int_0^{-t} dt_1 \int_0^t dt_2 \langle \xi(t_1)\xi(t_2)\rangle$$

$$= -\langle \xi^2 \rangle \int_0^{-t} dt_1 \int_0^t dt_2 \Phi_0(t_1 - t_2).$$

Note that $\Phi_0(t)$ is the Mori correlation function, which is independent of the sign of time, since it depends on the modulus of time. Let us carry out the calculation in the super-ohmic case where we assign to this correlation function the analytical form:

$$\Phi_0(t) = \left(\frac{T}{T+|t|}\right)^s. \tag{8.82}$$

Using this analytical form, we obtain from the above correlation of the diffusion process:

$$- \langle \xi^2 \rangle \int_0^{-t} dt_1 \int_0^t dt_2 \Phi_0(t_1 - t_2)$$

$$= \frac{\langle \xi^2 \rangle T^s}{1-s} \int_0^{-t} dt_1 \int_0^t dt_2 \frac{d}{dt_2}(T + |t_1 - t_2|)^{1-s}.$$

Take into account that we are differentiating with respect to the modulus of a negative time, that forces us to recover the positive sign. We have now to work on:

$$\frac{\langle \xi^2 \rangle T^s}{1-s} \int_0^{-t} dt_1 \{ [T + |t_1 - t|]^{1-s} - [T + |t_1|]^{1-s} \}$$

$$= -\frac{\langle \xi^2 \rangle T^s}{(1-s)(2-s)} \int_0^{-t} dt_1 \frac{d}{dt_1} \{ [T + |t_1 - t|]^{2-s} - [T + |t_1|]^{2-s} \},$$

changing sign again due to the time differentiation with respect to the modulus of negative times. Going to the asymptotic time limit we obtain:

$$\langle X(-t)X(t) \rangle = -\frac{\langle \xi^2 \rangle T^s}{(2H-1)2H} [|-2t|^{2H} - 2|t|^{2H}]. \qquad (8.83)$$

Finally, considering the fact that the second moment $\langle X^2(t) \rangle$ has the same overall factor as the correlation function we obtain Eq. (8.80), leading us to the conclusion that Mori theory can be used to generate the Mandelbrot FBM.

Note that the authors of [127] used the dynamical approach to FBM to obtain:

$$\langle X(t_1)X(t_2) \rangle = \langle X^2(|t_1|) \rangle + \langle X^2(|t_2|) \rangle - \langle X^2(|t_1 - t_2|) \rangle,$$

which can be used to derive the well-known FBM relation

$$\langle X(t_1)X(t_2) \rangle \propto |t_1|^{2H} + |t_2|^{2H} - |t_1 - t_2|^{2H},$$

yielding the infinite memory of Eq. (8.80).

8.3.1   FBM and $1/f$-noise

It is important to notice that Mandelbrot FBM has attracted significant attention as a possible way of explaining the origin of $1/f$-noise. For instance, the authors of the recent publication [170] in their interesting work on the biological auto-luminesence adopt the formula:

$$P(f) \propto \frac{1}{f^{2H+1}}, \qquad (8.84)$$

namely, establishing FBM $1/f$-noise under the condition $H \to 0$, corresponding to sub-diffusion. It is important to notice that these authors limit themselves to the study of the short-time rather than the long-time limit. In the long-time limit we use Eq. (8.82). It is important to recall that the correlation function is stationary. This allows us to evaluate the power spectrum $S_p(\omega)$ using the Wiener–Khintchine theorem [337]:

$$S_p(\omega) = \int_{-\infty}^{+\infty} e^{-i\omega t'} \Phi_0(t')dt'.$$

Using only the large-scale time limit of Eq. (8.82), and the change of variables $y = \omega t$, we obtain:

$$S_p(\omega) \propto \frac{1}{\omega^{1-s}} \int_0^{\infty} \frac{\cos y}{y^s} dy.$$

Note that the integral coincides with that in Eq. (8.65), leading to the conclusion:

$$S_p(\omega) \propto \frac{1}{\omega^{\beta}}.$$

The spectral index is given by:

$$\beta = 1 - s = 2H - 1, \tag{8.85}$$

rather than:

$$\beta = 1 - 2H,$$

thereby producing $1/f$-noise for $H \to 1$, namely, the condition of super-diffusion in the vicinity of the ideal ballistic limit.

It is important to note that when the $1/f$-noise is generated by crucial events, we subsequently show:

$$\beta = 3 - \mu \tag{8.86}$$

where μ, defined in Eq. (8.24), has the value $\mu = 2$.

8.3.2 Quantum mechanics and cognition

In the second act of this chapter we discuss crucial events as a possible signature of cognition. We also discuss the use of crucial events versus quantum coherence. To reduce the load on the readers in understanding that discussion, it is convenient to review here the arguments used by some advocates of quantum mechanics as a source of cognition.

The conviction that quantum mechanics is an essential ingredient of cognition is widely shared [289, 38] and the experimental work on bio-photons [435] contributed to the conjecture that consciousness has a quantum mechanical origin [49]. The psychological experiments of Tversky and Kahneman [540] are also interpreted using quantum logic and quantum probability [623].

The main problem with the adoption of quantum mechanics in this more general context is that, in spite of the phenomenal successes of the theory in the domain of physics, its realism is still the subject of debate [336]. The Copenhagen interpretation requires the settlement of the measurement issue and this unsettled problem is addressed either by generalizing the theory with nonlinear stochastic corrections [69] or by resting on decoherence theory [632], which interprets quantum collapse as an environment-induced de-phasing processes not requiring any extension of the current quantum theory.

In recent years, some researchers have adopted a theoretical interpretation of quantum mechanics not requiring the quantum wave function to collapse [172, 181, 273, 463]. These theories do not conflict with the existence of crucial events that we discuss in the next few sections, wherein we propose self-organized temporal criticality (SOTC) to be the origin of crucial events. As we shall see, this requires a bottom-up approach to organization, while the authors of [172, 181] adopt the principle of top-down causality.

8.4   Act II: Turbulence

Let us now discuss another source of $1/f$-noise that is often confused with the FBM $1/f$-noise. This second form of $1/f$-noise is generated by crucial events. Using turbulence theory we define crucial events as short-time intervals characterized by irregular behavior. These turbulent intervals are separated by extended time regions called *laminar regions* of regular motion, leading us to interpret the turbulent intervals as single events that we call crucial events. The complexity of the crucial events is temporal. We do not pay attention to their intensity, but we focus on the time intervals between consecutive turbulent intervals. The waiting-time PDFs $\psi(t)$ contain the temporal properties of the laminar regions. They have the structure of IPL, with an IPL index μ ranging from 1 to ∞ and the corresponding events are considered to be crucial when $\mu < 3$. Kaplan *et al.* [283], studying the EEGs of the human brain found short intervals of abrupt transitions from regular to a fast irregular behavior, and called them Rapid Transition Processes (RTPs). The time interval between consecutive RTPs were proven [30] to be separated by extended regions of regular behavior with a temporal complexity $\mu = 2$. For this reason RTPs are considered to be crucial events.

Let us fill the laminar regions with the symbols $+1$ and -1, using a fair coin-tossing prescription. We obtain a time series similar to those produced by DNA sequences assigning the symbol $+1$ to purines and the symbol -1 to pyrimidines. The DNA statistical analysis generates temporally nonsimple survival probabilities identical to those of turbulent processes. This leads us to conclude that the "dynamics" of DNA sequences are also driven by crucial events. We shall show that the analysis of DNA sequences done in 1994 [11] can be used to establish a clear distinction between FBM $1/f$-noise and $1/f$-noise generated by crucial events.

8.4.1   Procaccia and Schuster map

A tutorial approach to turbulence can be found in the book of Schuster [481] and it is based on the map of Procaccia and Schuster [441]:

$$x(n + 1) = x(n) + const|x(n)|^z, \qquad (8.87)$$

where z is a real number larger than 1. Here we discuss an idealized version of this map in the continuous time representation:

$$\frac{dX}{dt} = \alpha X^z. \qquad (8.88)$$

A crucial event is a condition of strong chaos lasting for a very short time. The purpose of the map is to establish how much time we have to wait before we encounter another crucial event. The occurrence of a crucial event corresponds to the coordinate X reaching the value $X = 1$ moving from an initial condition $0 < X(0) < 1$. When the coordinate X reaches the value 1 a crucial event occurs and a new initial condition $X(0)$ is selected with equal probability in the interval $[0, 1]$.

We showed in Sec. 5.2.2 that by introducing a uniform PDF of initial states over the interval $[0, 1]$ we obtain for the waiting-time PDF $\psi(t)$ the following hyperbolic expression:

$$\psi(t) = (\mu - 1)\frac{T^{\mu-1}}{(T + t)^\mu}, \qquad (8.89)$$

corresponding to the survival probability:

$$\Psi(t) = \left(\frac{T}{t + T}\right)^{\mu-1}. \qquad (8.90)$$

The time T and IPL index μ are expressed in terms of the model parameters in Sec. 5.2.2 and are not repeated here.

From survival probabilities to stationary correlation functions

It has to be stressed that the survival probability is not a stationary correlation function, however, it may be confused with a stationary correlation function if $\mu > 2$. For clarity, we repeat in an informal way some of the discussion presented earlier. In fact, in this case the laminar regions have a mean value that can be easily evaluated using Eq. (8.89). We note that the evaluation of the mean time multiplies the waiting-time PDF by τ turning the slow decay $1/\tau^\mu$ into the slower decay $1/\tau^{\mu-1}$, which is still integrable if $\mu - 1 > 1$, namely $\mu > 2$. The analytical evaluation of $\langle \tau \rangle$ yields:

$$\langle \tau \rangle = \frac{T}{\mu - 2}. \tag{8.91}$$

In this case the stationary correlation function $\Phi_{eq}(t)$ exists and, as done by the authors of [211], can be evaluated adopting the following procedure. Let us imagine that the laminar region are filled with $+1$ and -1, according to the tossing of a fair coin. We move along the time series with a window of size t. We have to look for laminar regions of size $t' > t$. In fact, the correlation function is evaluated by multiplying the sign at t, the end of the window, by the sign at the beginning of the window, and if the window is not imbedded in a larger laminar region, the probability that the product of these signs being positive is equal to the probability that it is negative, generating a vanishing contribution. The probability of finding a laminar region t' is $t'\psi(t')$. In the case where $t' = t$ there is only one position of the window to use for the calculation. If the window $t' > t$, the statistical weight of that window is $(t' - t)/t'$. Therefore, we get

$$\Phi_{eq}(t) = \frac{1}{\langle \tau \rangle} \int_t^\infty dt'(t' - t)\psi(t'). \tag{8.92}$$

To find the analytical expression for $\Phi_{eq}(t)$, let us differentiate Eq. (8.92) twice with respect to time and use Eqs. (8.91) and (8.89). We obtain:

$$\ddot{\Phi}_{eq} = \frac{(\mu - 2)(\mu - 1)T^{\mu-2}}{(t + T)^\mu},$$

which is the second time derivative of

$$\Phi_{eq}(t) = \left(\frac{T}{T + t} \right)^{\mu-2}. \tag{8.93}$$

8.4.2   DNA sequences

Can we interpret the result of Eq. (8.93) as corresponding to $\Phi_0(t)$ of the Mori theory? This is the assumption made by Allegrini *et al.* [11]. In

1995 these authors studied DNA sequences and converted them into 'time series', $\xi(t)$, with discrete times corresponding to the sequence sites, with ξ being equal to $+1$ for purines and to -1 for pyrimidines. They adopted the OP to evaluate the correlation function, which corresponds to selecting as initial conditions all the sites with $\xi = 1$. The initial size of the observation window is 1, thereby assigning the value 1 to the survival probability. Increasing the size of the mobile window has the effect of including some negative values within the mobile window, thereby making the survival probability decrease. The observation of coding DNA sequences lead them to noticing that the survival probability drops to small but non-zero values in the long times of the tails, observed in a log–log representation, yielding IPL behavior. They used FBM theory to interpret the long-time behavior, leading to a satisfactory theory, but not completely satisfactory agreement between theory and experimental observation. This disagreement was explained using the arguments that lead us to Eq. (8.93). To facilitate the reader's understanding of this subtle point, we jump back in time, to the 1994 article by Trefán *et al.* [530].

Trefán *et al.* [530] studied the case where a diffusion process is generated by crucial events, with laminar regions extending in time with a waiting-time PDF with an IPL index satisfying the condition $2 < \mu < 3$. This is a special condition where in the long-time limit a stationary correlation function exists, and generates the scaling index δ:

$$\delta = \frac{1}{(\mu - 1)}. \tag{8.94}$$

When $\mu < 2$, Eq. (8.94) generates a scaling parameter δ larger than the ballistic condition. Notice that the DNA model forces us to define a crucial event as an abrupt change from the adoption of the purine prescription $\xi = 1$ to a new regime where the purine and pyrimidine prescriptions are adopted with equal probability. This condition leads to a deviation from the Lévy central limit theorem where at any time step a fluctuation in ξ of different intensity may occur. The dynamical Lévy process [530] cannot exceed the maximal scaling $\delta = 1$. The condition $\mu < 2$ was studied in [106, 308] and is discussed subsequently. Here, let us focus on Eqs. (8.94) and (8.93) and assume that a stationary correlation function, that is not integrable, with $\mu - 2 < 1$, namely $\mu < 3$ is equivalent to FBM theory. We have seen that:

$$s = 2 - 2H, \tag{8.95}$$

which is Eq. (8.60), written here again to facilitate the argument. Let us further assume that $\Phi_0(t)$ of Eq. (8.93) is equivalent to FBM, and let us replace the scaling index δ with the Hurst notation H such that:

$$\mu - 2 = s. \tag{8.96}$$

This is equivalent to assuming that FBM is compatible with the existence of crucial events, a misleading assumption causing a great deal of confusion in the literature of anomalous diffusion. Inserting Eq. (8.95) into Eq. (8.96) we obtain:

$$H = \frac{4 - \mu}{2}. \tag{8.97}$$

Although Eqs. (8.97) and (8.94) are different analytical expressions, for $\mu = 2$ they both yield ballistic scaling and for $\mu = 3$, ordinary scaling. It is difficult to realize that they are quite different analytical prescriptions, corresponding to the action of infinite memory, with no crucial events, Eq. (8.97), to be contrasted with the important contribution of crucial events to both non-stationary memory and anomalous diffusion, Eq. (8.94). Allegrini et al. [11] did not stress the difference between the two predictions, but our theoretical arguments, based on the adoption of the map of Eq. (8.87), should have forced us to conclude that the FBM hypothesis is not a proper theoretical approach for interpreting DNA sequences. But at the time we did not have this insight.

However, we [11] did propose a theory to understand the apparently erratic nature of coding DNA sequences that we shall discuss more deeply in Act III of this chapter. Following what was suggested in [11] let us consider two time series, $\xi_1(t)$ and $\xi_2(t)$. The first is obtained by implementing the idealized map of Eq. (8.87). The second is obtained by implementing the algorithm described by Cakir et al. [127] to generate either the stationary correlation function of Eq. (8.66), for $s < 1$, or the stationary correlation function of Eq. (8.64), for $s > 1$. We then turn the time series into a diffusion process. According to [11] we expect to obtain the following scaling:

$$X(t) \propto \epsilon t^\delta + (1 - \epsilon)t^H. \tag{8.98}$$

In the work of [11] the Hurst index is $H = 1/2$, because the sequence $\xi_2(t)$ is obtained tossing a fair coin. We use FBM to generate this random condition. Rather than using this extreme limit of randomness, we can use a more realistic sequence corresponding to the adoption of an exponential correlation function:

$$\Phi_0(t) = e^{-\gamma t}.$$

Since this process does not carry the information of interest, it must ultimately be removed to make the crucial events with scaling index δ visible.

In this earlier publication [11] we generated the first time series using the Procaccia and Schuster's map and mistakenly denoted the anomalous scaling δ of the first time series with the symbol H thereby contributing to the widely shared, but misleading, conviction that all forms of anomalous diffusion are equivalent to FBM. They are not.

In the final part of this chapter, we discuss recent experimental results, of neurophysiological and biological interest, where both the first time series and the second time series generate anomalous diffusion.

8.4.3   Renewal processes

In the first few pages of his 1967 book on renewal events [155], Sir David Roxbee Cox defined the occurrence of failures, as renewal events, in a form that can be used to shed light into what crucial events are all about. We repeat the discussion from an earlier chapter to make the presentation given here self-contained. Imagine that a living organism is a perfectly working machine when it is brand new. Unfortunately this machine has a failure rate $g(t)$, with t being the age of the machine. It is expected that $g(t)$ increase with increasing t. Let us imagine that when a failure occurs, a team of engineers/doctors instantaneously brings back the machine to its original condition. Thus, a failure event coincides with a rejuvenation event. In this scenario what is the time interval between consecutive failures?

According to Cox:

$$g(t) = \frac{\psi(t)}{\Psi(t)},$$

where $\psi(t)$ is the waiting-time PDF offering information about the time interval between consecutive failures, and $\Psi(t)$ is the survival probability, the probability that no failure occurs at a time interval t from the last rejuvenation event. We know that $\psi(t) = -\dot{\Psi}(t)$. Thus, a simple quadrature yields the formal expression for the survival probability:

$$\Psi(t) = e^{-\int_0^t dt' g(t')dt'}. \tag{8.99}$$

In the simplest case the failure rate is constant, $g(t) = r_0$, so we obtain:

$$\Psi(t) = e^{-r_0 t},$$

namely a Poisson process.

Let us imagine an ideal condition where $g(t)$ rather than increasing as a function of time, it surprisingly decreases. If the decrease is too fast, the survival probability does not tend to zero for $t \to \infty$. We have to remain at the border between integrable and non-integrable $g(t)$:

$$g(t) = \frac{r_0}{1 + r_1 t}. \tag{8.100}$$

In this case, the integration of Eq. (8.99) yields:

$$\Psi(t) = \left(\frac{T}{t+T}\right)^{\mu-1},\tag{8.101}$$

which is identical to Eq. (8.90), with $T \equiv \frac{1}{r_1}$ and the IPL index $\mu \equiv 1 + \frac{r_0}{r_1}$.

This equivalence sheds light into the meaning of crucial events. Crucial events are renewal and their occurrence can be interpreted as a rejuvenation of the nonsimple system. They should not be confused with organization collapses. The renewal events make it possible for the system to adapt itself to a changing environment, free from the burden of the infinite memory burden of FBM.

8.4.4   Aging of the crucial events

FBM is characterized by stationary correlation functions. Crucial events are mistaken as a form of FBM because in the case $\mu > 2$ they produce the stationary correlation function of Eq. (8.93), one that is slower than the brand new survival probability. How can we explain this aging effect?

We assume that the process generating crucial events with the dynamic model of Eq. (8.88) starts at time $t = -\infty$, and begin the observation of crucial events at time $t = 0$. We perceive the first crucial event at a time $t > 0$. The probability of finding a laminar region of length τ^* is $\tau^*\psi(\tau*)/\langle\tau\rangle$. The probability density of observing the first change of laminar phase after a time t, being in a laminar region of length τ^*, is $\Theta(\tau^* - t)/\tau^*$. Consequently:

$$\psi_\infty(t) = \frac{1}{\langle\tau\rangle} \int_0^\infty d\tau^* \tau^* \psi(\tau^*) \frac{1}{\tau^*} \Theta(\tau^* - t) = \frac{1}{\langle\tau\rangle}\psi(t).$$

We conclude that:

$$\psi_\infty(t) = (\mu - 2)\frac{T^{\mu-2}}{(t+T)^{\mu-1}}.$$

This is the waiting-time PDF corresponding to the survival probability of Eq. (8.93).

8.5   Act III: Joint Action

This third act is a synthesis of the results of the research contained in two recent papers [82, 270]. Like any good three-act play, the third act clarifies any confusion and resolves any conflict that may have arisen in the first two acts. We hope.

The first paper we consider here at some length is by Jelinek *et al.* [270] and examines the heart rate time series of patients affected by diabetes-induced autonomic neuropathy of varying severity. These authors find that the progression of cardiac autonomic neuropathy (CAN) shifts the IPL index μ for the time interval PDF of the HRV time series from the border with perennial variability ($\mu = 2$) toward the border with Gaussian statistics ($\mu = 3$). The degree of shift provides a novel, sensitive measure for assessing disease progression. We observe that the analysis of brain dynamics done by Allegrini *et al.* [28] reveals that the brain of healthy patients in the awake state is characterized by $\mu = 2$ and consequently by ideal $1/f$-noise. We have seen that the condition of ideal $1/f$-noise can be realized in two different ways, through Eq. (8.85), for FBM, and through Eq. (8.86), when $1/f$-noise is generated by crucial events. The authors of [270] made their statistical analysis using the method of diffusion entropy analysis (DEA) modified with stripes (MDEA) [157] which has the important property of filtering out the FBM contribution to anomalous diffusion. Thus, they confirmed the observation made by Allegrini *et al.* [28] a decade earlier that the healthy brain operates at $\mu = 2$ if we make the plausible conjecture that a form of CME exists between the brain and the heart dynamics.

On the basis of these results, we would be led to reach the conclusion that the progress of CAN has the end effect of moving the brain from the state of maximal cognition, corresponding to $\mu = 2$, to the border with ordinary statistical physics, which is expected to be incompatible with life and cognition.

However, the results of Jelinek *et al.* [270] are more complicated and suggest that biological processes may, in addition to crucial events, host FBM nonsimplicity. In fact, Jelinek *et al.* [270] also analyzed their data adopting the method of Multiscale Entropy (ME) [152, 153]. The ME method analyzes nonsimple data at different time scales and makes it possible to split physiological data into two groups, the processes with ME virtually independent of the time scale, called $1/f$-processes, and the processes where ME decreases upon increase of time scale, called white noise processes. Quite surprisingly the progress of CAN does not turn heartbeat dynamics from the condition of $1/f$-noise to the condition of white noise. Jelinek *et al.* [270] prove that in the final stage of CAN, the variability of heartbeats is strongly reduced as is evidenced by μ moving to the Gaussian condition $\mu \geq 3$. However, the transition from $\mu < 3$ to $\mu > 3$, according to Procaccia and Schuster [441] should be interpreted as a transition from $1/f$ to white noise. However, this interpretation is not supported by the ME analysis of the HRV data. This is a consequence of the fact that Procaccia and Schuster refer to an interpretation of nonsimplicity based on crucial events. The transition from the non-Gaussian to the Gaussian condition is incompatible with the $1/f$-noise of Procaccia and Schuster, but it is not incompatible with the FBM $1/f$-noise. In fact, we saw that the Mori approach to FBM led us to adopt the Gaussian statistical probability hypothesized by Mandelbrot.

Let us turn our attention to the paper of Benfatto et $al.$ [82]. This second paper is based on the study of photon emissions from germinating seeds, using an experimental technique designed to detect photons of extremely low intensity when the signal/noise ratio is small. As we mentioned earlier, the conviction that quantum mechanics is an essential ingredient of cognition is widely shared [38, 289] and the experimental work on bio-photons [435] contributed to the conjecture that consciousness is of a quantum mechanical origin [49]. Benfatto et $al.$ [82] analyze the dark count signal in the absence of germinating seeds, as well as the photon emission during the germination process. The updated version of DEA used by Benfatto et $al.$ [82] is designed to determine if the signal's nonsimplicity is generated by either non-ergodic crucial events with a non-stationary correlation function, or by the infinite memory of a stationary but non-integrable correlation function, or by a mixture of both processes. These investigators find that the dark count yields ordinary scaling, thereby showing that no nonsimplicity of either kind may occur in the absence of any seeds in the chamber.

In the presence of seeds in the chamber anomalous scaling emerges, reminiscent of that found in neurophysiological processes. However, this time series is a mixture of both FBM and crucial event processes and with the progress of germination the non-ergodic component tends to vanish and the nonsimplicity is dominated by the stationary infinite memory of FBM. To be more precise, Benfatto et $al.$ process the data using DEA both with and without stripes. In the second case they find that the scaling with the progress of the germination process over time remains with a scaling index larger than 0.7, which, according to Eq. (8.97) would correspond to $\mu = 2.6$. Such a value would correspond to a condition of dynamical nonsimplicity compatible with cognition, being significantly smaller than the border value $\mu = 3$. However, the analysis done with stripes (MDEA) shows that the actual value of μ with the progress of germination over time rapidly approaches the border value $\mu = 3$, indicating that when the seeds generate roots the dynamical nonsimplicity observed is that of Mandelbrot's FBM. In other words, there are no crucial events with $\mu = 2.6$, since the filtered value of μ is close to the value $\mu = 3$, and dynamical nonsimplicity is due to $H \geq 0.7$.

For this reason, we are led to adopt the scaling perspective of Eq. (8.98), with ϵ depending on either the progress of CAN [270] or the progress of germination [82], thereby moving from the ideal condition $\epsilon = 1$ for pure crucial events to the condition $\epsilon = 0$ for pure non-crucial events.

3.5.1 Joint action of crucial events and FBM

It is always of value to determine where in the past the ideas emerged which led to the present understanding of a difficult problem and to examine how they shaped what we now believe to be true. We draw your attention to the work of Roncaglia *et al.* [457]. This paper was written in 1995 and was strongly influenced by the growing interest in those years in the potential application of classical chaos to quantum phenomena. Five years earlier Chirikov had been an invited speaker to the *International School on Order, Chaos and Patterns* [140] and his lectures attracted the attention of researchers in the field of dynamical chaos to a remarkably simple dynamical system, called the kicked rotor. The following two-dimensional map describes the dynamics of the kicked rotor:

$$p_{n+1} = p_n + K \sin \theta_n \qquad (8.102)$$

$$\theta_{n+1} = \theta_n + p_{n+1}, mod(2\pi).$$

If we focus our attention on the discrete kicked variable n, called x, we obtain the time series, $x_1, x_2, ..., x_i$, where the diffusional coordinate x yields the ordinary scaling:

$$x_n \propto \sqrt{n}. \qquad (8.103)$$

We assign to the control parameter K a value necessary to generate full deterministic chaos. Chirikov in his lectures [140] pointed out the connection with criticality and renormalization group theory, observing that at the edge of chaos a fractal phase space may emerge. The standard map has a Hamiltonian origin, making the resulting dynamical chaos compatible with the big question: "What would be the behavior of a quantum system that in the classical case is chaotic?" A remarkably interesting illustration of this issue can be found in the 1990 review paper by Felix Izrailev [266]. The work of Izrailev affords a clear picture of the origin of the quantum manifestations of chaos, with special emphasis on the discovery of quantum suppression of classical chaos [136].

Roncaglia *et al.* [457] decided to adopt the proposal made by Hillery *et al.* [254] to write a quantum mechanical Liouville equation that in the classical limit should become identical to the classical Liouville equation, given that the correspondence principle holds true. Using the Wigner formalism for a quantum particle moving under the influence of a potential $V(q)$ these authors found the following equation of motion for the Wigner quasi-probability $\rho_W(t)$ [457]:

$$\frac{\partial}{\partial t} \rho_W(q, p, t) = [L_{class} + L_Q] \rho_W(q, p, t), \qquad (8.104)$$

where L_{class} is the classical and L_Q is the quantum Liouvillian:

$$L_Q = \sum_{r=3}^{\infty} \left(\frac{\partial^r}{\partial q^r} V(q,t) \right) \left(\frac{\hbar}{2i} \right)^{r-1} \frac{1}{r!} \frac{\partial^r}{\partial p^r}. \qquad (8.105)$$

The quantum Liouville equation is very attractive because it shows that when the potential is linear and the derivatives of third order and higher vanish, the time evolution of the Wigner quasi-probability becomes indistinguishable from the classical Liouville equation. On the other hand, when the potential is nonlinear, it can generate chaotic dynamics. We are especially interested in the case of weak chaos [393, 541].

Weak chaos can be defined using the physics of turbulence and the Procaccia–Schuster map of Eq. (8.87). The idealized version of this map given by Eq. (8.88) coincides with the idealized version of the popular Manneville map [355] and illustrated in [19]. The Manneville map is also known as the Pomeau–Manneville map [434]. At $z = 1.5$, $\mu = 3$, the white noise regime ends and we enter a region characterized by aging. The aging becomes perennial at $z \geq 2$ where the Lyapunov coefficient vanishes thereby creating weak chaos according to [393]. Herein we adopt a definition of weak chaos that is not limited to $z > 2$, where the Lyapunov coefficient vanishes and include the region $1.5 < z < 2$, where the Lyapunov coefficient is still finite, but aging is already significantly important so as to establish a difference between Eqs. (8.94) and (8.97). According to Nee [393], the Pomeau–Manneville map is important in biology because "it generates: intermittency, aka sporadic behavior, and a declining hazard."

Nee [393] defines the 'aka sporadic behavior' of Pomeau–Manneville map in the following way: if the resetting generates x very close to the origin of the interval $[0, 1]$ the system moves very slowly away from the origin. This is a property well-reproduced by the idealized map of Eq. (8.88). In fact, Allegrini *et al.* [16] proved that as a result of this property the equilibrium PDF of x becomes proportional to $1/x^{z-1}$, which for $z > 2$ becomes not integrable as $x \to 0$. This property led some scientists to claim that this ergodicity breaking generates non-normalizable PDFs [7, 295, 296]. We prefer to denote these processes as characterized by perennial aging, in accordance with the numerical treatment of the Procaccia, Schuster, Pomeau–Manneville idealized prescription of Eq. (8.88).

The declining hazard of Nee [393] is a property of the renewal nature of these processes that was stressed earlier using the Cox formalism by means of Eq. (8.100). It is in fact surprising to realize that the survival probability generated by the intermittency prescription of Eq. (8.88) is a consequence of the fact that the risk of failure tends to decrease rather than increase upon aging.

It is remarkable that according to Vandermeer [541] weak chaos can be used to shed light into the phenomenon of bounded rationality, establishing a bridge between Turing and Simon. Vandermeer adopts weak chaos as a tool to span the gap separating the nomothetic and idiographic perspective, two opposite branches of modern psychology. He also applies these intriguing ideas to ecology thereby establishing a new form of agriculture that he defines as *agroecology*. This is an attractive use of weak chaos that seems to be connected with the popular issue of determining the foundation of cognition. A foundation that does not require the arguments of quantum probability developed by Yearsley and Busemeyer [623].

These are remarkably interesting observations, establishing that weak chaos, discovered in the field of turbulence is a powerful tool allowing us to contribute to the progress in areas of biology [393] and to the foundation of agroecology [541]. How do we explain the results of the two recent papers [82, 270]? We do not yet have a well-founded theory with which to explore all these interconnections. However, in the absence of such a theory the results of [457] allow us to make reasonable conjectures based on the quantum behavior of a system that in the classical case generates weak chaos.

Roncaglia *et al.* [457] addressed this intriguing question using the pseudo-probability of Hillery *et al.* [254] to evaluate the energy transmitted by the kicks to the rotator. They did the calculation using both the classical and quantum Liouvillians. Their investigations found that the mean energy per kick increases quickly in time until the rotor reached a regime of stochastic fluctuations around a constant value, which is identical in both the quantum and classical cases consistent with the predictions of Zurek and Paz [633]. This rapid increase persists for a time t_χ:

$$t_\chi = \frac{1}{\lambda}\ln\left(\frac{1}{\hbar}\right), \tag{8.106}$$

where λ is the Lyapunov exponent for the kicked rotor. It is important to notice that they did the calculation for $2 < \mu < 3$, where the Lyapunov coefficient is still finite. In this regime of rapid increase, the quantum and the classical conditions generate indistinguishable results. After this time, however, the quantum fluctuations become clearly different.

Of course, the classical fluctuations host the crucial events, even if the condition $2 < \mu < 3$ may lead to the incorrect impression that the fluctuations are a form of Mandelbrot's FBM process. On the basis of the two recent papers [82, 270] we make the plausible conjecture that the quantum fluctuations may host both crucial events and FBM non-crucial events. It is also plausible that the concentration ϵ of non-crucial events depends on the time interval from the system preparation, namely, a dependence on the number of kicks. We notice that according to Roncaglia *et al.* [457] the regime of the fluctuations around a common value stops at the time t_B predicted by Izrailev [266] on the basis of the theory of quantum suppression of classical chaos.

These remarks lead us to conjecture that new calculation done using the method of DEA and MDEA, should lead to the discovery that the dynamical behavior of quantum systems in the classical condition of weak chaos may host both crucial and non-crucial events. The non-crucial events may be a sign of quantum mechanics in a condition where a nonlinear potential makes the quantum behavior distinguishable from the classical behavior. The study of quantum chaos is still a subject of active research work, see, for instance, the recent work [328], and for this reason we hope that our remarks may attract the attention of some researchers to assess if the combination of crucial events and non-crucial events of quantum origin is the correct explanation of the results of the two recent papers [82, 270].

Before moving to illustrate another kind of conjectured origin of crucial events, let us make some additional remarks about the joint action of crucial and non-crucial events of quantum origin, based on the papers [108, 109]. Both papers study quantum systems that in the classical limit would be driven by weak chaos. They explore the predictions of Zurek and Paz [633] on the transition from the quantum to the classical regime, stressing that the condition of weak chaos may have the effect of maintaining the strange quantum properties, the superposition of two states with different eigenvalues, but for a macroscopic time. If experimental research fails to find this macroscopic quantum regime, Bonci *et al.* [108] argue that this should force the investigators to look for a theoretical approach different from decoherence theory, to explain the emergence of our classical world. Quantum tunneling is a property generated by the existence of states that are a superposition of different sites and consequently, for times larger than t_B, these articles predict the phenomenon of *chaos assisted tunneling*. This kind of tunneling is a form of irregular coherence that, however, may be connected to the origin of non-crucial events interpreted as a form of FBM. Chaos assisted tunneling, as well as other forms of quantum manifestations of chaos, are still the subject of active research work, see, for example, [268]. We hope that the conjectures made here to interpret the interesting results of the two papers [82, 270] may attract the attention of researchers to assess by how much these conjectures miss the mark.

8.5.2 Dynamical origin of CTRW

We have seen that CTRW generates results that may be mistaken as a form of FBM. We have also noticed that FBM can be derived from the quantum formalism of Mori [388] supplemented by the quantum arguments of Weiss [568]. As far as CTRW is concerned, in the recent literature there exists interest to derive CTRW from Langevin equations with multiplicative rather than merely additive stochastic forces [626]. This is a topic of increasing interest because, for example, the important issue of climate change is addressed by some investigators adopting the formalism of multiplicative fluctuations. We invite the study of the important work of Bianucci *et al.* [97], which is devoted to the study of climate change and to large-scale ocean dynamics, making the assumption that the key tool for explaining these processes is the perturbation through a multiplicative fast chaotic forcing.

The present book is devoted to crucial events and for this reason the main question that we have to address is whether or not multiplicative fluctuations may be one of the origins of crucial events. Since we have discussed the close connection between crucial events and CTRW, we may formulate the question in a slightly indirect form, as to whether or not multiplicative fluctuations may be the origin of CTRW.

In 1982 Grigolini [229] focused his attention on the study of the Kubo's stochastic oscillator:

$$\frac{dX(t)}{dt} = i\omega(t)X(t), \tag{8.107}$$

where the frequency fluctuations are independent of the motion of the variable of interest $X(t)$. Assuming that there exists a coupling between $X(t)$ and $\omega(t)$, Grigolini turned Eq. (8.107) into the Mori GLE that we have discussed in detail. The GLE is based on stationary correlation functions and as consequence can be used to explain the origin of FBM, being, however, incompatible with the generation of crucial $1/f$-noise. On the basis of these remarks we would be led to rule out multiplicative fluctuations as a possible origin of crucial events. However, if we pay attention to the title of the interesting work of Zanette and Manrubia [626] involving multiplicative processes with resets, we find that the mechanism of resets may change our conclusion.

The idealized equation of Eq. (8.88) is based on resets. When the coordinate $X(t)$ moving from an initial condition $X(0)$ in the interval $[0,1]$ reaches the value $X = 1$, we reset a new initial condition. The resetting condition is judiciously realized using the proper nonlinear maps, but the key point is that we have to assume that the new initial condition is selected with equal probability within the interval $[0,1]$.

Zanette and Manrubia [626] prompts us to go through the resetting literature and one paper they reviewed is of special interest due to the simple form of multiplicative fluctuations adopted by Takayasu *et al.* [525].

These authors discuss the following multiplicative process:

$$x(t+1) = b(t)x(t) + f(t), \qquad\qquad (8.108)$$

where $f(t)$ is a random additive noise and $b(t)$ is a non-negative stochastic coefficient which means dissipation for $b(t) < 1$ and amplification for $b(t) > 1$. If we compare Eq. (8.108) to Eq. (8.107) and do not pay attention to the use of complex number in Eq. (8.107) we notice that the main difference is the lack of an additive stochastic force in Eq. (8.107). This additive fluctuation is, in fact, the source of resetting. We might conclude that these authors had stumbled on the origin of crucial events. They did not. Let us see why.

They prove that Eq. (8.108) yields the steady solution

$$\langle X^2 \rangle = \frac{\langle f^2 \rangle}{1 - \langle b^2 \rangle},$$

which could lead one to conclude that it is possible to get a steady solution only in the case $\langle b^2 \rangle < 1$. This is not so. This is a consequence of assuming that Gaussian statistics, resting on finite second moments, is an inviable condition of statistical physics. This is also not so. It is possible to find equilibrium process with diverging values of $\langle X^3 \rangle$. The authors find that a stable but non-Gaussian equilibrium is obtained also in the case $\langle b^2 \rangle > 1$ provided the condition $\langle \ln b \rangle < 0$ is adopted.

So why did they did not find crucial events? The reason is because they maintain the condition that a stable and normalizable equilibrium distribution exists. We have seen that for the existence of crucial events we have either to abandon either the normalization condition [108, 109, 268] or the condition that a stable equilibrium condition exists [16]. Therefore, we think it would be interesting to study the simple model of Eq. (8.108) without requiring the existence of the condition of stable equilibrium.

8.5.3   Final remarks

We noticed that the generation of crucial events depends on the suitable choice of the control parameter K. There is a connection between the control parameter K of Eq. (8.102) and the control parameter K of the DMM model. The phenomena of phase transitions are generated by the judicious selection of the control parameter K. The work of Turalska and West based on the DMM shows that criticality, namely the judicious choice of K, generates crucial events. The work of Mahmoodi *et al.* [344] goes beyond DMM and demonstrates that the evolutionary game theory approach to social organization can be made compatible with criticality and crucial events by connecting the choice of K to the societal benefits of the decisions made by the society's members. If we adopt the view of Vandermeer [541] we may reach the conclusion that to understand the origin of bounded rationality we must invoke critical dynamics. However, the recent results of [82, 270] suggest that criticality should be supplemented by coherence. We hope that the conjectures made in this last chapter triggers research work to reach a deeper understanding of crucial events, after all is said and done, these events are crucial.

We now draw the curtain on our three-act play and hope that we may share insight into the nature of crucial events with you in the future.

Appendix A

Time Series Information

Bialek *et al.* [92] did the calculations presented in the first chapter by considering a one-dimensional chain of (two-state clocks) Ising spins, with a Hamiltonian:

$$H_S = -\sum_{i,j} J_{ij} s_i s_j \tag{A.1}$$

where spin j has the possible values $s_j = \pm 1$ and the interaction strength with spin i, J_{ij}. The interaction coefficient can be short-range, restricted to nearest neighbor interactions, but variable; long-range, that is, i and j do not need to be nearest neighbors, but variable; a constant. The equilibrium PDF is given by the Boltzmann distribution:

$$P(\mathbf{s}) \propto \exp[-H_S], \tag{A.2}$$

with the interaction strength normalized, such that the temperature is given by $k_B \Theta = 1$.

Consider two data streams A and B. The joint PDF $P(A, B)$ can be expressed in terms of the conditional PDF $P(A|B)$ as:

$$P(A, B) = P(A|B)P(B), \tag{A.3}$$

where $P(B)$ is the single process PDF. The predictive information associated with the two data streams is:

$$\mathcal{I}_p(A, B) = \left\langle \log_2 \left(\frac{P(A|B)}{P(A)} \right) \right\rangle$$

$$= \left\langle \log_2 \left(\frac{P(A, B)}{P(A)P(B)} \right) \right\rangle, \tag{A.4}$$

285

where the second equation is obtained by substituting from Eq. (A.3) for the conditional PDF and the brackets denote an average over the joint PDF $P(A, B)$. Equation (A.4) can be expressed in terms of entropies

$$\mathcal{I}_p(A, B) = -\langle \log_2 P(A) \rangle - \langle \log_2 P(B) \rangle + \langle \log_2 P(A, B) \rangle. \qquad \text{(A.5)}$$

Now assume the two data streams to be the past B and the future A of a single time series $Q(t)$, extending over the interval $-L \leq t \leq L'$ with $t = 0$ marking the present. Equation (A.5) can be expressed in terms of entropies labeled with the extensive variable, which for time series is the duration of the sequence. The entropy for the past depends only on the observation window L:

$$S(L) = -\langle \log_2 P(B) \rangle, \qquad \text{(A.6)}$$

the entropy for the future depends on the observation window L':

$$S(L') = -\langle \log_2 P(A) \rangle, \qquad \text{(A.7)}$$

and the total entropy depends on the overall observation window $L + L'$:

$$S(L + L') = -\langle \log_2 P(A, B) \rangle. \qquad \text{(A.8)}$$

Substituting these three expressions for entropy into the appropriate places in the equation for the predictive information yields:

$$\mathcal{I}_p(L, L') = S(L) + S(L') - S(L + L'), \qquad \text{(A.9)}$$

which is also the mutual information for the past and future of the time series.

It is possible to realize that because entropy is extensive, consequently the information given by Eq. (3.42) ought to be zero asymptotically. However, we also know that there is a non-extensive residual to the entropy so that:

$$S(L) = S_0 L + S_1(L) \qquad \text{(A.10)}$$

as a direct generalization of the discrete case given by Eq. (3.42) and subsequently:

$$\lim_{L \to \infty} \frac{S_1(L)}{L} = 0. \qquad \text{(A.11)}$$

Equation (A.11) indicates that the residual grows more slowly in time than linearly, it is therefore subextensive. Bialek et $al.$ [92] prove that the predictive information:

$$\mathcal{I}_p(L, L') \geq 0, \qquad \text{(A.12)}$$

and it is subextensive.

A.0.1 Examples of predictive information

Consider the scaling PDF for a time series $Q(t)$ with $P(q,t)dq$ as the probability that the dynamic variable has a value in the interval $(q, q + dxq)$ at time t:

$$P(q,t) = \frac{1}{t^\delta} F\left(\frac{q}{t^\delta}\right), \tag{A.13}$$

where $F(\cdot)$ is an unknown analytic function. Substituting this value for the PDF at the time t into the definition of the Wiener/Shannon entropy yields:

$$S(t) = \overline{S} + \delta \log_2 t \tag{A.14}$$

where the constant term is defined by the integral over the scaling variable $z = q/t^\delta$:

$$\overline{S} = -\int F(z) \log_2 F(z) dz \tag{A.15}$$

and the PDF is properly normalized:

$$\int P(q,t)dq = \int F(z)dz = 1.$$

Inserting the expression for the entropy obtained using the scaled PDF Eq. (A.14) into that for the predictive information given by Eq. (A.9), and simplifying, yields:

$$\mathcal{I}_p(L, L') = \delta \log_2 \left(\frac{LL'}{L + L'}\right). \tag{A.16}$$

Thus, in the infinitely extended future $L' \to \infty$, we can measure the information about the entire future from the past time of length L:

$$\mathcal{I}_p(L) = \lim_{L' \to \infty} \delta \log_2 \left(\frac{LL'}{L + L'}\right) = -\delta \log_2 L. \tag{A.17}$$

Note that Eq. (A.17) is the subextensive entropy:

$$S_1(L) = -\delta \log_2 L, \tag{A.18}$$

which for the Gaussian PDF has $\delta = 0.5$ and for the Lévy PDF has $\delta = 1/\mu$ and the Lévy index is in the interval $0 < \mu \leq 2$. Furthermore, if the past and future time series are of equal length $L = L'$, it is clear from Eq. (A.16) that:

$$\mathcal{I}_p(L, L) = -\delta \log_2(2) = -\delta, \tag{A.19}$$

and the predictive information is incorporated into the constant scaling index δ.

Bialek *et al.* [92] emphasize that the information measure over a time L is given by $S(L)$, which asymptotically has the extensive form $S_0 L$, but the predictive information increases subextensively as $S_1(L)$. Consequently, the total information garnered over the infinite past swamps the predictive information:

$$\lim_{L \to \infty} \frac{\text{predictive information}}{\text{total information}} = \lim_{L \to \infty} \frac{\mathcal{I}_p(L)}{S(L)} = 0 \qquad (A.20)$$

from which it is clear that most of the information gathered has no value. Here we are using value in the sense of being able to predict the future; if the information cannot be used in making predictions, it has no practical value.

Appendix B

Dynamic Linear Response Theory

Average response; stationary case: $2 < \mu_S < 3$.

In the stationary regime the waiting-time PDF $\psi_S(t)$, you will recall, has a finite mean value τ_S. Therefore a finite time scale t_C exists, such that for $t > t' > t_C$ the following approximation, corresponding to reaching the stationary condition, is valid [21]:

$$\psi_S(t, t') \simeq \frac{1}{\tau_S} \int_t^\infty ds \psi_S(s - t') = \frac{\Psi_S(t - t')}{\tau_S}, \qquad (B.1)$$

where

$$\tau_S = \frac{T_S}{\mu_S - 2},$$

is the mean value of $\psi_S(t)$. Equation (B.1) is exact for $t_C \to \infty$. We can also use Eq. (6.1) for the response in the case when the interaction is turned on at a later time t_C that we assume to be so large as to satisfy the approximation Eq. (B.1) using the following procedure.

We introduce into Eq. (6.1) an effective perturbation which is turned on at time $t = 0$, but is zero until t_C:

$$\xi_P^{eff}(t) = \Theta(t - t_C)\xi_P(t - t_C). \qquad (B.2)$$

In this scheme we can assume that the "real" process $\xi_P(t)$ is prepared at time $t = t_C$ in a new condition. The time t_C therefore, corresponds to the age of the system S, the age being the time the system S has been evolving prior to the onset of the perturbation from system P. The average response $\langle\sigma(t)\rangle$ of S in this case reads

$$\langle\sigma(t)\rangle = \epsilon \int_0^t dt'\psi_S(t,t')\langle\Theta(t'-t_C)\xi_P(t'-t_C)\rangle$$

$$= \epsilon \int_0^t dt'\psi_S(t,t')\Psi_P(t'-t_C), \qquad (B.3)$$

and the approximation Eq. (B.1) can be used to replace $\psi_S(t,t')$. With the substitution $\tau = t - t_C$ and after renaming $t - t_C$ as t back again, the average response of the system S of age t_C reads

$$\frac{\langle\sigma(t)\rangle|_{t_C}}{\epsilon} = \frac{1}{T_S} \int_0^t d\tau\Psi_S(t-\tau)\Psi_P(\tau), \qquad (B.4)$$

which becomes exact when $t_C \to \infty$. The asymptotic behavior of Eq. (B.4) is easily obtained in the Laplace domain:

$$\langle\hat{\sigma}(u)\rangle|_{t_C} = \frac{\epsilon}{T_S} \frac{(1-\widehat{\psi}_S(u))}{u} \frac{(1-\widehat{\psi}_P(u))}{u}, \qquad (B.5)$$

which we examine in the asymptotic regime $u \to 0$, that is, the limit $t \gg T_S, T_P$. For the hyperbolic form of the waiting-time PDFs we can write schematically

$$\widehat{\psi}(u) \simeq 1 - \tau u + \Gamma(\mu-2)u^{\mu-1}, \quad \mu > 2, \qquad (B.6)$$

and

$$\widehat{\psi}(u) \simeq 1 + \Gamma(\mu-2)u^{\mu-1}, \quad \mu < 2. \qquad (B.7)$$

Input–output cross-correlations

Non-stationary case I: $\mu_S < 2$, $\mu_P < 2$.

In the non-stationary case considered here, we use Eq. (6.25) along with the general expression of the waiting-time PDF for responder S $\psi_S(t,\tau)$ and the survival probability for stimulator P $\Psi_P(t,\tau)$ as obtained using Eqs. (6.7) and (6.9), respectively. The details for obtaining the general expression can be found elsewhere [44], which in the asymptotic limit yields:

$$\zeta_D(\mu_S, \mu_P) = \lim_{t\to\infty} \Phi(t) = -\frac{\sin(\pi\mu_P)}{\pi} \frac{\Gamma(\mu_P + \mu_S - 1)}{(\mu_P-1)\Gamma(\mu_P+1)\Gamma(\mu_S-1)}$$

$$\times F(\{\mu_P - 1, \mu_P - 1, \mu_P + \mu_S - 1\}, \{\mu_P, \mu_P + 1\}, 1), \quad (B.8)$$

where $F(\cdot)$ is the generalized confluent hypergeometric function. The functional form for the asymptotic input–output correlation function is given by region I of the cross-correlation cube in Fig. 1.3.

Non-stationary case II: $\mu_S < 2$, $\mu_P > 2$.

In the case where S is in a non-stationary regime, but the stimulator P is stationary, we can assume that, when the interaction is turned on, the stimulator has already reached a stationary condition. Consequently, in the input–output correlation function given by Eq. (6.25) the waiting-time PDF $\psi_S(t, \tau)$ is again given by Eq. (6.7), but the survival probability is stationary $\Psi_P(t, \tau) = \overline{\Psi}_P(t - \tau)$, and is given by the functional form of Eq. (6.23), with the labeling changed from S to P. Thus, we insert

$$\overline{\Psi}_P(t) = \left(\frac{T_P}{t + T_P} \right)^{\mu_P - 2} \tag{B.9}$$

into Eq. (6.25).

The IPL index $\mu_P - 2$ reflects the stationary condition realized with the preparation of the perturbation P at a time $t_P = -\infty$. In this case, we obtain

$$\lim_{t \to \infty} \Phi(t) = 0. \tag{B.10}$$

In the time asymptotic limit, the system S turns out to be independent of P in spite of the fact that at $t = 0$ we switch on the S–P interaction. This is depicted in region II of the input–output cube in Fig. 1.4.

$$x = \frac{\Delta T}{T_P}, b = 2 - \mu_P \text{ when } T_P < T_S, \tag{B.11}$$

and

$$x = \frac{\Delta T}{T_S}, b = 3 - \mu_S \text{ when } T_P > T_S. \tag{B.12}$$

When the two T-parameter values are the same, the Beta function reduces to

$$\lim_{\Delta T \to 0} B(x, \mu_S + \mu_P - 4, b) = \frac{\mu_S - 2}{\mu_S + \mu_P - 4}. \tag{B.13}$$

These values of the input–output correlation function are depicted in region IV of Fig. 1.4.

Stationary case I: $\mu_S > 2$, $\mu_P > 2$.

In the situation, where both the perturber and the perturbed are stationary, we again use Eq. (6.25) and assume that, when the interaction is turned on, the perturbation P has already reached a stationary condition so that $\Psi_P(t,\tau) = \overline{\Psi}_P(t-\tau)$, with $\overline{\Psi}_P(t)$ given by Eq. (B.9). For $\mu_S > 2$ a finite mean time τ_s of the waiting-time PDF $\psi_S(t)$ exists. Therefore, a finite time scale $t_C \propto \tau_S$ exists such that for $t > t_C$, Eq. (B.1) can be used again to approximate the waiting-time PDF $\psi_S(t,\tau)$. With these substitutions, the expression for the cross-correlation function becomes asymptotically equal to [44]:

$$
\Phi(t) = \int_0^t d\tau \frac{\Psi_S(t-\tau)}{\tau_s} \Psi_P(t,\tau)
$$

$$
= \overline{\Psi}_S(0)\Psi_P(t,t) - \overline{\Psi}_S(t)\Psi_P(t,0) - \int_0^t d\tau \overline{\Psi}_S(t-\tau)\frac{d\Psi_P(t,\tau)}{d\tau}.
$$

$$(B.14)$$

In the asymptotic limit, this expression reduces to the analytic form

$$
\Phi_\infty = \lim_{t\to\infty} \Phi(t) = 1 - (\mu_P - 2)T_P^{\mu_P-2}T_S^{\mu_S-2}\Delta T^{4-\mu_S-\mu_P}B(x, \mu_S+\mu_P-4, b),
$$

$$(B.15)$$

where $\Delta T = |T_P - T_S|$, $B(x, a, b)$ is the incomplete Beta function with

$$
x = \frac{\Delta T}{T_P}, b = 2 - \mu_P \text{ when } T_P < T_S, \tag{B.16}
$$

and

$$
x = \frac{\Delta T}{T_S}, b = 3 - \mu_S \text{ when } T_P > T_S. \tag{B.17}
$$

When the two T parameter values are the same, the Beta function reduces to

$$
\lim_{\Delta T\to 0} B(x, \mu_S+\mu_P-4, b) = \frac{\mu_S - 2}{\mu_S + \mu_P - 4}. \tag{B.18}
$$

These values of the input–output correlation function are depicted in region IV of Fig. 1.3.

Stationary case II: $\mu_S > 2$, $\mu_P < 2$.

In this case, where the perturbing system P is no longer stationary, again a finite time scale t_C can be found such that the approximation in Eq. (B.1) is valid. Thus, again the input–output correlation function given by Eq. (6.25) can be rewritten:

$$\Phi(t) \simeq \int_0^{t_c} d\tau \psi_S(t-\tau) \Psi_P(t,\tau) + \int_{t_c}^t d\tau \frac{d\overline{\Psi}_S(t-\tau)}{d\tau} \Psi_P(t,\tau). \qquad \text{(B.19)}$$

Again, in the asymptotic limit $t \to \infty$, the first term in Eq. (B.19) vanishes and after integrating the second term by parts, one obtains:

$$\Phi(t) \simeq \overline{\Psi}_S(0)\Psi_P(t,t) - \overline{\Psi}_S(t-t_C)\Psi_P(t,t_C) - \int_{t_C}^t d\tau \overline{\Psi}_S(t-\tau)\frac{d\Psi_P(t,\tau)}{d\tau}.$$
$$\text{(B.20)}$$

In the asymptotic limit the second term on the RHS trivially vanishes and also the third term can be shown to vanish [44]. The only remaining contribution in Eq. (B.20) is given by the first term, which is exactly one. It follows:

$$\Phi_\infty = 1, \qquad \text{(B.21)}$$

which is depicted in region III of Fig. 1.3.

Appendix C

Quantum Nomenclature

The well-known Schrödinger equation in Dirac notation for the wave function and the operator Hamiltonian is:

$$\frac{d}{dt}|\psi(t)\rangle = -\frac{i}{\hbar}\widehat{H}|\psi(t)\rangle, \tag{C.1}$$

and affords the time evolution of a generic ket $|\psi\rangle$ with unitary time evolution:

$$|\psi(t)\rangle = e^{-\frac{i\widehat{H}}{\hbar}t}|\psi\rangle. \tag{C.2}$$

We make frequent use of the density matrix:

$$\rho(t) = \sum_j p_j |\psi_i(t)\rangle\langle\psi_j(t)|, \tag{C.3}$$

to establish a connection with the literature of statistical physics. The weight factors p_j of the projection operator onto the state $|\psi_i\rangle$ imply the probability of that state being occupied and properly normalized:

$$\sum_j p_j = 1. \tag{C.4}$$

The quantum density matrix serves the purpose of generating an equation of motion which plays the same role in the microscopic domain that the classical Liouville equation plays in the macroscopic domain. In classical mechanics, the solution to the Liouville equation represents the time evolution of the phase space trajectory density function (TDF). In the special case of chaotic trajectories or of regular dynamics under the influence of random perturbation, the trajectories are erratic, while the time evolution of the TDF may be regular, thereby providing information about the PDF.

In the quantum mechanical case each ket $|\psi_i(t)\rangle$ represents a trajectory with a time evolution given, according to Eq. (C.2), by the equation:

$$|\psi_j(t)\rangle = e^{-\frac{i\widehat{H}}{\hbar}t}|\psi_j\rangle, \tag{C.5}$$

corresponding to trajectories with different initial conditions.

Thus, the time evolution of Eq. (C.3) can be expressed as:

$$\rho(t) = \sum_j p_j e^{-\frac{i\widehat{H}}{\hbar}t}|\psi_j\rangle\langle\psi_j|e^{\frac{i\widehat{H}}{\hbar}t}, \tag{C.6}$$

whose dynamics satisfy the equation of motion:

$$\frac{d}{dt}\rho(t) = -\frac{i}{\hbar}[\widehat{H}, \rho(t)] \equiv -\mathcal{L}\rho(t), \tag{C.7}$$

which is known as quantum Liouville equation, with super-operator $\mathcal{L}$ given by the commutator with the Hamiltonian operator.

It is interesting to notice that Heisenberg describes quantum mechanics using an equation of motion where observables change in time while the wave function remains constant. However, the mean values in the Heisenberg representation are identical to the mean values in the Schrödinger representation. Let us consider a generic observable A such that:

$$\langle A(t)\rangle = \langle\psi(t)|A|\psi(t)\rangle = \langle\psi|e^{\frac{i\widehat{H}}{\hbar}t}Ae^{-\frac{i\widehat{H}}{\hbar}t}|\psi\rangle, \tag{C.8}$$

making the observable A change in time according to the prescription:

$$A(t) = e^{\frac{i\widehat{H}}{\hbar}t}Ae^{-\frac{i\widehat{H}}{\hbar}t}, \tag{C.9}$$

yielding the equation of motion in terms of the quantum Liouville operator:

$$\frac{d}{dt}A(t) = \mathcal{L}A(t), \tag{C.10}$$

which is similar to the quantum Liouville equation for the TDF given by Eq. (C.7).

C.1   Mori Formalism

Let us now focus our attention on $k = 0$ and apply P_0 to both sides of Eq. (B.10):

$$\frac{d}{dt}P_0|f_0(t)\rangle = P_0\Gamma_0|f_0(t)\rangle, \tag{C.11}$$

and inserting unity in the form of Eq. (8.4) into this expression yields:

$$\frac{d}{dt}P_0|f_0(t)\rangle = P_0\Gamma_0 P_0|f_0(t)\rangle + P_0\Gamma_0 Q_0|f_0(t)\rangle. \tag{C.12}$$

Following the same procedure but this time starting with the multiplication by Q_0 yields:

$$\frac{d}{dt}Q_0|f_0(t)\rangle = Q_0\Gamma_0 P_0|f_0(t)\rangle + Q_0\Gamma_0 Q_0|f_0(t)\rangle, \tag{C.13}$$

whose solution can be expressed by the quadrature:

$$Q_0|f_0(t)\rangle = e^{Q_0\Gamma_0 t}Q_0|f_0\rangle + \int_0^t dt' e^{Q_0\Gamma_0(t-t')}Q_0\Gamma_0 P_0|f_0(t')\rangle. \tag{C.14}$$

To prove that Eq. (C.14) is the exact solution of Eq. (C.13), let us differentiate the left-hand side (LHS) with respect to t. Consequently the time derivative of the first term on the RHS of Eq. (C.14) and with respect to t within the integral of the second term on the RHS of the same equation has the effect of producing $Q_0\Gamma_0 Q_0|f_0(t)\rangle$, namely the second term on the RHS of Eq. (C.13). We have to also differentiate with respect to the upper time t of the time integral on the RHS of Eq. (C.14). This has the effect of generating the first term on the RHS of Eq. (C.14).

We note that as a consequence of $Q_0|f_0\rangle = 0$ Eq. (C.14) reduces to:

$$Q_0|f_0(t)\rangle = \int_0^t dt' e^{Q_0\Gamma_0(t-t')}Q_0\Gamma_0 P_0|f_0(t')\rangle. \tag{C.15}$$

Let us insert Eq. (8.3) into the RHS of Eq. (C.15) to obtain:

$$Q_0|f_0(t)\rangle = \int_0^t dt' e^{Q_0\Gamma_0(t-t')}Q_0\Gamma_0|f_0\rangle \Phi_0(t'). \tag{C.16}$$

Note that the equilibrium correlation functions can be generally defined:

$$\Phi_k(t) \equiv \frac{\langle \widetilde{f}_k|f_k(t)\rangle}{\langle \widetilde{f}_k|f_k\rangle}, \tag{C.17}$$

where the time-dependent wave function of Schrödinger is:

$$|f_k(t)\rangle = e^{\Gamma_k t}|f_k\rangle, \tag{C.18}$$

and the $k = 0$ equilibrium correlation function appears in Eq. (C.16).

It is a significant utility of the Mori formalism that only the state $|f_0\rangle$ is driven by the whole operator Γ_0. The state $|f_1\rangle$ is driven by Γ_1 and this makes it possible for the state $|f_0\rangle$ to perceive the state $|f_1\rangle$ as a representation of its environment with dynamical properties independent of the dynamics of the system of interest. The same property applies to the state f_k, whose dynamics depend on $|f_{k+1}\rangle$, the latter having a time evolution independent of that of $|f_k\rangle$. The adoption of this equilibrium correlation function leads us to write the projection off the initial state in the form given by Eq. (C.16). On the basis of Eq. (C.18) we write Eq. (C.16) more simply:

$$Q_0|f_0(t)\rangle = \int_0^t dt' \Phi_0(t')|f_1(t - t')\rangle. \tag{C.19}$$

We are now in a position to write the equation of motion for the part of the system in which we are interested, namely, Eq. (C.12). By inserting Eq. (C.19) into the second term of Eq. (C.12) we obtain:

$$\frac{d}{dt} P_0|f_0(t)\rangle = P_0\Gamma_0 P_0|f_0(t)\rangle + \int_0^t dt' P_0\Gamma_0|f_1(t - t')\rangle \Phi_0(t'). \tag{C.20}$$

Using the definition of Eq. (8.3) we write the first term on the RHS of Eq. (C.20) as follows:

$$P_0\Gamma_0 P_0|f_0(t)\rangle = |f_0\rangle \frac{\langle f_0|\Gamma_0|f_0\rangle}{\langle f_0|f_0\rangle} \frac{\langle f_0|f_0(t)\rangle}{\langle f_0|f_0\rangle}. \tag{C.21}$$

We define in general

$$\lambda_k \equiv \frac{\langle \widetilde{f}_k|\Gamma_k|f_k\rangle}{\langle \widetilde{f}_k|f_k\rangle}. \tag{C.22}$$

Using this definition and that of Eq. (C.17) as well, we rewrite Eq. (C.21) in the following more attractive form:

$$P_0\Gamma_0 P_0|f_0(t)\rangle = |f_0\rangle \lambda_0 \Phi_0(t). \tag{C.23}$$

Let us now work on the second term on the RHS of Eq. (C.20) again using the definition given by Eq. (8.3) to obtain:

$$\int_0^t dt' P_0\Gamma_0|f_1(t - t')\rangle \Phi_0(t') = |f_0\rangle \int_0^t dt' \frac{\langle f_0|\Gamma_0|f_1(t - t')\rangle \Phi_0(t')}{\langle f_0|f_0\rangle}. \tag{C.24}$$

To make the correlation function $\Phi_1(t)$ emerge we use $P_0 + Q_0 = 1$ and notice that due to the orthogonality of the states $P_0|f_1(t - s)\rangle = 0$, so that explicitly introducing the correlation function $\Phi_1(t)$ after some algebra, we rewrite this equation as:

$$|f_0\rangle \int_0^t dt' \frac{\langle f_0|\Gamma_0|f_1(t - t')\rangle \Phi_0(t')}{\langle f_0|f_0\rangle} = -|f_0\rangle \Delta_1^2 \int_0^t dt' \Phi_1(t - t') \Phi_0(t'),$$

$$\tag{C.25}$$

based on the definition:

$$\Delta_1^2 = -\frac{\langle \widetilde{f}_1 | f_1 \rangle}{\langle f_0 | f_0 \rangle} > 0, \tag{C.26}$$

which we adopt so as to assign to Δ_1 a real value in the Hermitian case. In conclusion Eq. (C.20) becomes

$$\frac{d}{dt}\Phi_0(t) = \lambda_0 \Phi_0(t) - \Delta_1^2 \int_0^t dt' \Phi_1(t - t')\Phi_0(t'). \tag{C.27}$$

Tri-diagonal representation

It is interesting to notice that using the Mori prescription we find for Γ_0 expressed in the basis set of the normalized states $|f_k\rangle$, the following tri-diagonal form

$$\Gamma_0 = \begin{pmatrix} \lambda_0 & \Delta_1 & 0 & 0 & 0 & 0 & \dots \\ \Delta_1 & \lambda_1 & \Delta_2 & 0 & 0 & 0 & \dots \\ 0 & \Delta_2 & \lambda_2 & \Delta_2 & 0 & 0 & \dots \\ 0 & 0 & \Delta_3 & \lambda_3 & \Delta_3 & 0 & \dots \\ 0 & 0 & 0 & \Delta_4 & \lambda_4 & \Delta_4 & \dots \\ 0 & 0 & 0 & 0 & \Delta_5 & \lambda_5 & \dots \\ \dots & \dots & \dots & \dots & \dots & \dots & \dots \end{pmatrix} \tag{C.28}$$

which is a Hermitian matrix.

C.2   Conditional Probability

Let us return to the entanglement issue discussed in Sec. 8.1.2. Let us imagine that the quantum system a is a quantum oscillator and the quantum system b is a second quantum oscillator. In both cases, let us consider only the ground state $|0\rangle\rangle$ and the first excited state $|1\rangle$. Consider the ket:

$$|\psi_T\rangle = \frac{1}{\sqrt{2}}(|10\rangle + |01\rangle), \tag{C.29}$$

a ket that we can also write as:

$$|\psi_T\rangle = c_{10}|10\rangle + c_{01}|01\rangle, \tag{C.30}$$

where the coefficients are:

$$c_{10} = c_{01} = \frac{1}{\sqrt{2}}. \tag{C.31}$$

If the oscillator b plays the role of the environment of oscillator a and we evaluate the reduced density matrix of the oscillator a we obtain, as shown in Sec. 8.1.2:

$$\sigma = \frac{1}{2}(|0\rangle\langle 0| + |1\rangle\langle 1|) \tag{C.32}$$

in full agreement with the Copenhagen interpretation of quantum mechanics. The total density matrix ρ reads:

$$\rho = |c_{10}|^2|10\rangle\langle 10| + c_{10}c_{01}^*|10\rangle\langle 01| + c_{01}c_{10}^*|01\rangle\langle 10|$$
$$+ |c_{01}|^2|01\rangle\langle 01|. \tag{C.33}$$

We note that while $1/2$ of the reduced density matrix of Eq. (C.32) is a probability, the expansion coefficient $|c_{10}|^2$, for instance, is not a probability, but a conditional probability. Let us make a jump back in time to when quantum mechanics was not yet known. With the help of Wikipedia let us go back to Thomas Bayes.

Thomas Bayes (1701–1761) was an English statistician, philosopher and Presbyterian minister who is known for formulating a specific case of the theorem that today bears his name: Bayes' theorem. Bayes never published what would become his most famous accomplishment; his notes were edited and published after his death.

Suppose that we are interested in evaluating the probability that the event A occurs under the condition that the event B occurs. This conditional probability is defined by the symbol $p(A|B)$ and Bayes' theorem may be expressed as:

$$p(A|B)p(B) = p(B|A)p(A) = p(A \cup B). \tag{C.34}$$

The theorem is interpreted to mean that the probability for A and B to be true at the same time can be obtained by assuming that B is true with probability $p(B)$ multiplied by the conditional probability that A is true if B is true. The same result is obtained by assuming that A is true with probability $p(A)$ and multiplying $p(A)$ by the conditional probability that B is true, under the assumption that A is true. Consequently, this equation expresses the probability that both A and B are true.

Bayes' theorem reads for the conditional probability:

$$p(A|B) = \frac{p(B|A)p(A)}{p(B)}. \tag{C.35}$$

To relate this result to our problem, we need to define A as the oscillator a in the state $|1\rangle$ under the condition that the oscillator b is in the state $|0\rangle$.

The event B is the oscillator b in the state $|0\rangle$. We can make the following association:

$$p(A|B) = |c_{10}|^2 = \frac{1}{2}, \tag{C.36}$$

thereby expressing the quantum coupling coefficients with the conditional probability of Bayes. On the other hand, to evaluate the probability of finding the oscillator a in the state $|0\rangle$ under the condition that the oscillator b is in the state $|1\rangle$, we use:

$$p(A'|B') = \frac{p(B'|A')p(A')}{p(B')}, \tag{C.37}$$

where A' is the oscillator a in the state $|0\rangle$ and B' is the oscillator b in the state $|1\rangle$. We find:

$$p(A'|B') = |c_{01}|^2 = \frac{1}{2}. \tag{C.38}$$

In conclusion, we know $p(A|B)$ and $p(A'|B')$, but we do not yet know $p(A)$ and $p(A')$.

Let us return to Eq. (C.35). Due to the symmetry of the problem we should have the equality $p(B|A) = p(A|B)$, which yields:

$$p(A) = p(B), \tag{C.39}$$

and from Eq. (C.37) we have:

$$p(A') = p(B'). \tag{C.40}$$

On the other hand, the symmetry of the problem also forces us to set $p(B'|A') = p(A'|B')$. Let us divide the conditional probability given by Eq. (C.35) by the conditional probability given by Eq. (C.37) to obtain:

$$\frac{p(A|B)}{p(A'|B')} = \frac{p(B|A)p(A)}{p(B)} \frac{p(B')}{p(B'|A')p(A')}. \tag{C.41}$$

Using $p(A|B) = p(A'|B')$ and $p(B'|A') = p(A'|B')$ we obtain:

$$\frac{p(A)}{p(B)} = \frac{p(A')}{p(B')}, \tag{C.42}$$

which is equivalent to:

$$\frac{p(A)}{p(A')} = \frac{p(B)}{p(B')}, \tag{C.43}$$

from which we conclude that the answer

$$p(A) = p(A') = \frac{1}{2}, \tag{C.44}$$

does not violate Bayes' theorem, but it is not derived from it. In conclusion, we cannot use Bayes' theorem to prove that the conditional probability is a probability. The use of Bayes statistics in the cognition literature (see, for instance, [289]) must be examined with attention.

C.3   Quantum Harmonic Oscillator

In Appendix C.1, we explained how to build up a tri-diagonal matrix representation for an observable A moving from a ket $|f_0\rangle$, which is not an eigenstate of the observable A. In this section, we prove that in the case when the time evolution operator Γ_0 is the Hamiltonian operator $\widehat{H}$, namely, we are using the Schrödinger equation, the Mori prescription recovers the results of quantum mechanical textbooks. We adopt the treatment illustrated in Chapter XII of the Messiah's textbook.

C.3.1   Eigenstates and eigenvalues of the quantum harmonic oscillator

The Hamiltonian of a one-dimensional quantum oscillator of mass m is given by:

$$\mathcal{H} = \frac{1}{2m}(\widehat{p}^2 + m^2\omega^2\widehat{q}^2), \tag{C.45}$$

the displacement operator $\widehat{q}$ and the momentum operator $\widehat{p}$ being connected through the commutation relation:

$$[\widehat{q}, \widehat{p}] = i\hbar. \tag{C.46}$$

It is convenient to use the dimensionless representation:

$$\mathcal{H} = \widehat{H}\hbar\omega, \widehat{q} = \left(\frac{\hbar}{m\omega}\right)^{\frac{1}{2}} \widehat{Q} \text{ and } \widehat{p} = (m\hbar\omega)^{\frac{1}{2}}\widehat{P}, \tag{C.47}$$

where by construction $\widehat{H}$, $\widehat{Q}$ and $\widehat{P}$ are dimensionless operators. The Hamiltonian operator $\widehat{H}$ reads:

$$\widehat{H} = \frac{1}{2}(\widehat{P}^2 + \widehat{Q}^2) \tag{C.48}$$

and the commutator between $\widehat{Q}$ and $\widehat{P}$ becomes:

$$[\widehat{Q}, \widehat{P}] = i. \tag{C.49}$$

Creation and annihilation operators

We define the annihilation operator:

$$a = \frac{1}{\sqrt{2}}(\widehat{Q} + i\widehat{P}) \tag{C.50}$$

and creation operator:

$$a^+ = \frac{1}{\sqrt{2}}(\widehat{Q} - i\widehat{P}). \tag{C.51}$$

These operators do not commute and their commutation relation is:

$$[a, a^+] = 1. \tag{C.52}$$

These relations may be inverted by summing and subtracting Eqs. (C.50) and (C.51) to obtain the displacement operator:

$$\widehat{Q} = \frac{1}{\sqrt{2}}(a + a^+) \tag{C.53}$$

and the momentum operator:

$$\widehat{P} = -\frac{i}{\sqrt{2}}(a - a^+). \tag{C.54}$$

Thus, the Hamiltonian given by Eq. (C.48) becomes:

$$\widehat{H} = \frac{1}{2}(aa^+ + a^+a). \tag{C.55}$$

We define also the number operator $\widehat{N}$:

$$\widehat{N} \equiv a^+a. \tag{C.56}$$

It is important to notice that this operator is Hermitian. In fact, taking its complex conjugate yields the tautology:

$$\widehat{N}^+ = (a^+a)^+ = a^+(a^+)^+ = a^+a = \widehat{N}. \tag{C.57}$$

Let us study the product:
$$\widehat{N}a = a^+aa \tag{C.58}$$

along with:
$$\widehat{N}a^+ = a^+aa^+. \tag{C.59}$$

Equation (C.52) can be rearranged to obtain:

$$a^+a = aa^+ - 1 \tag{C.60}$$

as well as:

$$aa^+ = a^+a + 1. \tag{C.61}$$

Inserting Eq. (C.60) into Eq. (C.58) yields:

$$\begin{aligned} \widehat{N}a &= (a^+a)a = aa^+a - a \\ &= a(a^+a - 1) = a(\widehat{N} - 1). \end{aligned} \tag{C.62}$$

Similarly, inserting Eq. (C.61) into Eq. (C.59) yields:

$$\begin{aligned} \widehat{N}a^+ &= a^+(aa^+) = a^+a^+a + a^+ \\ &= a^+(a^+a + 1) = a^+(\widehat{N} + 1). \end{aligned} \tag{C.63}$$

The eigenvalues of $\widehat{N}$ are integer numbers

If $|\nu\rangle$ is an eigenvector of $\widehat{N}$, and ν the corresponding eigenvalue, then, necessarily $\nu \geq 0$. This is a consequence of the properties of scalar products. In fact, let us define $|w\rangle = a|\nu\rangle$. We know that $\langle w|w\rangle \geq 0$ and $\widehat{N}|\nu\rangle = \nu|\nu\rangle$ so the scalar product reads:

$$\begin{aligned} \langle w|w\rangle &= \langle \nu|a^+a|\nu\rangle = \langle \nu|\widehat{N}|\nu\rangle \\ &= \nu\langle \nu|\nu\rangle = \nu \geq 0. \end{aligned} \tag{C.64}$$

Let us now prove that the eigenvalues ν are integer numbers. Use Eqs. (C.62) and (C.63) to define:

$$\widehat{N}a|\nu\rangle = a(\widehat{N} - 1)|\nu\rangle = a(\nu - 1)|\nu\rangle = (\nu - 1)a|\nu\rangle,$$

along with:

$$\widehat{N}a^+|\nu\rangle = a^+(\widehat{N} + 1)|\nu\rangle = a^+(\nu + 1)|\nu\rangle = (\nu + 1)a^+|\nu\rangle.$$

Thus, if $|\nu\rangle$ is an eigenvector of $\widehat{N}$ with eigenvalue ν, then $a|\nu\rangle$ is an eigenvector of $\widehat{N}$ with eigenvalue $\nu - 1$. By induction it is then a simple matter to prove:

$$\widehat{N}a^n|\nu\rangle = (\nu - n)|\nu\rangle. \tag{C.65}$$

If ν is an integer number, for instance n, then we obtain the eigenvalue 0, which is a legitimate result. If, on the other hand, ν is an integer number but less than n, we would get a negative eigenvalue, which is in conflict with Eq. (C.64).

The creation and annihilation operators act on the eigenstates $|n\rangle$ with integer eigenvalues as follows:

$$a^+|n\rangle = (n + 1)^{\frac{1}{2}}|n + 1\rangle \tag{C.66}$$

and

$$a|n\rangle = (n)^{\frac{1}{2}}|n-1\rangle. \tag{C.67}$$

In fact, combining Eqs. (C.66) and (C.67), we obtain:

$$\langle n|a^+a|n\rangle = n\langle n-1|n-1\rangle = n \tag{C.68}$$

in accordance with the operation of the number operator:

$$\langle n|\widehat{N}|n\rangle = n. \tag{C.69}$$

In the same way, we obtain:

$$\langle n|aa^+|n\rangle = (n+1)\langle n+1|n+1\rangle = n+1. \tag{C.70}$$

Using Eq. (C.61), the LHS of Eq. (C.70) becomes

$$\langle n|(a^+a+1)|n\rangle = n+1, \tag{C.71}$$

in accordance with

$$\langle n|(\widehat{N}+1)|n\rangle = \langle n|\widehat{N}|n| + \langle n|n\rangle = n+1. \tag{C.72}$$

Matrix representation of $\widehat{N}$, $\widehat{H}$, a and a^+

It is straightforward to show that in this dimensionless representation we have the Hamiltonian:

$$\widehat{H} = \widehat{N} + \frac{1}{2}. \tag{C.73}$$

The number operator $\widehat{N}$ has the following eigenstates and eigenvalues:

$$\widehat{N}|n\rangle = n|n\rangle, \tag{C.74}$$

where $n = 0, 1, 2, ...$, yielding the following diagonal matrix representation:

$$\widehat{N} = \begin{pmatrix} 0 & 0 & 0 & 0 & 0 & 0 & \cdots \\ 0 & 1 & 0 & 0 & 0 & 0 & \cdots \\ 0 & 0 & 2 & 0 & 0 & 0 & \cdots \\ 0 & 0 & 0 & 3 & 0 & 0 & \cdots \\ 0 & 0 & 0 & 0 & 4 & 0 & \cdots \\ 0 & 0 & 0 & 0 & 0 & 5 & \cdots \\ \cdots & \cdots & \cdots & \cdots & \cdots & \cdots & \cdots \end{pmatrix}.$$

The operator $\widehat{H}$ has the following eigenstates and eigenvalues

$$\widehat{H}|n\rangle = \left(n+\frac{1}{2}\right)|n\rangle, \tag{C.75}$$

where $n = 0, 1, 2, ...$, yielding the following diagonal matrix representation:

$$\hat{H} = \begin{pmatrix} \frac{1}{2} & 0 & 0 & 0 & 0 & 0 & ... \\ 0 & \frac{3}{2} & 0 & 0 & 0 & 0 & ... \\ 0 & 0 & \frac{5}{2} & 0 & 0 & 0 & ... \\ 0 & 0 & 0 & \frac{7}{2} & 0 & 0 & ... \\ 0 & 0 & 0 & 0 & \frac{9}{2} & 0 & ... \\ 0 & 0 & 0 & 0 & 0 & \frac{11}{2} & ... \\ ... & ... & ... & ... & ... & ... & ... \end{pmatrix}.$$

The creation operator a^+ of Eq. (C.66) generates the off-diagonal matrix:

$$a^+ = \begin{pmatrix} 0 & 0 & 0 & 0 & 0 & 0 & ... \\ \sqrt{1} & 0 & 0 & 0 & 0 & 0 & ... \\ 0 & \sqrt{2} & 0 & 0 & 0 & 0 & ... \\ 0 & 0 & \sqrt{3} & 0 & 0 & 0 & ... \\ 0 & 0 & 0 & \sqrt{4} & 0 & 0 & ... \\ 0 & 0 & 0 & 0 & \sqrt{5} & 0 & ... \\ ... & ... & ... & ... & ... & ... & ... \end{pmatrix}. \qquad \text{(C.76)}$$

The annihilation operator a of Eq. (C.67) generates the off-diagonal matrix:

$$a = \begin{pmatrix} 0 & \sqrt{1} & 0 & 0 & 0 & 0 & ... \\ 0 & 0 & \sqrt{2} & 0 & 0 & 0 & ... \\ 0 & 0 & 0 & \sqrt{3} & 0 & 0 & ... \\ 0 & 0 & 0 & 0 & \sqrt{4} & 0 & ... \\ 0 & 0 & 0 & 0 & 0 & \sqrt{5} & ... \\ 0 & 0 & 0 & 0 & 0 & 0 & ... \\ ... & ... & ... & ... & ... & ... & ... \end{pmatrix}. \qquad \text{(C.77)}$$

Bibliography

[1] Abney D.H., A. Paxton, R. Dale and C.T. Kello, "Complexity matching of dyadic conversation", *J. Exp. Psychol.: Gen.* **143**, 2304 (2014).

[2] Abney D.H., C.T. Kello and A.S. Warlaumont, "Production and convergence of multiple scale clustering in speech", *Ecol. Psychol.* **27**, 222 (2015).

[3] Abney D.H., A.S. Warlaumont, A.S. Oller, D.K. Wallot and C.T. Kello, "The multiscale clustering of infant vocalization bouts", unpublished (2015).

[4] Aburn M.J., C.A. Holmes, J.A. Roberts, T.W. Boonstra and M. Breakspear, "Critical fluctuations in cortical models near instability", *Front. Physiol.* **3**, 331 (2012).

[5] Adamic L.A. and B.A. Huberman, *Glottometrics* **3**, 143 (2002).

[6] Adrian R., *The Analyst; or Mathematical Museum, vol.* **1**, 93–109 (1809).

[7] Akimoto T., "Distributional Response to Biases in Deterministic Superdiffusion", *Phys. Rev. Lett.* **108**, 164101 (2012).

[8] Akin O.C., P. Paradisi and P. Grigolini, *Physica A* **371**, 157–170 (2006).

[9] Akselrod S., D. Gordon, F.A. Ubel, P.C. Shannon, A.L. Borger and R.J. Cohen, *Science* **213**, 220 (1981).

[10] Albert R. and A.-L. Barabási, *Rev. Mod. Phys.* **74**, 48 (2002).

[11] Allegrini P., M. Barbi, P. Grigolini and B.J. West, "Dynamical model for DNA sequences", *Phys. Rev. E* **52**, 5281 (1995).

[12] Allegrini P., P. Grigolini and B.J. West, *Phys. Rev. E* **54**, 4760 (1996).

[13] Allegrini P., M. Buiatti, P. Grigolini and B.J. West, *Phys. Rev. E* **57**, 4558 (1998).

[14] Allegrini P., P. Grigolini, P. Hamilton, L. Palatella and G. Raffaelli, "Memory beyond memory in heart beating, a sign of healthy physiological condition", *Phys. Rev. E* **65**, 041926 (2002).

[15] Allegrini P., J. Bellazzini, G. Bramanti, M. Ignaccolo, P. Grigolini and J. Yang, *Phys. Rev. E* **66**, 015101(R) (2002).

[16] Allegrini P., G. Aquino, P. Grigolini, L. Palatella and A. Rosa, "Generalized master equation via aging continuous-time random walks", *Phys. Rev. E* **68**, 056123 (2003).

[17] Allegrini P., R. Balocchi, S. Chillemi, P. Grigolini, P. Hamilton, R. Maestri, L. Palatella and G. Raffaelli, *Phys. Rev. E* **67**, 062901 (2003).

[18] Allegrini P., G. Aquino, P. Grigolini, L. Palatella and A. Rosa, *Phys. Rev. E* **68**, 056123 (2003).

[19] Allegrini P., V. Benci, P. Grigolini, P. Hamilton, M. Ignaccolo, G. Menconi, L. Palatella, G. Raffaelli, N. Scafetta, M. Virgilio and J. Yang, *Chaos Soliton. Fract.* **15**, 517 (2003).

[20] Allegrini P., P. Grigolini, L. Palatella and B.J. West, *Phys. Rev. E* **70**, 046118 (2004).

[21] Allegrini P., G. Aquino, P. Grigolini, L. Palatella, A. Rosa and B.J. West, *Phys. Rev. E* **71**, 066109 (2005).

[22] Allegrini P., F. Barbi, P. Grigolini and P. Paradisi, *Phys. Rev. E* **73**, 046136 (2006).

[23] Allegrini P., P. Grigolini and B.J. West, *Phys. Rev. Lett.* **99**, 010603 (2007a).

[24] Allegrini P., F. Barbi, P. Grigolini and P. Paradisi, *Chaos Soliton. Fract.* **34**, 11–18 (2007b).

[25] Allegrini P., M. Bologna, P. Grigolini and B.J. West, *Phys. Rev. Lett.* **99**, 010603 (2007c).

[26] Allegrini P., M. Bologna, P. Grigolino and M. Lukovic, arXiv:cond-mat/0608341v1 [cond-mat.stat-mech].

[27] Allegrini P., M. Bologna, L. Fronzoni, P. Grigolini and L. Silvestri, *Phys. Rev. Lett.* **103**, 030602 (2009).

[28] Allegrini P., D. Menicucci, R. Bedini, L. Fronzoni, A. Gemignani, P. Grigolini, B.J. West and P. Paradisi, "Spontaneous brain activity as a source of ideal 1/f noise", *Phys. Rev. E* **80**, 061914 (2009).

[29] Allegrini P., D. Menicucci, R. Bedini, A. Gemignani and P. Paradisi, "Complex intermittency blurred by noise: Theory and application to neural dynamics", *Phys. Rev. E* **82**, 015103 (2010).

[30] Allegrini P., P. Paradisi, D. Menicucci and A. Gemignani, "Fractal complexity in spontaneous EEG metastable-state transitions: new vistas on integrated neural dynamics", *Front. Physiol.* **1**, 128 (2010). *doi: 10.3389/ fphys.2010.00128*

[31] Allegrini P., P. Paradisi, D. Menicucci, M. Laurino, R. Bedini, A. Piarulli and A. Gemignani, *Chaos Soliton. Fract.* **55**, 32 (2013).

[32] Allegrini P., P. Paradisi, D. Menicucci, M. Laurino, A. Piarulli and A. Gemignani, "Self-organized dynamical complexity in human wakefulness and sleep: Different critical brain-activity feedback for conscious and unconscious states", *Phys. Rev. E* **92**, 032808 (2015).

[33] Almurad Z.M.H., C. Roume and D. Delignières, "Complexity matching in side-by-side walking", *Hum. Mov. Sci.* **54**, 125 (2017).

[34] Almurad Z.M.H., C. Roume and D. Delignières, "Complexity Matching: Restoring the Complexity of Locomotion in Older People Through Arm-in-Arm Walking", *Front. Physiol.* **9**, 1766 (2018). doi: 10.3389/fphys.2018.01766 (2018).

[35] Alvarez-Ramirez J., C. Ibarra-Valdez, E. Rodriguez and L. Dagdug, *Physica A* **387**, 281 (2008).

[36] Anderson P.W., "More is different", *Science* **177**, 393 (1971).

[37] Arecchi F.T. and F. Lisi, *Phys. Rev. Lett.* **49**, 34 (1982).

[38] Asano M., I. Basieva, A. Khrennikov, M. Ohya, Y. Tanaka and I. Yamato, "Quantum Information Biology: From Information Interpretation of Quantum Mechanics to Applications in Molecular Biology and Cognitive Psychology", *Found. Phys.* **45**, 1362 (2015).

[39] Ascolani G., M. Bologna and P. Grigolini, "Subordination to periodic processes and synchronization", *Physica A* **388**, 2727 (2009).

[40] Altemeier W.A., S. McKinney and R.W. Glenny, "Fractal nature of regional ventilation distribution", *J. Appl. Physiol.* **88**, 1551–1557 (2000).

[41] Attanasi A., A. Cavagna, L. Del Castello, I. Giardina, S. Melillo, L. Parisi, O. Pohl, B. Rossaro, E. Shen, E. Silvestri and M. Viale, *Phys. Rev. Lett.* **113**, 238102 (2014).

[42] Aquino G., P. Grigolini and B.J. West, *Europhys. Lett.* **80**, 10002 (2007).

[43] Aquino G., M. Bologna, P. Grigolini and B.J. West, "Beyond the death of linear response theory: criticality of the 1/-noise condition", *Phys. Rev. Lett.* **105**, 040601 (2010).

[44] Aquino G., M. Bologna, P. Grigolini and B.J. West, *Phys. Rev. E* **83**, 051130 (2011).

[45] Babloyantz A., in *Dimensions and Entropies in Chaotic Systems*, ed. G. Mayer-Kress, Springer-Verlag, Berlin (1986).

[46] Babloyantz A. and A. Destexhe, in *Proceed. Int. Conf. on Neural Networks*, San Diego (1987).

[47] Baddeley R., L.F. Abbott, M.C.A. Booth, F. Sengpiel, T. Freeman, E.A. Wakeman and E.T. Rolls, *Proc. R. Soc. Lond.* **264**, 1775 (1997).

[48] Bahcall S., *Loonshots: How to Nurture the Crazy Ideas that Win Wars, Cure Diseases, and Transform Industries*, St. Martin's Press, New York (2019).

[49] Baipai R., *Implications of Biophotons to Consciousness*, International Institute of Biophysics, Neuss, researchgate.net (2003).

[50] Bak P., C. Tang and K. Wiesenfeld, *Phys. Rev. Lett.* **59**, 381 (1987).

[51] Bak P., *How Nature Works: The Science of Self-organized Criticality*, Copernicus, Springer-Verlag (1996).

[52] Baiesi M., M. Paczusky and A.L. Stella, *Phys. Rev. Lett.* **96**, 05103 (2006).

[53] Bales M.E. and S.B. Johnson, *J. Biomed. Inf.* **39**, 451 2006

[54] Ball D.A., *Information Theory*, 2nd Edition, Pitman, New York (1956).

[55] Bologna M., G. Ascolani and P. Grigolini, "Density approach to ballistic anomalous diffusion: An exact analytical treatment", *J. Math. Phys.* **51**, 043303 (2010).

[56] Bao J.-D., Y.-L. Song and Y.-Z. Zhuo, *Phys. Rev. E* **72**, 011113 (2005).

[57] Bao J.-D., Y.-Z. Zhuo, F.A. Oliveira and P. Haänggi, *Phys. Rev. E* **74**, 061111 (2006).

[58] Barabási A.-L. and R. Albert, "Emergence of scaling in random networks", *Science* **286**, 509–512 (1999).

[59] Barabási A.-L., *Linked*, Plume, New York (2003).

[60] Barabási A.-L., *Nature* **435**, 207 (2005).

[61] Barabási A.-L., *Bursts*, Penguin Group, New York (2010).

[62] Barbi F., M. Bologna and P. Grigolini, *Phys. Rev. Lett.* **95**, 220601 (2005).

[63] Bardou F., J.-P. Bouchaud, A. Aspect and C. Cohen-Tannoudji, *Lévy Statistics and Laser Cooling*, Cambridge University Press, Cambridge, U.K. (2002).

[64] Barenblatt G.I., *Scaling, Self-similiarity, and Intermediate Asymptotics: Dimensional Analysis and Intermediate Asymptotcs*, Cambridge University Press, Dec 12, 1996

[65] Barkai E. and R.J. Silbey, *J. Phys. Chem. B* **104**, 3866 (2000).

[66] Barkai E., *Phys. Rev. Lett.* **90**, 104101 (2003).

[67] Barrow-Green J., *Poincaré and the Three body Problem*, History of Mathematics Vol. 11, American Mathematical Society, London Mathematical Society, Providence, RI (1997).

[68] Basset A.B., *A Treatise on Hydrodynamics, Vol.* **2**, pp. 285–297, Deighton Bell, Cambridge, MA (1888).

[69] Bassi A., K. Lochan, S. Satin, T.P. Singh and H. Ulbricht, "Models of wave-function collapse, underlying theories, and experimental tests", *Rev. Mod. Phys.* **85**, 471 (2013).

[70] Bassingthwaighte J.B., L.S. Liebovitch and B.J. West, *Fractal Physiology*, Oxford University Press, Oxford UK (1994).

[71] Beck C. and E.G.D. Cohen, *Physica A* **322**, 267 (2003).

[72] Bédard C., H. Kröger and A. Destexhe, *Phys. Rev. Lett.* **97**, 118102 (2006).

[73] Bedeaux D., K. Lakatos and K. Shuler, *J. Math. Phys.* **12**, 2116 (1971).

[74] Beggs J.M. and D. Plenz, "Neuronal Avalanches in Neocortical Circuits", *J. Neurosci.* **23**(35) 11167–11177 (2003).

[75] Beggs J.M., *Nat. Phys.* **3**, 834 (2007).

[76] Beig M.T., A. Svenkeson, M. Bologna, B.J. West and P. Grigolini, *Phys. Rev. E* **91**, 012907 (2015).

[77] Bejan A., *The Physics of Life, The Evolution of Everything*, St. Martins Press, NY (2016).

[78] Bel G. and E. Barkai, *Phys. Rev. Lett.* **94**, 240602 (2005).

[79] Bel G. and E. Barkai, *Europhys. Lett.* **74**, 15 (2006a).

[80] Bel G. and E. Barkai, *J. Phys. C* **17**, S4287 (2006b).

[81] Bel G. and E. Barkai, *Phys. Rev. E* **73**, 016125 (2006c).

[82] Benfatto M., E. Pace, C. Curceanu, A. Scordo, A. Clozza, I. Davoli, M. Lucci, R. Francini, F. De Matteis, M. Grandi, R. Tuladhar and P. Grigolini, "Biophotons: low signal/noise ratio reveals crucial events", *BioRxiv*, 558353 (2020).

[83] Ben-Naim A., *A Farewell to Entropy: A Statistical Thermodynamics Based on Information*, World Scientific, Singapore (2008).

[84] Bennett M., M.F. Schatz, H. Rockwood and K. Wiesenfeld, "Huyguen's clocks", *Proc. Roy. Soc. A, Mathematical and Engineering Sciences* **458**, 563 (2002).

[85] Bennett C.H., *Int. J. Theor. Phys.* **21**, 905 (1982); *Sci. Am.* **257**, 108 (1987).

[86] Benzi R., A. Sutera and A. Vulpiani, *J. Phys. A* **14**, L453 (1981).

[87] Benzi R., *Nonlin. Process. Geophy.* **17**, 431–441 (2010).

[88] Bernoulli D., *Econometrica* **22** (1954), Translated from Latin to English by L. Sommer from *Comm. Acad. Sca. Imp. Pet. V*, 175 (1738).

[89] Bernoulli D., *Reflexions et Exlaircissementts sur les Nouvelles Vibrations des Cordes Exoises dans les Memoires de l'Academie*, Roy. Acad. Berlin, pp. 147–172 (1753).

[90] Berry M.V., *J. Phys. A: Math. Gen.* **12**, 781 (1979).

[91] von Bertalanffy L., "Quantitative laws in metabolism and growth", *Q. Rev. Biol.* **32**, 217–231 (1957).

[92] Bialek W., I. Nemenman and N. Tishby, *Neural Comput.* **13**, 2409–2463 (2001).

[93] Bianco S., P. Grigolini and P. Paradisi, *J. Chem. Phys.* **123**, 174704 (2005).

[94] Bianco S. and P. Grigolini, *Chaos Soliton. Fract.* **34**, 41 (2007).

[95] Bianco S., M. Ignaccolo, M.S. Rider, M. Ross, P. Winsor and P. Grigolini, *Phys. Rev. E* **75**, 061911 (2007).

[96] Bianco S., E. Geneston, P. Grigolini and M. Ignaccolo, *Physica A* **387**, 1387 (2008).

[97] Bianucci M., A. Capotondi, S. Merlino and R. Mannella, "Estimate of the average timing for strong El Nio events using the recharge oscillator model with a multiplicative perturbation", *Chaos* **28**, 103118 (2018).

[98] Bigerelle M. and A. Iost, *Chaos Soliton. Fract.* **11**, 2179 (2000).

[99] Billock V.A. and B.H. Tsou, *Proc. Natl. Acad. Sci.* **104**, 8490 (2006).

[100] Birkhoff G.D., *Aesthetic Measure*, Cambridge: Harvard University Press (1933).

[101] Boccaletti S. and D.L. Valladares, *Phys. Rev. E* **62**, 7497 (2000).

[102] Bohara G., D. Lambert, B.J. West and P. Grigolini, "Crucial events, randomness, and multifractality in heartbeats", *Phys. Rev. E* **96**, 062216 (2017).

[103] Bohara G., B.J. West and P. Grigolini, "Bridging Waves and Crucial Events in the Dynamics of the Brain", *Front. Physiol.* **9**, 1174 (2018). doi: 10.3389/fphys.2018.01174

[104] Bologna M., P. Grigolini and J. Riccardi, *Phys. Rev. E* **60**, 6435 (1999).

[105] Bologna M., A.A. Budini, F. Giraldi and P. Grigolini, "From power law intermittence to macroscopic coherent regime", *J. Chem. Phys.* **130**, 244106 (2009).

[106] Bologna M., G. Ascolani and P. Grigolini, "Density approach to ballistic anomalous diffusion: An exact analytical treatment", *J. Math. Phys.* **51**, 043303 (2010).

[107] Bologna M., B.J. West and P. Grigolini, "Renewal and memory origin of anomalous diffusion: A discussion of their joint action", *Phys. Rev. E* **88**, 062106 (2013).

[108] Bonci L., P. Grigolini, A. Laux and R. Roncaglia, "Anomalous diffusion and environment-induced quantum decoherence", *Phys. Rev. A* **54**, 112 (1996).

[109] Bonci L., A. Farusi, P. Grigolini and R. Roncaglia, "Tunneling rate fluctuations induced by nonlinear resonances: A quantitative treatment based on semiclassical arguments", *Phys. Rev. A* **58**, 5689 (1998).

[110] Boon J.P. and O. Decroly, *Chaos* **5**, 501 (1995).

[111] Boonstra T.W., B.J. He and A. Daffertshofer, "Scale-free dynamics and critical phenomena in cortical activity", *Front. Physiol.* **4**, 79 (2013).

[112] Botcharova M., L. Berthouze, M.J. Brookes, G.R. Barnes and S.F. Farmer, "Resting state MEG oscillations show long-range temporal correlations of phase synchrony that break down during finger movement", *Front. Physiol.* **6**, 183 (2015).

[113] Bouchaud J.P. and A. Georges, *Phys. Rep.* **195**, 127 (1990)

[114] Bouchaud J., *J. Phys.* **12**, 1705 (1992).

[115] Bradley R.T., R. McCraty, M. Atkinson, D. Tomasino, A. Daugherty and L. Arguelles, "Emotion Self-Regulation, Psychophysiological Coherence, and Test Anxiety: Results from an Experiment Using Electrophysiological Measures", *Appl. Psychophysiol. Biofeedback* **35**, 261 (2010).

[116] Brillouin L., *Science and Information Theory*, Academic Press, New York (1962).

[117] Brokmann X., J.-P. Hermier, G. Messin, P. Desbiolles, J.-P. Bouchaud and M. Dahan, *Phys. Rev. Lett.* **90**, 120601 (2003).

[118] Brooks D.R. and E.O. Wiley, *Evolution as Entropy: Toward a Unified Theory of Biology*, University of Chicago Press, Chicago (1986).

[119] Brown R., *Phys. Rev. Lett.* **81**, 4835 (1998).

[120] Buchman T.G., "Physiologic failure: multiple organ dysfunction syndrome," in *Complex Systems Science in BioMedicine*, Eds. T.S. Deisboeck and S.A. Kauffman, New York: Kluwer Academic/Plenum Publishers, pp. 631–640 (2006).

[121] Budini A.A., "Stochastic Representation of a Class of Non-Markovian Completely Positive Evolutions", *Phys. Rev. A* **69**, 042107 (2004).

[122] Budini A.A. and P. Grigolini, *Phys. Rev. A* **80**, 022103 (2009).

[123] Buiatti M., P. Grigolini and L. Palatella, *Physica A* **268**, 214 (1999).

[124] Buiatti M. and M. Buiatti, "Chance vs. Necessity in Living Systems: A False Antinomy", *Rivista di Biologia/Biology Forum* **101**, 29–66 (2008).

[125] Buiatti M., D. Papo, P.-M. Baudonniére and C. van Vreeswijk, *Neuroscience* **146**, 1400 (2007).

[126] Bush V., *Science, The Endless Frontier*, United States Government, Washington (1945); http//www.nsf.gov/od/lpa/nsf50/vbush1945.htm.

[127] Cakir R., P. Grigolini and A.A. Krokhin, *Phys. Rev. E* **74**, 021108 (2006).

[128] Calabrese P. and A. Gambassi, *J. Phys. A* **38**, R133 (2005).

[129] Calder III, W.A., *Size, Function and Life History*, Harvard University Press, Cambridge, MA (1984).

[130] Cannon W.B., *The Wisdom of the Body*, New York: W.W. Norton & Co. (1932).

[131] Cardy S., *Scaling and Renormalization in Statistical Physic*, Cambridge Lecture Notes in Physics, Cambridge University Press, Cambridge (UK) (1996).

[132] Castellana M., W. Bialek, A. Cavagna and I. Giardiana, "Entropic force in a non-equilibrium system: flocks of birds", *Phys. Rev. E* **93**, 052416 (2016).

[133] Câteau H. and A.D. Reyes, *Phys. Rev. Lett.* **96**, 058101 (2006).

[134] Carhart-Harris R.L., R. Leech, P.J. Hellyer, M. Shanahan, A. Feilding, E. Tagliazucchi, D.R. Chialvo and D. Nutt, *Front. Hum. Neurosci.* **8**, 1 (2014).

[135] Carroll T.L. and L.M. Pecora, *Phys. Rev. Lett.* **64**, 821 (1990).

[136] Casati G., B.V. Chirikov and F.M. Izrailev, Stochastic Behavior of a Quantum Pendulum under a Periodic Perturbation, *Lect. Notes Phys.* **93**, 334 (1979).

[137] Cavagna A., L. Di Carlo, I. Giardina, L. Grandinetti, T.S. Grigera and G. Pisegna, "Dynamical renormalization group approach to the collective behaviour of swarms", arXiv:1905.01227v1 (2019).

[138] Chaté H. and M.A. Munoz, *Physics* **7**, 120 (2014).

[139] Chatterjee A., A. Ghosh and B.K. Chakrabarti, "Universality of Citation Distributions for Academic Institutions and Journals", *PLOS One* **11**(1), e0146762 (2016).

[140] Chirikov B.V., *Patterns in Chaos,* revised text of the lectures given at the *International School on Order, Chaos and Patterns, Como* (1990).

[141] Coey C.A., A. Washburn, J. Hassebrock and Mi. J. Richardson, "Complexity matching effects in bimanual and interpersonal syncopated finger tapping", *Neurosci. Lett.* **616**, 204 (2016).

[142] Cocchi L., L.G. Gollo, A. Zalesky and M. Breakspear, "Criticality in the brain: A synthesis of neurobiology, models and cognition", *Prog. Neurobiol.* **158**, 132 (2017).

[143] Collins J.J. and C.J. De Luca, *Phys. Rev. Lett.* **73**, 764 (1994).

[144] Collins J.J., C.C. Chow and T.T. Imhoff, *Nature* **376**, 236 (1995).

[145] Collins J.J., C.C. Chow, A.C. Capela and T.T. Imhoff, *Phys. Rev. E* **54**, 5575 (1996a).

[146] Collins J.J., T.T. Imhoff and P. Grigg, *J. Neurophysiology* **76**, 642 (1996b).

[147] Compte A., *Phys. Rev. E* **53**, 4191 (1996).

[148] Condorcet, Marquis de, *https://www.goodreads.com/quotes/tag/calculus.*

[149] Contoyiannis Y.F., F.K. Diakonos and A. Malakis, "Intermittent Dynamics of Critical Fluctuations", *Phys. Rev. Lett.* **89**, 035701 (2002).

[150] Correll J., *J. Personality and Social Psychol.* **94**, 48 (2008).

[151] Cortes E., B.J. West and K. Lindenberg, *J. Chem. Phys.* **82**, 2708 (1985).

[152] Costa M., A.L. Goldberger and C.-K. Peng, "Multiscale entropy analysis of complex physiologic time series", *Phys. Rev. Lett.* **89**, 068102 (2002).

[153] Costa M., A.L. Goldberger and C.-K. Peng, "Multiscale entropy analysis of biological signals", *Phys. Rev. E* **71**, 021906 (2002).

[154] Couzin I., "Collective Minds", *Nature* **445**, 715 (2007).

[155] Cox D.R., *Renewal Theory*, Metheun & Co, London (1967).

[156] Crisanti A. and F. Ritort, *J. Phys. A: Math. Gen.* **36**, R181 (2003).

[157] Culbreth G., B.J. West and P. Grigolini, "Entropic Approach to the Detection of Crucial Events", *Entropy* **21**, 178 (2019).

[158] da Silva L., A.R.R. Papa and A.M.C. de Souza, *Phys. Lett. A* **242**, 343 (1998).

[159] Einstein A., "On the movement of small particles suspended in stationary liquids required by the molecular-kinetic theory of heat", *Ann. Phys.* **17**, 549–560 (1905), appearing in *The Collected Papers of Albert Einstein*, English translation by Anna Beck, Princeton U.P., Princeton, NJ, 1989, Vol. 2, pp. 123–134.

[160] de Sola Price D., *Little Science, Big Science*, Columbia University Press, NY (1963).

[161] Dacorogna M.M., R. Gencoy, U. Müller, R.B. Olsen and O.V. Pictet, *An Introduction to High Frequency Finance*, Academic Press, San Diego, CA, (2001).

[162] de Arcangelis L. and H.J. Herrmann, "Activity-dependent neuronal model on complex networks", *Front. Physiol.* **3**, 62 (2012).

[163] Dehghani N., N.G. Hatsopoulos, Z.D. Haga, R.A. Parker, B. Greger, E. Halgren, S.S. Cash and A. Destexhe, "Avalanche analysis from multielect rode ensemble recordings in cat, monkey, and human cerebral cortex during wakefulness and sleep", *Front. Physiol.* **3**, 1–18, (2012).

[164] De Jaegher H., E. Di Paolo and S. Gallagher, "Can social interaction constitute social cognition?", *Trends Cogn. Sci.* **14**, 441 (2010).

[165] Deligniéres D., L. Lemoine and K. Torre, *Hum. Mov. Sci.* **23**, 87 (2004).

[166] Deligniéres D., K. Torre and L. Lemaine, *Acta Psychoogica* **127**, 382 (2008).

[167] Deligniéres D. and K. Torre, *J. Appl. Physiol.* **106**, 1272–1279 (2009).

[168] Deligniéres D., Z.M.H. Almurad, C. Roume and V. Marmelat, "Multifractal signatures of complexity matching", *Exp. Brain Res.* **234**, 2773 (2016).

[169] Demetrius L., S. Legendre and P. Harremoes, "Evolutionary entropy: a predictor of body size, metabolic rate and maximal life span", *Bull. Math. Biol.* **71**, 800–818 (2009).

[170] Dlask M., J. Kukal, M. Poplová, P. Sovka and M. Cifra, "Short-time fractal analysis of biological autoluminescence", *PLOS One* **14**, e0214427 (2019).

[171] Doob J.L., *Ann. Math.* **43**, 351 (1942).

[172] Drossel B. and G. Ellis, "Contextual Wavefunction collapse: an integrated theory of quantum measurement", *New J. Phys.* **20**, 113025 (2018).

[173] Dumas G., J. Nadel, R. Soussignan, J. Martinerie and L. Garnero, "Inter-Brain Synchronization during Social Interaction", *PLOS One* **5**, e12166 (2010).

[174] Dumas G. and H. Scott, *The KAM Story - A Friendly Introduction to the Content, History, and Significance of Classical Kolmogorov–Arnold–Moser Theory*, World Scientific Publishing (2014).

[175] Dunbar R.I.M., "Neocortex size as a constraint on group size in primates", *J. Hum. Evol.* **20**, 469–493 (1991).

[176] Dunbar R.I.M., "Neocortex size as a constraint on group size in primates", *J. Hum. Evol.* **22**, 469 (1992).

[177] Dunbar R.I.M., "Coevolution of neocortical size, group size and language in humans", *Behav. Brain Sci.* **16**, 681–735 (1993).

[178] Dunbar R.I.M., "The Social Brain Hypothesis", *Evol. Anthropol.* **6**, 178 (1998).

[179] Dutta P. and P.M. Horn, *Rev. Mod. Phys.* **53**, 497 (1981).

[180] Einstein A., *Ann. Physik* **17**, 549 (1905).

[181] Ellis G.F.R., "Top-down causation and quantum physics", *Proc. Natl. Acad. Sci.* **46**, 11661 (2018).

[182] Faloutsos M., P. Faloutsos and C. Faloutsos, *Proc. ACM Sigcomm* (1999).

[183] Fechner G.T., *Elemente der Psychophysik*, Leipzig: Breitkopf and Hartel (1860).

[184] Feder J., *Fractals*, Plenum Press, New York (1988).

[185] Feldmann A., A.C. Gilbert, W. Willenger and T.G. Kurtz, *Comput. Commun. Rev.* **28**, 1 (1998).

[186] Feller W., *An Introduction to Probability Theory and Its Applications*, Vol. II, 2nd ed. Wiley, New York (1971).

[187] Femat R. and G. Solis-Perales, *Phys. Lett. A* **262**, 50 (1999).

[188] Feng J. and P. Zhang, *Phys. Rev. E* **63**, 051902 (2001).

[189] Ferrario M. and P. Grigolini, *J. Math. Phys.* **20**, 2567 (1979).

[190] Ferris T., *Coming of Age in the Milky Way*, Harper Collins, NY (1988).

[191] Fine J.M., A.D. Likens, E.L. Amazeen and P.G. Amazeen, "Emergent Complexity Matching in Interpersonal Coordination: Local Dynamics and Global Variability", *J. Exp. Psychol.: Human Perception and Performance* **41**, 723 (2015).

[192] Flomembon O., K. Velonia, D. Loos, S. Masuo, M. Cotlet, Y. Engelborghs, J. Hofkens, A.E. Rowan, R.J.M. Nolte, M. van der Auweraer, F.C. de Schyver and J. Klafter, *Proc. Natl. Acad. Sci. USA* **102**, 2368 (2005).

[193] Ford G.W., M. Kac and P. Mazur, *J. Math. Phys.* **6**, 504 (1965).

[194] Forsythe A., M. Nadal, N. Sheehy, C.J. Cela-Conde and M. Sawey, *Brit. J. Psych.* **102**, 49–70 (2011).

[195] Fowler H. and W.E. Leland, *IEEE J. Selected Areas Commun.* **9**, 1139 (1991).

[196] Fox R., *J. Math. Phys.* **18**, 2331 (1977).

[197] Fraiman D. and D.R. Chialvo. "What kind of noise is brain noise: anomalous scaling behavior of the resting brain activity fluctuations." *Front. Physiol.* **3**, 307 (2012).

[198] Freeman W.J., *Mass Action in the Nervous System, Chapter 7*, Academic Press, New York, pp. 489 (1975).

[199] Freeman W.J., *Biol. Cybern.* **56**, 139–150 (1987).

[200] Frederick S., G. Loewenstein and T. O'Donoghue, *J. Econ. Lit.* **40**, 351 (2002).

[201] Freud S., *Beyond the Pleasure Principle*, trans. C.J.M. Hubback, London, Vienna: Intl. Psycho-Analytical (1922).

[202] Froese T., H. Iizuka and T. Ikegami, "Embodied social interaction constitutes social cognition in pairs of humans: a minimalist virtual reality experiment", *Sci. Rep.* **4**, 3672 (2014).

[203] Fronczak A., P. Fronczak and J.A. Holyst, "Thermodynamic fores, flows, and Onsager coefficients in complex networks", *Phys. Rev. E* **76**, 06116 (2007).

[204] Frost R., *The Road Not Taken and Other Poems*, Penguin Classics Deluxe Edition, NY (2015).

[205] Gallos L.K., M. Sigman and H.A. Makse, "The conundrum of functional brain networks: small-world efficiency or fractal modularity", *Front. Physiol.* **3**, 123 (2012).

[206] Gammaitoni L., P. Hänggi, P. Jung and F. Marchesoni, *Rev. Mod. Phys.* **70**, 223 (1998).

[207] Gammaitoni L., P. Hänggi, P. Jung and F. Marchesoni, *Eur. Phys. J. B* **69**, 1–3 (2009).

[208] Gard T., J.J. Noggle, C.L. Park, D.R. Vago and A. Wilson, "Potential self-regulatory mechanisms of yoga for psychological health", *Front. Human Neurosci.* **8**, 770 (2014).

[209] Gauss C.F., Theoria motus corporum coelestrium, Hamburg, 1809, In: Dover Eng. Trans. *Theory of Motion of Heavenly Bodies Moving about the Sun in Conic Sections*, New York, 1963.

[210] Gibbs J.W., *Collected Works*, Longmans, Green (1928).

[211] Geisel T., J. Nierwetberg and A. Zacherl, "Accelerated Diffusion in Josephson Junctions and Related Chaotic Systems", *Phys. Rev. Lett.* **54**, 616 (1985).

[212] Glass L., "Synchronization and rhythmic processes in physiology", *Nature* **410**, 277–284 (2001).

[213] Glass L. and M.C. Mackey, *From Clocks to Chaos; The Rhythms of Life*, Princeton University Press, Princeton, NJ (1988).

[214] Glöckle W.G. and T. Nonnenmacher, *J. Stat. Phys.* **71**, 741 (1993a).

[215] Glöckle W.G. and T. Nonnenmacher, *Rheol. Acta* **33**, 337 (1993b).

[216] Glöckle W.G. and T.F. Nonnenmacher, *Macromolecules* **24**, 6426 (1991).

[217] Godréche G. and J.M. Luck, *J. Stat. Phys.* **104**, 489 (2001).

[218] Goldberger A.L., V. Bhargava, B.J. West and A.J. Mandell, *Biophys. J.* **48**, 525 (1985).

[219] Goldberger A.L., D.R. Rigney and B.J. West, "Chaos, fractals and physiology", *Sci. Am.* **262**, 42–49 (1990).

[220] Goldberger A.L., L. Amaral, L. Glass, J.M. Hausdorff, P.Ch. Ivanov, R.G. Mark, J.E. Mietus, G.B. Moody, C.K. Peng and H.E. Stanley, "PhysioBank, PhysioToolkit, and PhysioNet: Components of a New Research Resource for Complex Physiologic Signals", *Circulation* **101**(23), e215–e220 (2000). *http://circ.ahajournals.org/content/101/23/e215.full, doi: 10.1161/01.CIR.101.23.e215.*

[221] Goldberger A.L., L.A.N. Amaral, J.M. Hausdorff, P.Ch. Ivanov, C.-K. Peng and H.E. Stanley, "Fractal dynamics in physiology: Alterations with disease and aging", *Proc. Natl. Acad. Sci.* **99**, 2466–2472 (2002).

[222] Gong P., A.R. Nikolaev and C. van Leeuwen, *Phys. Rev. E* **76**, 011904 (2007).

[223] Goychuck I. and P. Hänggi, *Phys. Rev. Lett.* **91**, 070601 (2003).

[224] Gradstein I.S. and I.M. Ryzhik, *Table of Integrals, Series, and Products*, Academic Press, San Diego (1980).

[225] Granovetter M., "The strength of weak ties", *American J. Sociol.* **78**, 1360–1380 (1973).

[226] Greene J.A. and J. Loscalzo, "Putting the Patient Back Together – Social Medicine, Network Medicine, and Limits of Reductionism", *New England J. Mrd.* **377**, 2493–2499 (2017).

[227] Greer C. and E. McLaughlin, *Brit. J. Criminol.* **50**, 1041–1059, (2010).

[228] Griffin L., D.J. West and B.J. West, "Random stride intervals with memory", *J. Biol. Phys.* **26**, 185–202 (2000).

[229] Grigolini P., "A generalized Langevin equation for dealing with non-additive fluctuations", *J. Stat. Phys.* **27**, 283 (1982).

[230] Grigolini P., "Theoretical Foundation", in *Advances in Chemical Physics, Memory Function Approaches to Stochastic Problems in Condensed Matter* (1985).

[231] Grigolini P., A. Rocco and B.J. West, *Phys. Rev. E* **59**, 2603 (1999).

[232] Grigolini P., L. Palatella and G. Raffaelli, "Asymmetric anomalous diffusion: an efficient way to detect memory in time series", *Fractals* **9**, 439 (2001).

[233] Grigolini P., D. Leddon and N. Scafetta, *Phys. Rev. E* **65**, 046203 (2002).

[234] Grigolini P., P. Allegrini and B.J. West, *Chaos Soliton. Fract.* **34**, 3–10 (2007).

[235] Grigolini P., *Chaos Soliton. Fract.* **81**, 575 (2015).

[236] Grigolini P., "Theoretical Foundation", in *Advances in Chemical Physics, Memory Function Approaches to Stochastic Problems in Condensed Matter* (1985).

[237] Gross G., J. Kowalski and B. Rhoades, in *Oscillations in Neural Systems*, edited by D. Levine, V. Brown and T. Shirey (Erlbaum, New York, 1999), Vol. 1, pp. 3–29.

[238] Hachinski K.V. and V. Hachinski, *J. Can. Med. Ass.* **151**, 293 (1994).

[239] Harte J., *Maximum Entropy and Ecology: A Theory of Abundance, Distribution and Energetics*, Oxford University Press, UK (2011).

[240] Hayano J., K. Kiyono, Z.R. Struzik, Y. Yamamoto, E. Watanabe, P.K. Stein, L.L. Watkins, J.A. Blumenthal and R.M. Carney, *Front. Physiol.* **2**, Article 65, 1 (2011).

[241] Hausdorff J.M., C.K. Peng, Z. Ladin, J.Y. Ladin, J.Y. Wei and A.L. Goldberger, *J. Appl. Physiol.* **78**, 349 (1995).

[242] Hausdorff J.M., L. Zemany, C.K. Peng and A.L. Goldberger, *J. Appl. Physiol.* **86**, 1040 (1999).

[243] Hawking S., *A Brief History of Time*, Bantam Dell Pub. Group, UK (1988).

[244] He B.J., J.M. Zempel, A.Z. Snyder and M.E. Raichle, "The Temporal Structures and Functional Significance of Scale-free Brain Activity", *Neuron* **66**, 353 (2010).

[245] M. Malik, *et al.*, "Heart rate variability", *European Heart J.* **17**, 354 (1996).

[246] Heinsalu E., M. Patriarca, I. Goychuk and P. Hänggi, *Phys. Rev. Lett.* **999**, 120602 (2007).

[247] Heinsalu E., M. Patriarca, I. Goychuk and P. Hänggi, *Phys. Rev. E* **79**, 041137 (2009).

[248] Helbing D. and S. Lämmer, "Supply and Production Networks: From the Bullwhip Effect to Business Cycles", cond-mat/0411486v1 (2004).

[249] Hennig T., P. Maass, J. Hayano and S. Heinrichs, *J. Biol. Phys.* **32**, 383 (2006).

[250] Henry B.I., T.A.M. Langlands and P. Straka, *Phys. Rev. Lett.* **105**, 170602 (2010).

[251] Heusner A.A., "Size and power in mammals", *J. Exp. Biol.* **160**, 25–54 (1991).

[252] Hidalgo J., J. Grilli, S. Suweis, M.A. Munoz, J.R. Banavar and A. Maritan, *Proc. Natl. Acad. Sci. USA* **111**, 10095 (2014).

[253] Hilfer R., Editor, *Applications of Fractional Calculus in Physics*, World Scientific Publishing, Singapore (2002).

[254] Hillery H., R.F. O'Connell, M.O. Scully and E.P. Wigner, "Distribution Functions In Physics: Fundamentals", *Phys. Rept.* **106**, 121 (1984).

[255] Hinomoto S., K. Shima and T. Tanji, *Neural Comput.* **15**, 2803 (2003).

[256] Holler J., K.H. Kendrick, M. Casillas and S.G. Levinson, Eds., *Turn-taking in Human Communicative Interactions, Front. Psychol. Res. Topics* **6**, 1919 (2016).

[257] Holloway G., "Entropic Forces in Geophysical Fluid Dynamics", *Entropy* **11**, 360 (2009).

[258] Hoop B. and C.-H. Peng, *J. Membrane Biol.* **177**, 177 (2000).

[259] M. Malik, *et al.*, HRV task force, "Heart rate variability", *European Heart J.* **17**, 354 (1996).

[260] Hsu K.J. and A. Hsu, *Proc. Natl. Acad. Sci.* **88**, 3507 (1991).

[261] Hughes B.D., *Random Walks and Random Environments*, Calderon Press, Oxford (1996).

[262] Ibáñez-Molina A.J., V. Lozano, M.F. Soriano, J.I. Aznarte, C.J. Gómez-Ariza and M.T. Bajo, "EEG Multiscale Complexity in Schizophrenia During Picture Naming", *Front. Physiol.* **9**, 1213 (2018).

[263] *AR4 Climate Change 2007: The Physical Science Basis, https://www.ipcc.ch/report/ar4/wg1/ https://www.brainyquote.com/ authors/jayz.*

[264] Ivanov P.C., M.G. Rosenblum, C.K. Peng, J. Meitus, S. Havlin and H.E. Stanley, "Scaling behavior of heartbeat intervals obtained by wavelet-based time-series analysis", *Nature* **383**, 323 (1996).

[265] Ivanov P.C., L.A. Amaral, A.L. Goldberger, *et al.*, "Multifractality in human heartbeat dynamics", *Nature* **399**, 461–465 (1999).

[266] Izrailev F.M., "Simple Models of Quantum Chaos: Spectrum and Eigenfunctions", *Phys. Rep.* **196**, 299 (1990).

[267] Jäger G. and R. van Rooij, *Synthese* **159**, 99 (2007).

[268] Jangid P., A.K. Chauhan and S. Wuster, "Tunneling rate fluctuations induced by nonlinear resonances: A quantitative treatment based on semiclassical arguments", *Phys. Rev. A* **102**, 043513 (2020).

[269] Jaynes E.T., "Information theory and statistical mechanics", *Phys. Rev.* **106**, 620 (1957); *ibid.* **108**, 171 (1957).

[270] Jelinek H.F., R. Tuladhar, G. Culbreth, G. Bohara, D. Cornforth, B.J. West and P. Grigolini, "Diffusion Entropy vs. Multiscale and Renyi Entropy to Detect Progression of Autonomic Neuropathy", *Front. Physiol.* **11**, 1759 (2020).

[271] Jensen H.J., *Self-Organized Criticality*, Cambridge Lecture Notes in Physics, Cambridge University Press, Cambridge UK (2000).

[272] Jennings H.D., P.Ch. Ivanov, A. de M. Martins, P.C. da Silva and G.M. Viswanathan, *Physica A* **336**, 585 (2004).

[273] Jeffery K.J. and C. Rovelli, "Transitions in Brain Evolution: Space, Time and Entropy", *Trends Neurosci.* **43**, 7 (2020).

[274] Jia T., D. Wang and B.K. Szymanski, "Quantifying patterns of research-interest evolution", *Nat. Hum. Behav.* **1**, 0078 (2017).

[275] Jizba P. and J. Korbel, "Multifractal diffusion entropy analysis: Optimal bin width of probability histograms", *Physica A* **413**, 438 (2014).

[276] Jordan K., J. Challis and K. Newell, "Long range correlations in the stride interval of running", *Gait Posture* **24**(1), 120–125 (2006).

[277] Jung P., A. Cornell-Bell, S. Madden and F. Moss, "Noise-Induced Spiral Waves in Astrocyte Syncytia Show Evidence of Self-Organized Criticality", *Am. Physiol. Soc.* **79**(2), 1098 (1998).

[278] Kadota H., K. Kudo and T. Ohtsuki, *Neurosci. Lett.* **370**, 97 (2004).

[279] Kahneman D., *Thinking, Fast and Slow*, Farrar, Straus and Grioux, New York (2011).

[280] Kalashyan A.K., M. Buiatti and P. Grigolini, *Chaos Soliton. Fract.* **39**, 895 (2009).

[281] Kang Y.-M. and Y.-L. Jiang, *J. Math. Phys.* (NY) **51**, 023301 (2010).

[282] Kaplan D.T. and M. Talajic, *Chaos* **1**, 251 (1991).

[283] Kaplan A.Y., A.A. Fingelkurts, A.A. Fingelkurts, B.S. Borisov and B.S. Darkhovsky, "Nonstationary nature of the brain activity as revealed by EEG/MEG: methodological, practical and conceptual challenges", *Signal Process.* **85**, 2190 (2005).

[234] Kello C.T., B.C. Beltz, J.G. Holden and G.C. Van Orden, *J. Exp. Psychol.* **136**, 551 (2007).

[285] Kelvin, Lord (Sir William Thomson), *Popular Lectures and Addresses, Vol.* **1**, MacMillan, NY (1883).

[286] Kenkre V.M., E.W. Montroll and M.F. Shlesinger, *J. Stat. Phys.* **9**, 45 (1973).

[287] Kenning P., H. Plassmann, M. Deppe, H. Kugel and W. Schwindt [Editorial of a special issue devoted to neuroeconomics, Münster, May 25–27 (2004)], *Brain Res. Bull.* **67**, 341 (2005).

[288] Khalsa M.K., J.M. Greiner-Ferris, S.G. Hofmann and S.B.S. Khalsa, "Yoga-Enhanced Cognitive Behavioral Therapy (Y-CBT) for Anxiety Management: A Pilot Study", *Clin. Psychol. Psychother.* **22**, 364 (2015).

[289] Khrennikov A., "Quantum-like modeling of cognition", *Front. Phys.* **3**(77), 77 (2015).

[290] Kinouchi O. and M. Copelli, *Nat. Phys.* **2**, 348 (2006).

[291] Kish L.B., G.P. Harmer and D. Abbott, *Fluc. Noise Lett.* **1**, L13–L19 (2001).

[292] Kobayashi M. and T. Musha, *IEEE Trans. Biomed. Eng.* **29**, 456 (1982).

[293] Kocarev L. and U. Parlitz, *Phys. Rev. Lett.* **76**, 1816 (1996).

[294] Koch R., *The 80/20 Principle: The Secret of Achieving More with Less*, Nicholas Brealey Publishing (1998).

[295] Korabel N. and E. Barkai, "Pesin-Type Identity for Intermittent Dynamics with a Zero Lyaponov Exponent", *Phys. Rev. Lett.* **102**, 050601 (2009).

[296] Korabel N. and E. Barkai, "Infinite Invariant Density Determines Statistics of Time Averages for Weak Chaos", *Phys. Rev. Lett.* **108**, 060604 (2012).

[297] Mahmoodi K., B.J. West and P. Grigolini, "Self-Organizing Complex Networks: individual versus global rules", *Front. Physiol.* **8**, 478 (2017).

[298] Mahmoodi K., B.J. West and P. Grigolini, "Self-Organized Temporal Criticality: bottom-up resilience versus top-down vulnerability", *Complexity* **2018**, Article 8139058 (2018).

[299] Mahmoodi K., P. Grigolini and B.J. West, "On social sensitivity to either zealot or independent minorities", *Chaos Soliton. Fract.* **110**, 185 (2018).

[300] Kramers H.A., *Physica* **7**, 284 (1940).

[301] Kubo R., *J. Phys. Soc. Jpn* **12**, 570 (1957).

[302] Kubo R., M. Toda and N. Hashitusume, *Statistical Physics*, Springer, Berlin (1985).

[303] Kuno M., D.P. Fromm, H.F. Hamann, A. Gallagher and D.J. Nesbitt, *J. Chem. Phys.* **115**, 1028 (2001).

[304] Kuno M., D.P. Fromim, S.R. Hohmson, A. Gallagher and D.J. Nesbitt, *Phys. Rev. B* **67**, 125304 (2003).

[305] Kuramoto Y., *Physica A* **126**, 128 (1981).

[306] Laloux L. and P. Le Doussal, *Phys. Rev. E* **57**, 6296 (1998).

[307] Lambert D., M. Bologna, R. Tuladhar, B.J. West and P. Grigolini, "Joint Action of Periodicity and Ergodicity Breaking Crucial Events: Spectrum Evaluation", unpublished.

[308] Lamperti J., "An Occupation Time Theorem for a Class of Stochastic Processes", *Trans. Am. Math. Soc.* **88**, 380 (1958).

[309] Landauer R., *Physica A* **194**, 551 (1993).

[310] Landauer R., *Science* **272**, 1914–1918 (1996a).

[311] Landauer R., *Phys. Lett. Sect. A* **217**, 188–193 (1996b).

[312] Langevin P., "Sur la theorie du mouvement brownien", *C. R. Acad. Sci. Paris* **146**, 530–533 (1908).

[313] Lannon R., F. Amini and T. Lewis, *A General Theory of Love*, Random House, New York (2000).

[314] Laskin N., I. Lambadaris, F.C. Harmantzis and M. Devetsikiotis, *Comp. Networks* **40**, 363–375 (2002).

[315] Layne S.P., G. Mayer-Kress and J. Holzfuss, in *Dimensions and Entropies in Chaotic Systems*, 246–256, ed. G. Mayer-Kress, Springer-Verlag, Berlin (1986).

[316] van de Leemput I., M. Wichers, S. Angélique, *et al.*, "Critical slowing down as early warning of the onset and termination of depression", *Proc. Natl. Acad. Sci.* **111**, 87–92 (2014).

[317] Leland W.E., M.S. Taqqu, W. Willinger and D.V. Wilson, *IEEE/ACM Trans. Networking* **2**, 1 (1994).

[318] Lemons D.S. and A. Gythiel, Paul Langevin's 1908 paper "On the Theory of Brownian Motion", *Am. J. Phys.* **65**, 11 (1997).

[319] Lee M.H., *Phys. Rev. Lett.* **49**, 1072 (1982).

[320] Levina A., J.M. Herrmann and T. Geisel, *Nat. Phys.* **3**, 857 (2007).

[321] Lewis M.D., *Behav. Brain Sci.* **28**, 169 (2005).

[322] Li T.Y. and J.A. Yorke, "Period Three Implies Chaos", *Am. Math. Month.* **82**, 985–992 (1975).

[323] Li F.A. and Q. Gu (Eds.), World Scientific, Singapore, pp. 1–46 (1992).

[324] Li M. and P. Vitányi, *An Introduction to Kolmogorov Complexity and Its Applications*, Springer, NY (1997).

[325] Li W. and D. Holste, *Phys. Rev. E* **71**, 041910 (2005).

[326] Lighthill J., "The recently recognized failure of predictability in Newtonian dynamics", *Proc. Roy. Soc. London A* **407**, 35 (1986).

[327] Lindenberg K. and B.J. West, *The Nonequilibrium Statistical Mechanics of Open and Closed Systems*, VCH Publishers, New York, NY (1990).

[328] Liu Q., H. Chen and W. Hai, "Chaos-assisted localization and delocalization of a particle in a driven optical lattice", *Chaos Soliton. Fract.* **122**, 80 (2019).

[329] Longo G. and M. Montévil, "The Inert vs. the Living State of Matter: Extended Criticality, Time Geometry, Anti-Entropy — An Overview", *Front. Physiol.* **3**, 39 (2012).

[330] Loscalzo J., A.-L. Barabasi and E.K. Silverman, "Network Medicine", pp. 1–16, *Complex Systems in Human Disease and Therapeutics*, Cambridge, MA, Harvard University Press, (2017).

331] Lotka A.J., *Elements of Physical Biology*, reprinted by Dover in 1956 as *Elements of Mathematical Biology*; *first published by The Williams and Wilkins Co. (1924)*.

332] Lotka A.J., *J. Wash. Sci.* **16**, 317 (1926).

333] Lowen S.B. and M.C. Teich, *Phys. Rev. E* **47**, 992 (1993).

334] Lovecchio E., P. Allegrini, E. Geneston, B.J. West and P. Grigolini, "From self-organized to extended criticality", *Front. Physiol.* **3**, 98 (2012).

335] Lowen S.B. and M.C. Teich, *Fractal-Based Point Processes*, John Wiley, New York (2005).

336] Luk R.W.P., "What Do We Mean by "True" in Scientific Realism?", *Found. Sci.* **25**, 845 (2020).

337] Lukovic M. and P. Grigolini, "Power spectra for both interrupted and perennial aging processes", *J. Chem. Phys.* **129**, 184102 (2008).

338] Lukovic M., F. Vanni, A. Svenkeson and P. Grigolini, "Transmission of information at criticality", *Physica A* **416**, 430 (2014).

339] Lutz E., *Phys. Rev. Lett.* **93**, 190602 (2004).

340] MacDonald N., *Trees and Networks in Biological Models*, John Wiley & Sons, NY (1983).

341] Machlup S., in *Sixth Int. Symp. on Noise in Physical Systems*, pp. 157–160, NBS (1977).

342] Mahmoodi K., B.J. West and P. Grigolini, "Self-Organizing Complex Networks: individual versus global rules", *Front. Physiol.* **8**, 478 (2017).

343] Mahmoodi K., B.J. West and P. Grigolini, "On The Dynamical Foundation of Multifractality", available at: http://arxiv.org/abs/1707.05988 (2017).

344] Mahmoodi K., B.J. West and P. Grigolini, "Self-Organized Temporal Criticality: bottom-up resilience versus top-down vulnerability", *Complexity* **2018**, ID 8139058 (2018).

345] Mahmoodi K., P. Grigolini and B.J. West, "On social sensitivity to either zealot or independent minorities", *Chaos Soliton. Fract.* **110**, 185 (2018).

[346] Mahmoodi K., B.J. West and P. Grigolini, "Complexity Matching and Requisite Variety", available at: http://arxiv.org/abs/1806.08808 (2018).

[347] Mafahim J.U., D. Lambert, M. Zare and P. Grigolini, *New J. Phys.* **17**, 015003 (2015).

[348] Magdziarz M., A. Weron and J. Klafter, *Phys. Rev. Lett.* **101**, 210601 (2008).

[349] Magin R.L., *Fractional Calculus in Bioengineering*, Begell House Pub., CA (2006).

[350] Mainardi F., *Chaos Soliton. Fract.* **7**, 1461 (1996).

[351] Malkiel B.G., *A Random Walk down Wall Street: The Time-Tested Strategy for Successful Investing* (eleventh Edition), W.W. Norton & Sons, NY (2016).

[352] Manaris B., J. Romero, P. Machado, D. Krehbiel, T. Hirzel, W. Pharr and R.B. Davis, *Comput. Music J.* **29**, 55 (2005).

[353] Mandelbrot B.B. and J.W. Van Ness, "Fractional Brownian motions, fractional noises and applications", *SIAM Rev.* **10**, 422 (1968).

[354] Mandelbrot B.B., *Fractals, Form and Chance*, W.F. Freeman, San Francisco, CA (1977).

[355] Manneville P., "Intermittency, self-similarity, and spectrm in dissaptive dynamical systems", *J. Physique* **41**, 1235–1243 (1980).

[356] Mantegna R.N. and H.E. Stanley, *An Introduction to Econophysics*, Cambridge University Press, Cambridge, UK (2000).

[357] Margolin G. and E. Barkai, *Phys. Rev. Lett.* **94**, 1 (2005).

[358] Margolin G. and E. Barkai, "Nonergodicity of a Time Series Obeying Lévy Statistic", *J. Stat. Phys.* **122**, 137 (2006).

[359] Marini U., B. Marconi, A. Puglisi, L. Rondoni and A. Vulpiani, *Phys. Rep.* **461**, 111 (2008).

[360] Marmelat V. and D.D. Deligniéres, "Strong anticipation: complexity matching in interpersonal coordination", *Exp. Brain. Res.* **222**, 137 (2012).

[361] Marx K. and F. Engels, *Manifesto of the Communist Party* (1848); *Marx/Engels Selected Works*, Vol. One, Progressive Publishers, Moscow, 1969, pp. 98–137.

[362] Maxwell J.C., *Theory of Heat*, originally published 1888, Dover, New York (2001).

[363] McDonald M.D. and D. Abbott, *PLoS Comp. Biol.* **5**, 1–9, e1000348 (2009).

[364] McMullen M., L. Girling, M.R. Graham and W.A. Mutch, "Biologically variable ventilation improves gas exchange and respiratory mechanics in a model of severe bronchospasm", *Critical Care Med.* **35**(7), 1749–1755 (2007).

[365] McNamara B., K. Wiesenfeld and R. Roy, *Phys. Rev. Lett.* **60**, 2626 (1988).

[366] Meakin P., *Fractals, Scaling and Growth far from Equilibrium*, Cambridge Nonlinear Science Series 5, Cambridge University Press, NY, NY (1998).

[367] Meerschaert M., E. Nane and P. Vellaisamy, *Elect. J. Prob.* **16**, 1600 (2011).

[368] Mega M.S., P. Allegrini, P. Grigolini, V. Latora, L. Palatella, A. Rapisarda and S. Vinciguerra, *Phys. Rev. Lett.* **90**, 188501 (2003).

[369] Metzler R., E. Barkai and J. Klafter, *Europhys. Lett.* **46**, 431 (1999a).

[370] Metzler R., E. Barkai and J. Klafter, *Phys. Rev. Lett.* **82**, 3563 (1999b).

[371] Metzler R., E. Barkai and J. Klafter, *Physica A* **266**, 343 (1999c).

[372] Metzler R., *Phys. Rev. E* **62**, 6233 (2000).

[373] Metzler R. and J. Klafter, *J. Phys. Chem. B* **104**, 3851 (2000b).

[374] Metzler R., J.-H. Jeon, A.G. Cherstvya and E. Barkai, *Phys. Chem. Chem. Phys.* **16**, 24128 (2014).

[375] Mikkelsen K.B. and T.E. Lund, "Sampling rate dependence of correlation at long time lags in BOLD fMRI measurements on humans and gel phantoms", *Front. Physiol.* **4**, 106 (2013).

[376] Mikosch T., S. Resnick, H. Rootzén and A. Stegeman, *Ann. Appl. Probab.* **12**, 23 (2002).

[377] Min W., G. Luo, B.J. Cherayil, S.C. Kou and X.S. Xie, *Phys. Rev. Lett.* **94**, 198302 (2005).

[378] Misra B. and E.C.G. Sudarshan, "The Zeno's paradox in quantum theory", *J. Math. Phys.* **18**, 756 (1977).

[379] Mohanty A.K. and A.V. Narayana Rao, *Phys. Rev. Lett.* **84**, 1832 (2000).

[380] Monto S., "Nested synchrony — a novel cross-scale interaction among neuronal oscillations", *Front. Physiol.* **3**, 384 (2012).

[381] Montroll E.W. and W.W. Badger, *Introduction to Quantitative Aspects of Social Phenomena*, Gordon and Breach, NY (1974).

[382] Montroll E.W. and G. Weiss, *J. Math. Phys.* **6**, 178 (1965).

[383] Montroll E.W. and B.J. West, in *Fluctuation Phenomena*, Eds. E.W. Montroll and J. Lebowitz, North-Holland (1979); Second edition (1986).

[384] Montroll E.W. and M.F. Shlesinger, *J. Stat. Phys.* **32**, 209 (1983).

[385] Monthus C. and J.-P. Bouchaud, *J. Phys. A* **29**, 3847 (1996).

[386] Mora T. and W. Bialek, "Are Biological Systems Poised at Criticality?", *J. Stat. Phys.* **144**, 268 (2011).

[387] Moran D.K., https://www.springboard.com/blog/41-shareable-data-quotes/.

[388] Mori H., "Transport, Coillective Motion, and Brownian Motion", *Prog. Theor. Physics* **33**, 423 (1965).

[389] Moss F., L.M. Ward and W.G. Sannita, *Clin. Neurophysiol.* **115**, 267–281 (2004).

[390] Musha T., in *Sixth Int. Sym. on Noise in Physical Systems*, eds. P.H.E. Meijer, R.D. Mountain and R.J. Soulen, pp. 142, Nat. Bureau of Standards (1981).

[391] Mutch W.A.C., S.H. Harm, G.R. Lefevre, M.R. Graham, L.G. Girling and S.E. Kowalski, "Biologically variable ventilation increases arterial oxygenation over that seen with positive end-expiratory pressure alone in a porcine model of acute respiratory distress syndrome", *Crit. Care Med.* **28**, 2457–2464 (2000).

[392] Néda Z., L. Varga and T.S. Biró, "Science and Facebook: The same popularity law!", *PLOS One* **12**(7): e0179656 (2017).

[393] Nee S., "Survival and weak chaos", *R. Soc. Open Sci.* **5**, 172181 (2018).

[394] Neumann R.M., *J. Chem. Phys.* **66**, 870 (1977).

[395] Newman M.E.J., *SIAM Rev.* **45**, 167 (2003).

[396] Nicolis C. and G. Nicolis, *Tellus* **33**, 225 (1981).

[397] Niemann N., H. Kantz, E. Barkai, "Fluctuations of 1/f-Noise and the Low-Frequency Cutoff Paradox", *Phys. Rev. Lett.* **110**, 140603 (2013).

[398] Nirmal M., B.O. Dabbousi, M.G. Bawendi, J.J. Macklin, J.K. Trautman, R.D. Harris and L.E. Brus, *Nature* **383**, 802 (1996).

[399] Nivala M., C.Y. Ko, M. Nivala, J.N. Weiss and Z. Qu, "Criticality in Intracellular Calcium Signaling in Cardiac Myocytes", *Biophys. J.* **102**, 2433 (2012).

[400] Noether E., "Invariante Variationsprobleme", Nachr. D. König. Gesellsch. D. Wiss. Zu Göttingen, Math-phys. Klasse, pp. 235–257 (1918).

[401] Norwich K.H., "Le Chatelier's principle in sensation and perception: fractal-like enfolding at different scales", *Front. Physiol.* **1**, 17 (2010).

[402] *Network Science*, National Research Council of the National Academies, Washington, DC. (2005). Available at: www.nap.edu.

[403] Nyikos L., L. Balazs and R. Schiller, *Fractals* **2**, 143 (1994).

[404] Oliveira J.G. and A.-L. Barabási, "Darwin and Einstein correspondence patterns", *Nature* **437**, 1251 (2005).

[405] Onsager L., *Phys. Rev.* **38**, 2265 (1931); *Phys. Rev.* **37**, 405 (1931).

[406] Oppenheim I., K.E. Shuler and G.H. Weiss, *Stochastic Processes in Chemical Physics: The Master Equation*, MT Press, Cambridge (1977).

[407] Ossowska M. and S. Ossowski, "The Science of Science", originally published in Polish as "Nauka o nauce" in *Nauka Polska* (Polish Science), **vol. XX** (1935), no. 3.

[408] Pais-Vieira M., M. Lebedev, C. Kunicki, J. Wang and M.A.L. Nicoliles, *Sci. Rep.* **3**, 1319 (2013).

[409] Papo D., "Time scales in cognitive neuroscience", *Front. Physiol.* **4**, 86 (2013).

[410] Papo D., M. Zanin, J.A. Pineda-Pardo, S. Boccaletti and J.M. Buldu, "Functional brain networks: great expectations, hard times and the big leap forward", *Philos. Trans. R. Soc. B* **369**, 20130525 (2014).

[411] Paradisi P., P. Allegrini, A. Gemignani, M. Laurino, D. Menicucci and A. Piarulli, "Scaling and intermittency of brain events as a manifestation of consciousness", *AIP Conf. Proc.* **151**, 1510 (2013).

[412] Paradisi P. and P. Allegrini, *Chaos Soliton. Fract.* **55**, 32 (2013).

[413] Paraschiv-Ionescu A., E. Buchser, B. Rutschmann and K. Aminian, *Phys. Rev. E* **77**, 021913 (2008).

[414] Pareto V., *Cours d'Economie Politique*, Lausanne (1897).

[415] Parmukkul P., A. Svenkedon, P. Grigoline, M. Bologna and B.J. West, "Complexity and the Fractional Calculus", *Adv. Math. Phys.* **2013**, 1 (2013). *http:/dx.doi.org/10.1153/2013/498789.*

[416] Pauli W., Festschrift zum 60 geburtstag A. Sommerfeld, Ed. S. Hirzel, Leipzig (1928).

[417] Paxson V. and S. Floyd, *IEEE/ACM Trans. Networking* **3**, 226–244 (1995).

[418] Pease A., K. Mahmoodi and B.J. West, "Complexity measures of music", *Chaos Soliton. Fract.* **108**, 82 (2018).

[419] Pelton M., D.G. Grier and P. Guyot-Sionnest, *Appl. Phys. Lett.* **85**, 819 (2004).

[420] Peng C.K., J. Mietus, J.M. Hausdorff, S. Havlin, H.G. Stanley and A.L. Goldberger, *Phys. Rev. Lett.* **70**, 1343 (1993).

[421] Peng C.-K., S. Havlin, H.E. Stanley and A.L. Goldberger, *Chaos* **5**, 1 (1995).

[422] Peng C.-K., J.E. Mietus, Y. Liu, G. Khalsa, P.S. Douglas, H. Benson and A.L. Goldberger, "Exaggerated heart rate oscillations during two meditation techniques", *Int. J. Cardiol.* **70**, 101 (1999).

[423] Peng C.-K., I.C. Henry, J.E. Mietus, J.M. Hausdorff, G. Khalsa, H. Benson and A.L. Goldberger, "Heart rate dynamics during three forms of meditation", *Int. J. Cardiol.* **95**, 19 (2004).

[424] Peng X., H. Yuan, W. Chen, T. Wang and L. Ding, *PLOS One*, April 17, pp. 1–24. https://doi.org/10.1371/journal.pone.0175354 (2017).

[425] Penna T.J.P., P.M.C. de Oliveira, J.C. Sartorelli, W.M. Goncalves and R.D. Pinto, "Long-range anti-correlation and non-Gaussian behavior of a leaky faucet", *Phys. Rev. E* **52**, R2168–R2171 (1995).

[426] Perrow C., *Normal Accidents: Living with High Risk Technology*, (1984).

[427] Piccinini N., D. Lambert, B.J. West, M. Bologna and P. Grigolini, *Phys. Rev. E* **93**, 062301 (2016).

[428] Pinto H.D., D. Escaff, U. Harbola, A. Rosa and K. Lindenberg, *Phys. Rev. E* **89**, 052143 (2014).

[429] Plenz D. and T.C. Thiagarajan, *Trends Neurosci.* **30**, 101 (2007).

[430] Plenz D., "The critical brain", *Physics* **6**, 47 (2013).

[431] Plenz D. and E. Niebur (editors), *Criticality in Neural Systems, Reviews of Nonlinear Dynamics and Complexity*, Wiley-VCH (2014).

[432] Pfurtscheller G., A.R. Schwerdtfeger, A. Seither-Preisler, C. Brunner, C.S. Aigner, J. Brito, M.P. Carmo and A. Andrade, "Brain-heart communication: Evidence for "central pacemaker" oscillations with a dominant frequency at 0.1 Hz in the cingulum", *Clin. Neurophysiol.* **128**, 183 (2017).

[433] Podlubny I., *Fractional Differential Equations*, Acadmic Press, San Diego (1999).

[434] Pomeau Y. and P. Manneville, "Intermittent transition to turbulence in dissipative dynamic systems", *Comm. Math. Phys.* **74**, 189–197 (1980).

[435] Popp F.A., "Some Essential Questions of Biophoton Research and Probable Answers", in *Recent Advances in Biophoton Research and its Applications*, F.A. Popp, K.H. Li and Q. Gu (Eds.), World Scientific, Singapore, pp. 1–46 (1992).

[436] Pottier N., "Aging properties of an anomalously diffusing particule", *Physica A* **317**, 371–382 (2003).

[437] Pramukkul P., A. Svenkeson, P. Grigolini, M. bologna and B.J. West, "Complexity and the Fractional Calculus", *Adv. Math. Phys.* **2013**, Article ID 498789 (2013).

[438] Pramukkul P., A. Svenkeson, B.J. West and P. Grigolini, *Europhys. Lett.* **111**, 58003 (2015).

[439] de S. Price D.J., "A general theory of bibliometric and other cumulative advantage processes", *J. Amer. Soc. Inform. Sci.* **27**, 292–306 (1976).

[440] de S. Price D.J., *Little Science, Big Science . . . and Beyond*, Columbia University Press, New York (1986).

[441] Procaccia I. and H. Schuster, "Functional renormalization-group theory of universal $1/f$-noise in dynamical systems", *Phys. Rev. A* **83**, 1210 (1983).

[442] Quetelet A., *Treatise on Man and the Development of his Faculties or Essay on Social Physics* (Sur l'homme et le developpement de ses faculties, ou essai de physique sociale), Publisher; First English edition, William and Robert Chambers (1835).

[443] Quastler H., Editor, *Information Theory in Biology*, University of Illinois Press, Urbana (1953).

[444] Rachlin H. and B.A. Jones, *J. Behav. Dec. Making* **21**, 29 (2008).

[445] van Raan A.F.J., "Measuring Science", H.F. Moed, *et al.*, Eds., *Handbook of Quantitative Science and Technolgy Research*, pp. 19–50, Kluwer Academic Publishers, Printed in the Netherlands (2004).

[446] van Raan A.F.J., "FTl TTRction: Conceptual and methodological problems in the ranking of universities by bibliometric methods", *Scientometrics* **62**, 133–143 (2005).

[447] Rao R.P.N., A. Stocco, M. Bryan, D. Sarma, T.M. Youngquist, J. Wu and C.S. Prat, *PLOS One* **9**, e111332 (2014).

[448] Rauscher F.H., G.L. Shaw and K.N. Ky, *Nature* (London) **365**, 611 (1993).

[449] *Studies in History and Philosophy of Science Part B: Studies in History and Philosophy of Modern Physics Volume* **36**, Issue 2, June 2005, Pages 245–273; Studies in History and Philosophy of Science

Part B: Studies in History and Philosophy of Modern Physics Boltzmann and Gibbs: An attempted reconciliation? Author links open overlay panel D.A. Lavis.

[450] Rebenshtok A. and E. Barkai, *Phys. Rev. Lett.* **999**, 210601 (2007).

[451] Redner S., "How popular in your paper? An empirical study of the citation distribution", *Eur. Phys. J. B* **4**, 131–134 (1998).

[452] Resnick S. and H. Rootzén, *Ann. Appl. Probab.* **10**, 753 (2000).

[453] Reyes-Manzano C.F., C. Lerma, J.C. Eheverria, M. Matinez-Lavin, L.A. Martinez-Martinez, O. Infanta and L. Guzmam-Vargas, "Multifractal analysis reveals decreased nonlinearity and stronger anticorrelatin in heart period fluctuation of fibromyagia patients", *Front. Physiol.* **9**, 1118 (2018).

[454] Rim S., I. Kim, P. Kang, Y.J. Park and C.M. Kim, *Phys. Rev. E* **66**, 015205 (2002).

[455] Rodríguez-Iturbe I. and A. Rinaldo, *Fractal River Basins: Chance and Self-Organization*, Cambridge University Press, Cambridge, UK (1997).

[456] Rodriguez E.G., N. Geoge, J.P. Lachaux, J. Martinerie, B. Renault and F. Varela, *Nature* **397**, 430 (1999).

[457] Roncaglia R., L. Bonci, B.J. West and P. Grigolini, "Anomalous diffusion and the correspondence principle", *Phys. Rev. E* **51**, 5524 (1995).

[458] Roos N., "Entropic forces in Brownian motion", *Am. J. Phys.* **22** (2013); doi:10.1119/1.4894381.

[459] Rosa E.R., E. Ott and M.H. Hess, *Phys. Rev. Lett.* **80**, 1642 (1998).

[460] Rosas A., D. Escaff, I.L.D. Pinto and K. Lindenberg, *J. Phys. A: Math. Theor.* **49**, 095001 (2016).

[461] Rosenblum M.G., A.S. Pikovsky and J. Kurths, *Phys. Rev. Lett.* **76**, 1804 (1996).

[462] Rosenblum M.G., A.S. Pikovsky and J. Kurths, *Phys. Rev. Lett.* **78**, 4193 (1997).

[463] Rovelli C., "An argument against the realistic interpretation of the wave function", *Found. Phys.* **46**, 1229 (2016).

[464] Rulkov N.F., M.M. Sushehik, L.S. Tsimring and H.D.I. Abarbanel, *Phys. Rev. E* **51**, 980 (1995).

[465] Sakai Y., S. Funahashi and S. Shinomoto, *Neural Networks* **12**, 1181 (1999).

[466] Sakai Y., *Neural Networks* **14**, 1145 (2001).

[467] Salinas E. and T.J. Seinowski, *Neural Comput.* **14**, 2111 (2002).

[468] Salingaros N.A. and B.J. West, "A universal rule for the distribution of sizes", *Environment and Planning B: Planning and Design* **26**, 909 (1999).

[469] Samko S.G., A.A. Kilbas and O.I. Marichev, *Fractional Integrals and Derivatives*, Gordon and Breach Science Publishers, Switzerland (1993).

[470] Saxe J.G., *The Poems of John Godfrey Saxe*, Osgood & Co., Boston (1872).

[471] Scalas E., *Chaos Soliton. Fract.* **34**, 33–40 (2007).

[472] Scafetta N., P. Hamilton and P. Grigolini, "The thermodynamics of social processes: the teen birth phenomenon", *Fractals* **9**, 193 (2001).

[473] Scafetta N. and B.J. West, "Solar flares intermittencies and the Earth's temperature anomalies", *Phys. Rev. Lett.* **90**, 248701 (2003).

[474] Scafetta N. and B.J. West, "Multiscaling Compartive Analysis of Time Series and a Discussion on "Earthquake Conversations" in California", *Phys. Rev. Lett.* **92**, 138501 (2004).

[475] Scafetta N., P. Grigolini, T. Imholt, J. Roberts and B.J. West, "Solar turbulence in earth's global and regional temperature anomalies", *Phys. Rev. E* **69**, 026303 (2004).

[476] Scafetta N., D. Marchi and B.J. West, *Chaos* **19**, 026108 (2009).

[477] Scher H. and E. Montroll, *Phys. Rev. B* **12**, 2455 (1975).

[478] Schmidt-Nielsen K., *Scaling, Why is Animal Size so Important?*, Cambridge University Press: Cambridge, UK (1984).

[479] Schneider W.R. and W. Wyss, *J. Math. Phys.* **30**, 134 (1989).

[480] Schrödinger E., *What is Life? The Physical Aspect of the Living Cell*, First pub. 1944, Cambridge University Press, UK (1967).

481] Schuster H.G., *Deterministic Chaos: An Introduction*, Second Revised Edition, VCH Publisher, New York (1988).

482] Scott J., *Social Network Analysis: A Handbook*, 2nd ed., Sage, London (2000).

483] Seshadri V. and B.J. West, *Proc. Natl. Acad. Sci.* **79**, 4501–4505 (1982).

484] Seuront L. and J.G. Mitchell, *J. Marine Syst.* **69**, 310 (2008).

485] Shadlen M.N. and W.T. Newsame, *J. Neurosci.* **18**, 3870 (1998).

486] Shahverdiev E.M., S. Sivaprakasam and K.A. Shore, *Phys. Lett. A* **292**, 320 (2002).

487] Shahverdiev E.M. and K.A. Shore, *Phys. Rev. E* **71**, 016201 (2005).

488] Shannon C.E., *Bell Sys. Tech. J.* **27**, 379–423; *ibid.* 623–656 (1948).

489] Shannon C.E. and W. Weaver, *The Mathematical Theory of Communication*, University of Illinois Press, Urbana, IL (1959).

490] Shannahoff-Khalsa D.S., "An Introduction to Kundalini Yoga Meditation Techniques That Are Specific for the Treatment of Psychiatric Disorders", *J. Alternat. Complem. Med.* **10**, 91 (2004).

491] Shimizu K.T., R.G. Neuhauser, C.A. Leatherdale, S.A. Empedocles, W.K. Woo and M.G. Bawendi, *Phys. Rev. B* **63**, 205316 (2001).

492] Shinomoto S. and Y. Tsubo, *Phys. Rev. E* **64**, 041910 (2001).

493] Shlesinger M.F. and B.D. Hughes, *Physica A* **109**, 597 (1981).

494] Shlesinger M.F., B.J. West and J. Klafter, *Phys. Rev. Lett.* **58**, 1100 (1987).

495] Shockley W., "On the Statistics of Individual Variations of Productivity in Research Laboratories", *Proc. IRE* **270** (1957).

496] Shushin A.I., *Phys. Rev. E* **78**, 051121 (2008).

497] Silvestri L., L. Fronzoni, P. Grigolini and P. Allegrini, *Phys. Rev. Lett.* **102**, 014502 (2009).

498] Skagerstam B.-S.K. and A. Hansen, *Europhys. Lett.* **72**, 513 (2005).

[499] Smith A., *An Inquiry into the Nature and Causes of the Wealth of Nations*, University of Chicago Press, Chicago, IL (1977); first published in 1776.

[500] Soberano E.K. and D.G. Kelty-Stephen, "Demystifying cognitive science: explaining cognition through network-based modeling", *Front. Physiol.* **6**, 88 (2015).

[501] Sokolov I.M., J. Klafter and A. Blumen, *Phys. Today* **49**, No. 2, 33 (1996).

[502] Sokolov I.M., *et al.*, *Physica A* **302**, 568 (2001).

[503] Sokolov I.M. and R. Metzler, *Phys. Rev. E* **67**, 010101(R) (2003).

[504] Sokolov I.M., *Phys. Rev. E* **73**, 067102 (2006a).

[505] Sokolov I.M. and J. Klafter, "Field-induced dispersion and subdiffusion", *Phys. Rev. Lett.* **97**, 140602 (2006).

[506] Sokolov I.M., "Statistical mechanics of entropic forces: disassembling a toy", *Eur. J. Phys.* **31**, 1353 (2010).

[507] Sornette D., *Phys. Rep.* **378**, 1–98 (2003).

[508] Stam C.J., *Clin. Neurophysiol.* **116**, 2266 (2005).

[509] Stam C.J. and J.C. Reijneveld, *Nonlinear Biomed. Phys.* **1**, 3 (2007).

[510] Stephen D.G. and J.A. Dixon, "Strong anticipation: Multifractal cascade dynamics modulate scaling in synchronization behaviors", *Chaos Soliton. Fract.* **44**, 160 (2011).

[511] Stephen D.G., J.R. Anastas and J.A. Dixon, "Scaling in cognitive performance reflects multiplicative multifractal cascade dynamics", *Front. Physiol.* **3**, 102 (2012).

[512] Stevens C.F. and A.M. Zador, *Nat. Neurosci.* **1**, 210 (1998).

[513] Strogatz S.H. and I. Stewart, *Sci. Am.* **269**, 102 (1993).

[514] Strogatz S., *SYNC*, Hyperion Books, New York (2003).

[515] Sturmberg J.P. and C.M. Martin, *Handbook of Systems and Complexity in Health*, Springer, New York (2013).

[516] Su Z.-Y. and T. Wu, *Physica A* **380**, 418 (2007).

[517] Suki B., A.M. Alencar, U. Frey, P.C. Ivanov, S.V. Buldyrev, A. Majumdar, H.E. Stanley, C. Dawson, G.S. Krenz and M. Mishima, "Fluctuations, noise and scaling in the cardiopulmonary system", *Fluct. Noise Lett.* **3**, R1–R25 (2003).

[518] Sumpter D.J.T., *Philos. Trans. R. Soc. B* **361**, 5 (2006).

[519] Sun Z., W. Xu and X. Yang, *Math. & Computer Modeling* (2008).

[520] Svenkeson A. and B.J. West, "Constructing a critical phase in a population of interacting two-state stochastic units", *Chaos Soliton. Fract.* **113**, 40 (2018).

[521] Szeto H.H., P.Y. Chang, J.A. Decena, Y. Chang, D. Wu and G. Dwyer, *Am. J. Physiol.* **262**, R141 (1992).

[522] Szilard L., *Z. Physik* **53**, 840 (1929).

[523] Tagliazucchi E., P. Balenzuela, D. Fraiman and D.R. Chialvo, "Criticality in large-scale brain fMRI dynamics unveiled by a novel point process analysis", *Front. Physiol.* **3**, 15 (2012).

[524] Takahashi T., K. Ikeda and T. Hasegawa, *Behav. Brain Funct.* **3**, 52 (2007).

[525] Takayasu H., A.-H. Sato and M. Takayasu, "Stable Infinite Variance Fluctuations in Randomly Amplified Langevin Systems", *Phys. Rev. Lett.* **79**, 966 (1997).

[526] Taylor R.P., A.P. Micolich and D. Jonas, *Nature* **399**, 422 (1999).

[527] Taylor R.P., A.P. Micolich and D. Jonas, *Leonardo* **35**, 203 (2002).

[528] Termsaithong T., M. Oku and K. Aihara, "Dynamical coherence patterns in neural field model at criticality", *Artif. Life Robot.* **17**, 75 (2012).

[529] Timme N.M., N.J. Marshall, N. Bennett, M. Ripp, E. Lautzenhiser and J.M. Beggs, "Criticality Maximizes Complexity in Neural Tissue", *Front. Physiol.* **7**, 425 (2016).

[530] Trefán G., E. Floriani, B.J. West and P. Grigolini, "Dynamical approach to anomalous diffusion: Response of Lévy processes to a perturbation", *Phys. Rev. E* **50**, 2564 (1994).

[531] Tsuruoka M., Y. Tsuruoka, R. Shibasaki and Y. Yasuoka, *Proceedings of the 29th Annu. Int. Conf. IEE EMBS Cité Internationale*, Lyon, France (2007).

[532] Tsuchiya M., A. Giuliani, M. Hashimoto, J. Erenpreisa and K. Yoshikawa, "Self-Organizing Global Gene Expression Regulated Through Criticality: Mechanism of the Cell-Fate Change", *PLOS One* doi: 10.1371/journal.pone.0167912 (2016).

[533] Tuladhar R., G. Bohara, P. Grigolini and B.J. West, "Meditation-induced coherence and crucial events", *Front. Physiol.* **9**:626 (2018).

[534] Tuladhar R., M. Bologna and P. Grigolini, "Non-Poisson renewal events and memory", *Phys. Rev. E* **96**, 042112 (2018).

[535] Turalska M., M. Lukovic, B.J. West and P. Grigolini, *Phys. Rev. E* **83**, 061142-1 (2009).

[536] Turalska M., B.J. West and P. Grigolini, *Phys. Rev. E* **80**, 021110-1 (2011).

[537] Turalska M., E. Geneston, B.J. West, P. Allegrini and P. Grigolini, "Cooperation-induced topological complexity: a promising road to fault tolerance and hebbian learning", *Front. Physiol.* **3**, 52 (2012).

[538] Turalska M. and B.J. West, "Fractional dynamics of individuals in complex networks", *Front. Phys.* **6**:110 (2018).

[539] Turing A.M., *Philos. Trans. R. Soc. London Ser. B* **327**, 37 (1952).

[540] Tversky A. and D. Kahneman, "Judgment under uncertainty: Heuristics and biases", *Science* **185**, 1124 (1974).

[541] Vandermeer J., "Confronting Complexity in Agroecology: Simple Models From Turing to Simon", *Front. Sustainable Food Syst.* **4**, 95 (2020).

[542] Valdez A.B. and E.R. Amazeen, *Exp. Brain Res.* DOI 10.1007/s00221-0088-1305-0 (2008).

[543] Vanni F., M. Lukovic and P. Grigolini, *Phys. Rev. Lett.* **107**, 078103 (2011).

[544] van Vreewijk C., *Neurocomputing* **38**, 417 (2001).

[545] Varela F., J.P. Lachauz, E. Rodriguez and J. Martinerie, *Nat. Rev. Neurosci.* **2**, 229 (2001).

[546] Vazques A., B. Rácz, A. Lukács, A.-L. Barabási, *Phys. Rev. Lett.* **98**, 158702 (2007).

[547] Veraart A., E.J. Faassen, V. Dakos, E.H. van Nes, M. Lürling and M. Scheffer, "Recovery rates reflect distance to a tipping point in a living system", *Nat.* **481**, 357–359 (2012).

[548] Vicsek T., A. Czirkok, E. Ben-Jacob, I. Cohen and O. Shochet, "Novel type of phase transition in a system of self-driven particles", *Phys. Rev. Lett.* **75**, 1226 (1995).

[549] Vierordt, *Uber das Gehen des Menchen in Gesunden und kranken Zustaenden nach Selbstregistrirender Methoden*, Tuebigen, Germany (1881).

[550] Vlad M.O., *Physica A* **184**, 290 (1992a).

[551] Vlad M.O., *Int. J. Mod. Phys. B* **6**, 419 (1992b).

[552] Vlad M.O., J. Ross and F.W. Schneider, *Phys. Rev. E* **62**, 1743 (2000).

[553] Voss R.F. and J. Clarke, *Nature* (London) **258**, 317 (1975).

[554] Voss R.F. and J. Clarke, *Phys. Rev.* **13**, 556 (1976).

[555] Voss R.F. and J. Clarke, *Sci. Am.* **238**, 16–32 (1978)

[556] Voss R.F., *Phys. Rev. Lett.* **68**, 3805 (1992).

[557] Voss R.F., *Fractals* **2**, 1 (1994).

[558] Voss R.F., *Phys. Rev. E* **61**, 5115 (2000).

[559] Wagenmakers E.-J., S. Farrell and R. Ratcli, *Psychonomic Bull. Rev.* **11**, 579 (2004).

[560] Wallin N.L., *Human Evolution* **15**, 199 (2000).

[561] Wallace R., *Int. J. Bifur. Chaos* **10**, 493–502 (2000).

[562] Wang D., "Habituation", in the *Handbook of Brain Theory and Neural Networks*, Ed. M.A. Arbib, pp. 441, MIT Press (1995).

[563] Watkins N.W., "On the continuing relevance of Mandelbrot's non-ergodic fractional renewal models of 1963 to 1967", *Eur. Phys. J. B* **90**, 241 (2017).

[564] Watts D.J., *Small Worlds*, Princeton University Press, Princeton, New Jersey (1999).

[565] Weibel E.R., *Symmorphosis*, Harvard University Press, Cambridge, MA (2000).

[566] Weishaupt D., V.D. Köchli and B. Marincek, *How does MRI Work? An Introduction to the Physics and Function of Magnetic Resonance Imaging*, Springer-Verlag, Berlin and Heidelberg (2003).

[567] Weiss G.H., *Aspects and Applications of the Random Walk*, North-Holland, Amsterdam (1994).

[568] Weiss U., *Quantum Dissipative Systems*, 2nd edition, World Scientific, Singapore (1999).

[569] Werner G., "Fractals in the nervous system: conceptual implications for theoretical neuroscience", *Front. Physiol.* **1**, 15 (2010).

[570] Werner G., "Letting the brain speak for itself", *Front. Physiol.* **2**, 60 (2011).

[571] Weron A., M. Magdziarz and K. Weron, *Phys. Rev. E* **77**, 036704 (2008).

[572] West B.J. and V. Seshadri, *Physica A* **113**, 203 (1982).

[573] West B.J., *An Essay on the Importance of Being Nonlinear*, Lect. Notes in Biomath., Ed. S. Levin, Springer-Verlag, NY (1985).

[574] West B.J., *Fractal Physiology and Chaos in Medicine*, Studies in Nonlinear Phenomena in Life Science Vol. 1, World Scientific, Singapore (1990); *ibid.* 2nd edition, Vol. 16 (2013).

[575] West B.J. and M. Shlesinger, "The Noise in Natural Phenomena", *Am. Sci.* **78**, 40 (1990).

[576] West B.J. and W. Deering, "Fractal physiology for physicists: Lévy statistics", *Phys. Rep.* **246**, 1–100 (1994).

[577] West B.J. and W. Deering, *The Lure of Modern Science*, Studies in Nonlinear Phenomena in Life Science Vol. 3, World Scientific, Singapore (1995).

[578] West B.J., P. Grigolini, R. Metzler and T.F. Nonnenmacher, *Phys. Rev. E* **55**, 99 (1997).

[579] West B.J. and L. Griffen, "Allometric control of human gait", *Fractals* **6**, 101 (1998).

580] West B.J., *Physiology, Promiscuity and Prophecy at the Millennium: A Tale of Tails*, Studies in Nonlinear Phenomena in Life Science Vol. 7, World Scientific, Singapore (1999).

581] West B.J., M. Bologna and P. Grigolini, *Physics of Fractal Operators*, Springer, New York (2003).

582] West B.J. and N. Scafetta, "A nonlinear model for human gait", *Phys. Rev. E* **67**, 051917 (2003).

583] West B.J., M. Latka, M. Glaubic-Latka and D. Latka, "Multifractality of cerebral blood flow", *Physica A* **318**, 453 (2003).

584] West B.J., L.A. Griffin, H.J. Frederick and R.E. Moon, "The independently fractal nature of respiration and heart rate during exercise under normobaric and hyperbaric conditions", *Respirat. Physiol. Neurobiol.* **145**, 219 (2005).

585] West B.J., *Where Medicine Went Wrong*, Studies in Nonlinear Phenomena in Life Science Vol. 11, World Scientific, New Jersey (2006).

586] West B.J. and P. Grigolini, *Phys. Rev. Lett.* **100**, 088501 (2008).

587] West B.J., E. Genesten and P. Grigolini, "Maximizing information exchange between complex networks", *Phys. Rep.* **468**, 1–99 (2008).

588] West B.J., "Control from an allometric perspective," in *Progress in Motor Control: A Multidisciplinary Perspective* (Advances in Experimental Medicine and Biology, Vol. 629, Ed. D. Sternad (New York: Springer), pp. 57–82 (2009).

589] West B.J., "The wisdom of the body; a contemporary view," *Front. Physiol.* **1**, 1 (2010).

590] West B.J., "Fractal physiology and the fractional calculus: a perspective", *Front. Physiol.* **1**, 12 (2010).

591] West B.J. and P. Grigolini, "Habituation and 1/f-noise", *Physica A* **389**, 5706 (2010).

592] West B.J. and N. Scafetta, *Disrupted Netwoks: From Physics to Climate Change*, Studies of Nonlinear Phenmena in the Life Science, Vol. 13, World Scientific, Singapore (2010).

593] West B.J. and P. Grigolini, *Complex Webs, Anticipating the Improbable*, Cambridge University Press, Cambridge, UK (2011).

[594] West B.J. and P. Grigolini, "The Principle of Complexity Management", in *Decision Making; a psychophysics application of Network Science*, Eds. P. Grigolini and B.J. West, World Scientific, Singapore (2011).

[595] West B.J. and D. West, "Stochastic ontogenetic growth model", *Europhys. Lett.* **97**, 48002 (2012).

[596] West B.J., *Fractal Physiology and Chaos in Medicine*, World Scientific, Singapore (2013).

[597] West B.J., M. Turalska and P. Grigolini, *Networks of Echoes: Imitation, Innovation and Invisible Leaders*, Springer, NY (2014).

[598] West B.J., "A mathematics for medicine: the network effect", *Front. Physiol.* **5**, 1–17 (2014).

[599] West B.J., M. Turalska and P. Grigolini, "Fractional calculus ties the microscopic and macroscopic scales of complex network dynamics", *New J. Phys.* **17**, 059003 (2015).

[600] West B.J., *Fractional Calculus View of Complexity, Tomorrow's Science*, CRC Press, Boca Raton, FL (2016).

[601] West B.J., *Simplifying Complexity: Life is Uncertain, Unfair and Unequal*, Benthem Books, Sharja, UAE (2016).

[602] West B.J., "Information force", *J. Theor. Comp. Sci.* **3**, 144 (2016).

[603] West B.J., *Nature's Patterns and the Fractional Calculus*, DeGruyter, Berlin (2017).

[604] West B.J. and M. Turalska, "Hypothetical control of heart rate variability", *Front. Physiol.* **10**, 1078 (2019).

[605] West B.J., K. Mahmoodi and P. Grigolini, *Empirical Paradox, Complexity Thinking and Generating New Kinds of Knowledge*, Cambridge Scholars Publishing, UK (2019).

[606] West B.J. and T.H. Nonnenmacher, "An ant in a gurge", *Phys. Lett. A* **278**, 255–259 (2001).

[607] West G.B., J.H. Brown and B.J. Enquest, "A general model for the origin of allometric scaling laws in biology", *Science* **276**, 122 (1997).

[608] West G.B., J.H. Brown and B.J. Enquest, "The fourth dimension of life: fractal geometry and allometric scaling of organisms", *Science* **284**, 1677 (1999).

[609] West G.B., J.H. Brown and B.J. Enquist, "A general model for ontogenetic growth", *Nature* **413**, 628–631 (2001).

[610] West G., *SCALE, The Universal Laws of Growth, Innovation, Sustainability, and the Pace of Life in Organisms, Cities, Economies and Companies*, Penguin Press, NY (2017).

[611] Wiener N., *Cybernetics*, MIT Press, Cambridge, MA (1948).

[612] Wiener N., "Time, Communication and the Nervous System", *Proc. New York Acad. Sci.* **50**, 197–220 (1948).

[613] Wiener N., *The Human Use of Human Beings*, Avon Books, NY (1950).

[614] Wiener N., *Nonlinear Problems in Random Theory*, MIT Press and John Wiley, New York (1958).

[615] Wiener N., *Generalized Harmonic Analysis and Tauberian Theorems*, MIT Press, Cambridge, MA (1964).

[616] Willinger W. and V. Paxson, *Notices Am. Math. Soc.* **45**, 961–970 (1998).

[617] Willis J.C., *Age and Area; A Study in Geographical Distribution and Origin of Species*, Cambridge University Press (1922).

[618] Wilson K.G., *Rev. Mod. Phys.* **55**, 583 (1975).

[619] Winfree A.T., *The Geometry of Biological Time*, Springer-Verlag, Berlin (1990).

[620] Wittmann M. and M.P. Paulus, *Trends in Cognitive Sciences* **12**, 7 (2007); M. Wittmann, D.S. Leland and M.P. Paulus, *Exp. Brain Res.* **179**, 1 (2007).

[621] Wood K., C. Van den Broeck, R. Kawai and K. Lindenberg, *Phys. Rev. Lett.* **96**, 145701 (2006).

[622] Yates F.E., "Order and Complexity in Dynamical Systems: Homeodynamics as a Generalized Mechanics for Biology", *Math. Comput. Model.* **19**, 49 (1994).

[623] Yearsley J.M. and J.R. Busemeyer, "Quantum cognition and decision theories: A Tutorial", *J. Math. Psychol.* **74**, 99 (2016).

[624] Yule G.U., *Proc. Roy. Soc. London* **213**, 403 (1924).

[625] Zaks M.A., E.J. Park, M.G. Rosenblum and J. Kurths, *Phys. Rev. Lett.* **82**, 4228 (1999).

[626] Zanette D.H. and S. Manrubia, "Fat tails and black swans: Exact results for multiplicative processes with resets", *Chaos* **30**, 033104 (2020).

[627] Zapata-Fonseca L., D. Dotov, R. Fossion and T. Froese, "Time-Series Analysis of Embodied Interaction: Movement Variability and Complexity Matching As Dyadic Properties", *Front. Psychol.* **7**, 1940 (2016).

[628] Zeng F.G., Q. Fu and R. Morse, *Brain Res. Interact.* **869**, 251 (2000).

[629] Zhan M., G.W. Wei and C.H. Lai, *Phys. Rev. E* **65**, 036202 (2002).

[630] Zipf G.K., *Human Behavior and the Principle of Least Effort*, Addison-Wesley (1949).

[631] Zumofen G. and J. Klafter, *Europhys. Lett.* **25**, 565 (1994).

[632] Zurek W.H., "Decoherence, eigenselection, and the quantum origins of the classical", *Rev. Mod. Phys.* **75**, 715 (2003).

[633] Zurek W.H. and J.P. Paz, "Decoherence, Chaos, and the Second Law", *Phys. Rev. Lett.* **72**, 2508 (1994).

Index

www.ingramcontent.com/pod-product-compliance
Lightning Source LLC
Chambersburg PA
CBHW050537190326
41458CB00007B/1814